OPTOELECTRONICS AND FIBRE OPTICS COMMUNICATION

OPTOELECTRONICS AND FIBRE OPTICS COMMUNICATION

(SECOND EDITION)

C K Sarkar
Professor of Electronics and Telecommunication Engineering
Jadavpur University, Kolkata

D C Sarkar
Former Professor of Physics
Kalyani University, West Bengal

Publishing Globally

NEW AGE INTERNATIONAL (P) LIMITED, PUBLISHERS
New Delhi • Bangalore • Chennai • Cochin • Guwahati
Hyderabad • Kolkata • Lucknow • Mumbai
Visit us at **www.newagepublishers.com**

Published by New Age International (P) Ltd., Publishers
First Edition : 2001
Second Edition: 2013

BRANCHES

- **Bangalore** 37/10, 8th Cross (Near Hanuman Temple), Azad Nagar, Chamarajpet, Bangalore-560 018
  Tel.: (080) 26756823, Telefax: 26756820, **E-mail: bangalore@newagepublishers.com**
- **Chennai** 26, Damodaran Street, T. Nagar, Chennai-600 017
  Tel.: (044) 24353401, Telefax: 24351463, **E-mail: chennai@newagepublishers.com**
- **Cochin** CC-39/1016, Carrier Station Road, Ernakulam South, Cochin-682 016
  Tel.: (0484) 2377004, Telefax: 4051303, **E-mail: cochin@newagepublishers.com**
- **Guwahati** Hemsen Complex, Mohd. Shah Road, Paltan Bazar, Near Starline Hotel, Guwahati-781 008
  Tel.: (0361) 2513881, Telefax: 2543669, **E-mail: guwahati@newagepublishers.com**
- **Hyderabad** 105, 1st Floor, Madhiray Kaveri Tower, 3-2-19, Azam Jahi Road, Nimboliadda, Hyderabad-500 027
  Tel.: (040) 24652456, Telefax: 24652457, **E-mail: hyderabad@newagepublishers.com**
- **Kolkata** RDB Chambers (Formerly Lotus Cinema) 106A, 1st Floor, S.N. Banerjee Road, Kolkata-700 014
  Tel.: (033) 22273773, Telefax: 22275247, **E-mail: kolkata@newagepublishers.com**
- **Lucknow** 16-A, Jopling Road, Lucknow-226 001
  Tel.: (0522) 2209578, 4045297, Telefax: 2204098, **E-mail: lucknow@newagepublishers.com**
- **Mumbai** 142C, Victor House, Ground Floor, N.M. Joshi Marg, Lower Parel, Mumbai-400 013
  Tel.: (022) 24927869, Telefax: 24915415, **E-mail: mumbai@newagepublishers.com**
- **New Delhi** 22, Golden House, Daryaganj, New Delhi-110 002
  Tel.: (011) 23262368, 23262370, Telefax: 43551305, **E-mail: sales@newagepublishers.com**

ISBN: 978-81-224-3457-6

C-12-08-6470

Printed in India at Nisha Enterprises,

Typeset at Goswami Associates, Delhi

PUBLISHING FOR ONE WORLD

NEW AGE INTERNATIONAL (P) LIMITED, PUBLISHERS
7/30 A, Daryaganj, New Delhi-110002
Visit us at **www.newagepublishers.com**

The author dedicates the book to his parents

The author dedicates the book to his parents

Preface

Recently optoelectronics and fiber communications have become an integral part of M.Sc. (Physics and Electronics Science) and B.E. and M.E. courses on Electronics and Communication Engineering of different universities. Although a large number of books on the subject are available, these are generally too difficult to be followed by beginners who need an easier treatment of the subject. The book explains the fundamentals of the subject in a simple and easy to understand manner. To fulfill this requirement, the first edition of the book was launched successfully in 2001.The book became popular with students and teachers. Considering the fast development of the subject of optical communication, the book needed a detailed revision satisfying the syllabi of different institutions. A revised edition of the book is presented by incorporating some advanced topics such as modulation techniques, electro optic effect, optoelectronic integrated circuits etc.

There are altogether twelve chapters in the book. *Chapter-1* starts with a short introduction to optical communications, the size and types of optical fiber. A brief description of an optical fiber link is also presented. This chapter also highlights the advantages of optical fiber relative to other transmission media. It includes a brief description of the development of optical communication system mentioning different wavelengths, optical sources and detector used. Following the description of optical fiber structure, two methods are used to show that a fiber guides light. The first approach described in the chapter utilized geometrical ray optics concept of light reflection and refraction to provide the determination of numerical aperture for meridional and skew rays.

Chapter-2 utilizes the second approach which treats light as an electromagnetic wave which propagates along the optical fiber waveguide. The analysis for a stepindex fiber has been done by solving Maxwell's field equations with a brief description of M.E. subject to the cylindrical boundary condition of the fiber. This chapter starts with the solution of Maxwell's field equations in cylindrical coordinate system (r, ϕ, x). The expressions for the different electric and magnetic field components in the core and cladding regions have been determined. From the boundary conditions at core-cladding interface, the conditions imposed on the phase constant 'β' for the propagation of the electromagnetic wave through the core is presented.

The lower and upper limits of the Bessal function in terms of the radius of the core for the propagation of an electromagnetic wave in a particular mode through the core have been discussed. The relationship between the V-number and the corresponding number of modes M has also been determined.

This chapter also deals with the amounts of energy flow through the core and cladding of the fiber with the help of Poynting vector.

Chapter-3 discusses the signal degradation arising from the attenuation and dispersion mechanisms in the fiber. It starts with the attenuation of the signal by different scattering mechanism, such as Rayleigh, Brillouin and Raman scattering. It also points out the threshold value of the signal for the nonlinearities to occur due to Raman and Brillouin scattering. The wavelength spectrum of losses due to absorption has been illustrated and the method of minimizing these losses has been discussed. Radiative losses in signal due to microbending and macrobending have also been discussed.

Different types of dispersion in the signal such as intermodal, material and waveguide dispersions are also discussed in this chapter and the effect of dispersion and attenuation on bit rate has been illustrated.

Chapter-4 starts with the requirements of optical sources used in fiber communications. In order to transmit data through optical fiber waveguides, the optical source compatible with the fiber size and transmission properties is required and a light emitting diode (LED) or laser diode is suitable for the purpose. Chapter-4 deals with the LED. In order to understand the LED, one must have a good idea about the operation of a *p-n* junction diode. The operation of a *p-n* junction diode has been discussed on a physical basis as well as on the basis of electron energy, band structure and Fermi level. The theory relating to electro luminescence has been developed. This chapter also discusses direct bandgap and indirect bandgap semiconductors as well as the basic theory of a DH (Double Heterostrucure) light emitting diodes along with carrier and optical energy confinements. Different LED structures, such as surface emitter, edge emitter, super luminescent LED have been illustrated and discussed. The lens coupling to the fiber, such as the use of spherical ended fiber and spherical microlens, have been illustrated.

A few drive circuits for modulating LED's have been presented explaining the drive operation of the circuits. Differences between optical and electrical modulation, bandwidths have been explained.

Chapter-5 starts with Einstein relationships between absorption spontaneous emission and stimulated emission showing that without population inversion at ordinary temperature stimulated emission is negligible. Some hints have also been given regarding some non-semiconductor lasers (such as ruby and He-Ne lasers). The threshold condition for lasing has been developed. The theory of confinement of optical power and carriers in DH laser structure has been discussed with proper illustration. The layer structures, materials and the operating characteristics of few semiconductor lasers such as Fabry Perot homojunction, stripe geometry laser, DFB and DBR lasers have been discussed with appropriate diagrams.

When optical power emerges out of the fiber, it must be converted into a electrical signal. The function is accomplished by a photodetector. Here the discussion on the *p-n* photodiode is followed by *p-i-n* photodiode and avalanche photodiode and these two types of photodetectors are used exclusively in fiberoptic systems. Chatper-6 is devoted to the photodetector systems. Noise generated in these systems has also been examined in detail. Finally, in Chapter-6,Solar cell and its operation and applications are discussed.

Chapter-7 starts with fiber production methods such as double crucible method, vapour phase axial deposition method, chemical vapour deposition, fiber drawing and coating method. This chapter describes the various materials used to make fiber and illustrates how optical fibers are incorporated in cable structure. Different types of typical fibers are illustrated and described in this chapter. A detailed discussion of splices and connectors with uses and operation is also included. Power launching and coupling of optical fibers have been included.

Chapter-8, devoted to optical fiber receivers, starts with optial fiber digital data link for digital transmission. It also discussed the different types of front ends of a receiver and the noise produced in a receiver. It presents the block diagram of an optical receiver explaining the function of each block. The introduction to optoelectronic integrated circuits (OEIC) is also presented.

Chapter-9 discusses the techniques for the measurement of different fiber parameters. The parameters measured are attenuation measurement by cut back technique and OTDR technique, dispersion measurement in both frequency and time domains, measurement of refractive index profile by using Mach Zehnander microscope, numerical aperture and core diameter measurement.

Chapter-10 is devoted to the use of fiber sensor, for measurement velocity of flow of liquid, optical displacement sensor. Single mode optical fiber sensor for current measurement, single mode fiber interferometer sensor for an accurate measurement of pressure or temperature, rotation measurement by fiber microbending sensor, optical fiber flow meter optical fluid level detector, photoelastic photo sensor have all been discussed in this chapter.

Chapter-11 deals with multipoint multiwave transmission networks. Discussions of Link power budget, different WDM system, optical amplifiers and solitions are included in this chapter.

Finally, *Chapter-12* has been added in the revised version of the book. The chapter incorporates the Electro optic effect, The various modulation techniques used in optical communication have been described. An introduction to Orthogonally Frequency Division Multiplexing(OFDM) has been discussed.

In the book, a large number of solved numerical problems are given for an easier understanding of the subject. A considerable number of problems are given at the end of each chapter for practice and self test.

The authors are grateful to Dr. M.N. Roy, Professor to Telecommunication Department of Jadavpur University for going through the manuscript. The authors are also grateful to Prof. D.P. Bhattacharyya of Jadavpur University for encouraging us to publish a book on fiber communications. The authors consulted Prof. S.K. Sarkar of Calcutta University and Prof. B.N. Biswas and Prof. S. Roy of Burdwan University.

The author expresses his sincere gratitude to Prof. Ansuman Sarkar. Assistant Professor, Kalyani Government Engineering College, Kalyani (W.B.) for his help in the preparation of the revised version.

C.K. Sarkar
D.C. Sarkar

Contents

CHAPTER 1

Introduction to Fiber Communications

CHAPTER FOCUS

- *Introduction*
- *Optical fiber transmission link*
- *Development of optical fiber communications*
- *Classification of fibers and ray path*
- *Advantages of multimode fibers*
- *Advantages of the cladding*
- *Numerical aperture: Step index fiber*
- *Light collected by a fiber from diffuse source*
- *Ray optics representation*
- *Acceptance angle of the skew rays*
- *Advantages of optical fiber over the copper wire system*
- *Solved examples and review questions*

INTRODUCTION

The optical fiber is simply a cylindrical waveguide system operating at optical frequencies. It consists of a core at the centre and a cladding or sheath outside the core. The core is generally a cylindrical dielectric glass with a refractive index n_1 and the cladding is the second dielectric sheath or cover usually of glass with a lower refractive index n_2 as shown in Fig. 1.1(*a*). Sometime the cladding is surrounded by a buffer coating [Fig. 1.1(*b*)].

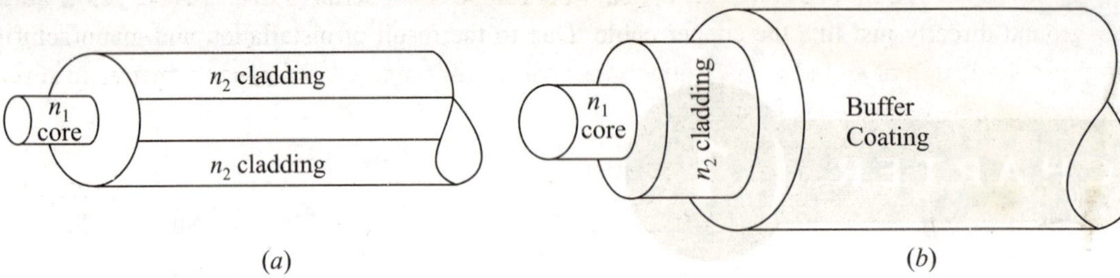

Fig. 1.1 (*a*) Optical fiber waveguide system with core of refractive index n_1 and a cladding of refractive index n_2 where n_1 is greater than n_2. (*b*) The cladding of the fiber is coated by a buffer coating

Lasers or light emitting diodes (LEDs) are capable of generating electromagnetic waves at optical frequencies which can easily be modulated by an electrical signal to be transmitted. If the light beam of the electromagnetic wave impinges on the system through the one end face of the fiber, it emerges out of the other face with a small attenuation. An avalanche photodiode (APD) or a semiconductor p-i-n diode is capable of detecting the signal. Therefore, the optical fiber is used for the transmission of optical signal in a manner identical to the propagation of the microwave signal through a cylindrical metallic waveguide.

1.1 OPTICAL FIBER TRANSMISSION LINK

Fig. 1.2 shows the elements of an optical fiber transmissions link.

1. A light source which may be a laser or a LED is capable of generating an optical signal at desired frequencies.
2. A drive circuit is a means for modulating the output light beam from the laser or LED. The envelope of the modulated beam has the same waveshape as the electrical signal to be transmitted.
3. A transmission medium is a cable providing mechanical and environmental protection to the hair-thin optical fibre. Then the optical signal is launched into the optical fibre which is contained inside the cable.
4. A receiver whose main component is a photodetector is capable of converting the received modulated wave back to the original signal which has the same waveshape as the optical wave envelope. An avalanche photodiode (APD) or a semiconductor p-i-n diode is used as a photodetector.
5. Electronic amplifier and signal restorer consisting of signal processor circuits, recover the original signal and present it in a suitable form.

1.2 DEVELOPMENT OF OPTICAL FIBER COMMUNICATIONS

The cabled optical fiber is one of the most important elements in an optical fibre link. The cable provides protection to the glass fibre during installation and service, and it may contain a copper wire for boostering the power of the repeater. The boostering (increasing) of power is needed for periodically amplifying and reshaping the signal when the link extends for a long distance. The cable generally contains many "hair thin" cylindrical glass fibers. Each glass fiber is an independent communication link channel just like a copper cable. The installation of the optical fiber can be either aerial or in duct under sea or buried under ground directly just like the copper cable. Due to the result of installation and manufacturing limitations, the length of an individual optical fiber cable will range several hundred metres to several kilometres for long distance applications.

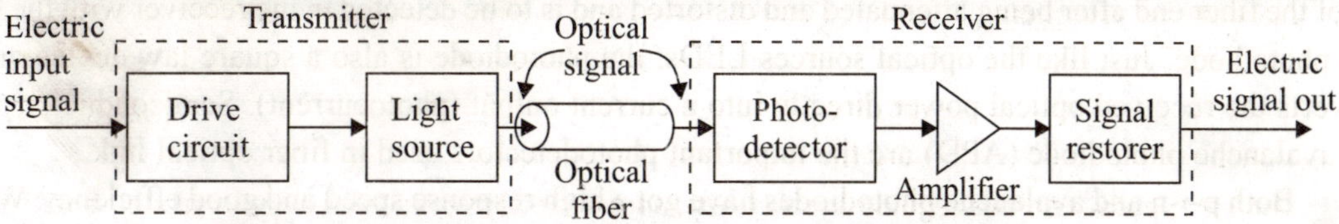

Fig. 1.2 Basic elements of an optical fiber transmission link

One of the important fiber characteristics is its attenuation as a function of wavelength of the propagating optical signal as shown in Fig. 1.3. The attenuation takes place due to the absorption and scattering of the optical wave. In the earlier days of fiber communications, there was an exclusive use of 0.8 to 0.9 μm band. Fibers manufactured at that time exhibited a local minima of attenuation in this region of the attenuation curve and at the same time, optical sources and photodetectors in this wavelength band were also available. By reducing the concentration of the hydroxyl ions and metallic ion impurities in the fiber materials fiber manufacturers were able to fabricate optical fiber wave guides with low loss in the 1.1 to 1.6 μm wavelength region. This spectral bandwidth is referred to as "long wavelength band" in fiber communications. In the wavelength of 1.3 μm a lot of interest, thus, developed, since this is the region of minimum distortion in pure silica fibers.

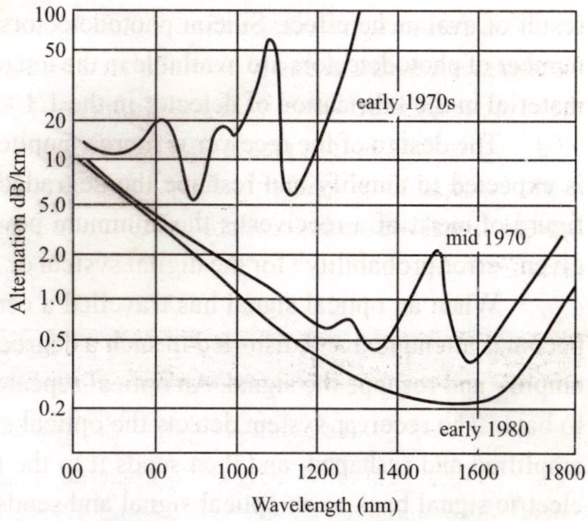

Fig. 1.3 Optical fiber attenuation as a function wavelength. Early fiber links were operated exclusively in the 800 to 900 nm range where there was a local attenuation minimum. Achievement of lower attenuation created interest in the longer wavelength operation.

When the cable is installed, a light source which is dimensionally compatible with the diameter of the fiber core is to inject optical wave into the fiber. Semiconductor light emitting diodes (LEDs) and laser diodes are found to be suitable transmitting sources for this purpose. The optical output of these devices can easily be modulated simple by varying the bias current. The electrical signal to be transmitted can be either in analog or in digital form. The drive current associated with the transmitter converts the electrical signal into the optical signal by varying the current through the light source. The envelope of the optical signal has the same waveform as that of the electrical signal. The optical source is a square law device and the linear variation in the drive current results in the corresponding linear change in the optical power of the light source.

When the wavelengths are in the range of 0.8 to 0.9 μm, the material of the light source is generally an alloy of GaAIAs. At the larger wavelengths ranging from 1.1 to 1.6 μm, an alloy of

InGaAsP is the principal optical source material. After the optical signal is launched in the fiber, the signal will progressively get attenuated and distorted with the increase in the length of the fiber. The attenuation takes place due to absorption and scattering, whereas the distortion in the shape of the signal occurs due to the dispersion mechanism in the fiber waveguide. The modulated optical power comes out of the fiber end after being attenuated and distorted and is to be detected in the receiver with the help of a photodiode. Just like the optical sources LEDs, the photodiode is also a square law device and it converts the received optical power directly into a current output (photocurrent). Semiconductor p-i-n and avalanche photodiode (APD) are the important photodetectors used in fiber optical link.

Both p-i-n and avalanche photodiodes have got a high response speed and good efficiency. When a low power optical signal is received, the use of an avalanche photodiode is preferred. The advantage of an avalanche photodiode is that it exhibits a greater sensitivity owing to the internal mechanism as a result of avalanche effect. Silicon photodetectors are used in the 0.8 to 0.9 μm wavelength region. A number of photodetectors are available in the longer wavelength region. An alloy InGaAs is the principal material in the fabrication of detector in the 1.1 to 1.6 μm wavelength region.

The design of the receiver is more complicated than that of the transmitter, because the receiver is expected to amplify and reshape the degraded signal received by the photodetector. The principal figure of merit of a receiver is the minimum power necessary at the desired data rate to attain either given "error probability" for the digital system or a specific "signal to noise ratio" for an analog system.

When an optical signal has travelled a certain distance along the length of the fiber, the signal becomes attenuated and distorted in such a degree that a "repeater" is needed in the transmission line to amplify and reshape the signal. An optical repeater consists of a receiver and a transmitter placed back to back. The receiver system detects the optical signal, converts it to an electrical signal which is then amplified and reshaped, and then sends it to the transmitter section. The transmitter then converts the electric signal back to an optical signal and sends it through the optical fiber waveguide.

1.3 CLASSIFICATION OF FIBERS AND RAY PATH

The basic structure of the optical fiber is given in Fig. 1.1. Variation in the composition of the core material gives rise to two types of optical fibers. These are (1) Step index fiber and (2) Graded index fiber.

1. Step Index Fiber

In the case of a step index fiber, the refractive index n_1 of the core material is uniform throughout and undergoes an abrupt change in step at the core cladding boundary as shown in Fig. 1.4 (*a*) and (*b*).

2. Graded Index Fiber

In the case of a graded index fiber the refractive index of the core is made to vary as a function of radial distance from the centre of the fiber as shown in Fig. 1.4(*c*). The use of graded index fiber is gradually diminishing. Both step index and graded index fibers can further be divided into singlemode and multimode classes. A singlemode fiber can sustain only "one mode" of propagation, whereas multimode fiber can contain a "large number of modes". A typical size of the singlemode and multimode fibers with ray path is illustrated in Fig. 1.4. The singlemode fibers is also referred to as the "monomode" fiber.

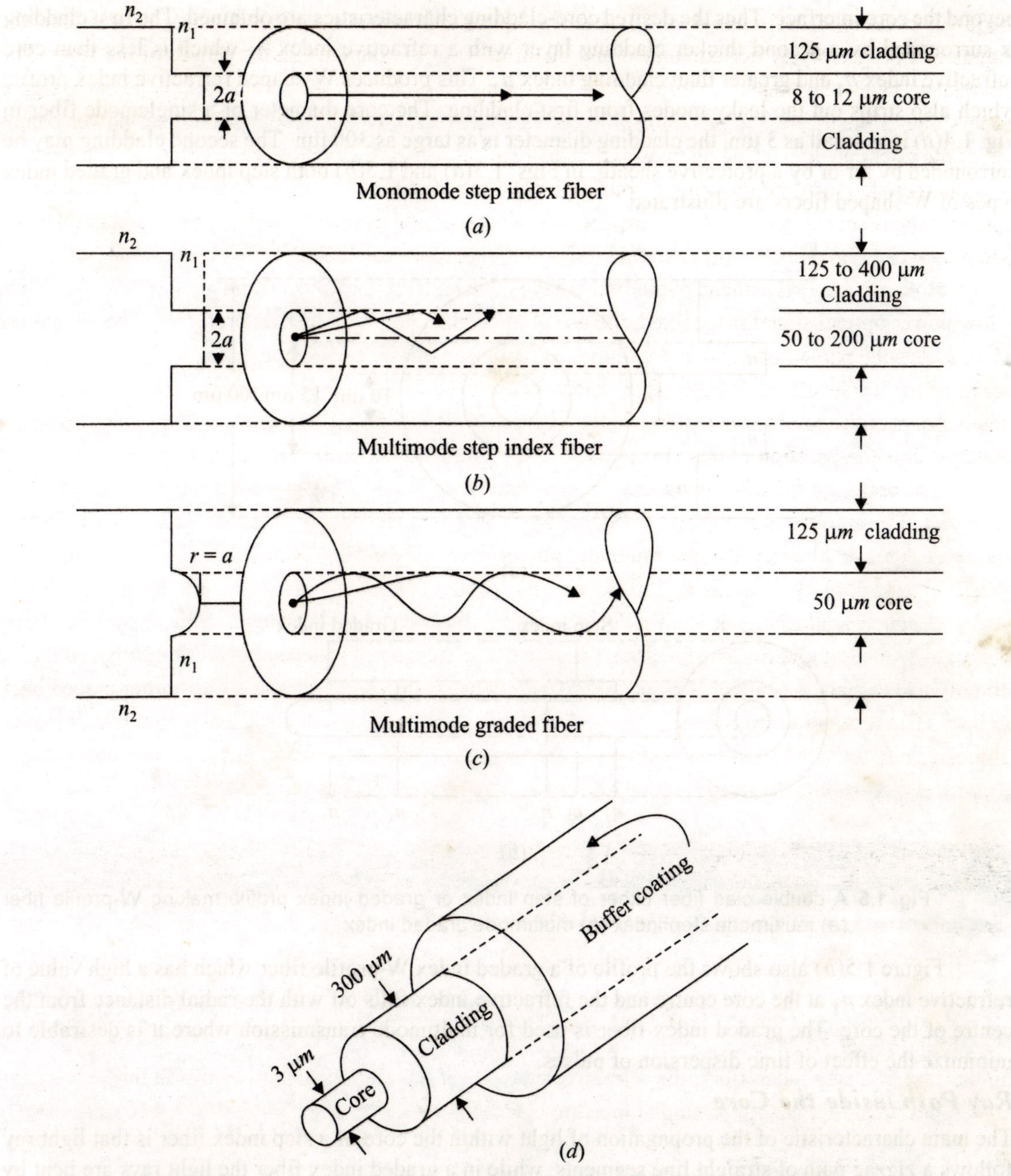

Fig. 1.4 Comparison of a singlemode and multimode step index fibers on the basis of size and ray path (*a*) Monomode step index (*b*) multimode step and (*c*) multimode graded index fibers (*d*) singlemode structure, *a*(core diameter) = 3 μm and *b*(cladding diameter) = 300 μm.

W-profile fibers

Fig. 1.5(*a*) shows the W-profile fiber. In this case the width of cladding with refractive index n_2 is made thick enough so that it encloses all of the light energy field-ends which extend for a short distance beyond the core interface. Thus the desired core-cladding characteristics are obtained. The first cladding is surrounded by a second thicker cladding layer with a refractive index n_3 which is less than core refractive index n_1 and greater than cladding index n_2. This produces W-shaped refractive index profile which also strips out the leaky modes from first cladding. The core diameter of a singlemode fiber in Fig. 1.4(*d*) is as small as 3 μm, the cladding diameter is as large as 300 μm. The second cladding may be surrounded by air or by a protective sheath. In Figs. 1.5(*a*) and 1.5(*b*) both step index and graded index types of W-shaped fibers are illustrated.

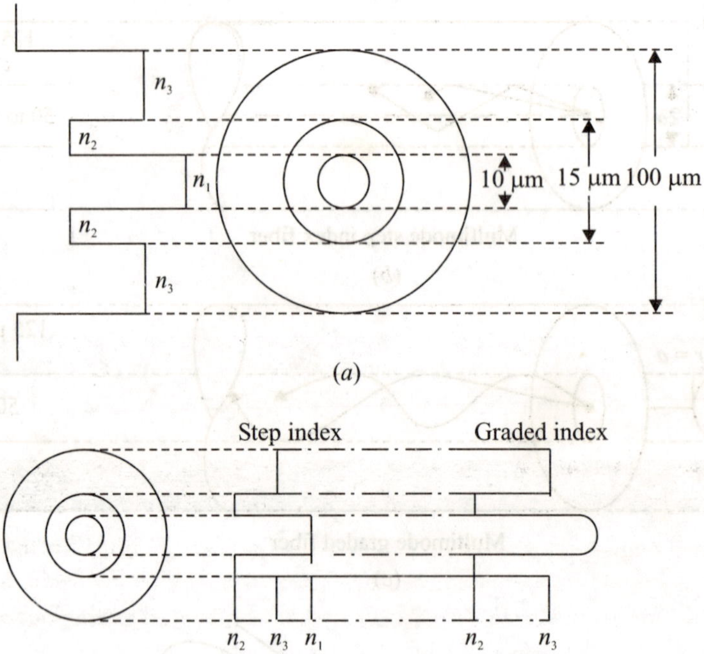

Fig. 1.5 A double clad fiber either of step index or graded index profile making W-profile fiber (*a*) multimode step index (*b*) multimode graded index

Figure 1.5(*b*) also shows the profile of a graded index W-profile fiber which has a high value of refractive index n_1 at the core centre and the refractive index falls off with the radial distance from the centre of the core. The graded index fiber is used for multimode transmission where it is desirable to minimize the effect of time dispersion of pulses.

Ray Path inside the Core

The main characteristic of the propagation of light within the core of a step index fiber is that light ray follows a zigzag path of straight line segments, while in a graded index fiber the light rays are bent by refraction so that they are curved inward toward the axis as shown in Fig. 1.4.

1.4 ADVANTAGES OF MULTIMODE FIBERS

Multimode fibers possess several advantages as compared to the single mode fibers. These are:

1. The larger radii of the multimode fibers make it easier to launch the optical power into the fiber.
2. The second advantage is that light can be launched in a multimode fiber with the help of a light

emitting diode source, whereas in a single mode fiber light can only be launched by a laser diode. Although the LEDs have lesser power output than the laser diodes, LEDs are easier to manufacture, less expensive, requires less complex electronic circuitry and have a longer life time than the laser diodes. Thus, LEDs are found to be more suitable in many applications.

The disadvantage of multimode fibers is that they undergo intermodal dispersion which distorts the shape of the propagating light pulses. This will be discussed in a later section.

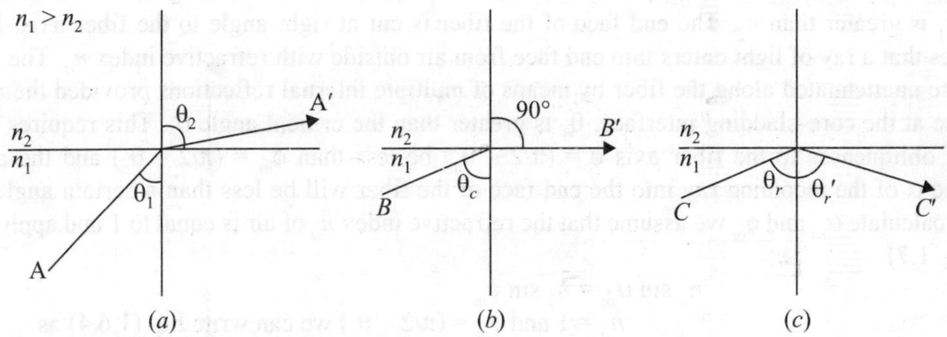

Fig. 1.6 Refraction and total internal reflection at the dielectric interface: (*a*) ray *AA'* is refracted according to Snell's law of refraction $n_1 \sin \theta_1 = n_2 \sin \theta_2$ (*b*) ray *BB'* is the critical ray $n_1 \sin \theta_c = n_2$, (*c*) ray *CC'* is totally reflected at the interface $\theta_r = \theta_r'$

1.5 ADVANTAGES OF THE CLADDING

The presence of the cladding is not absolutely essential for the optical fiber for the propagation of optical signals through the core of the optical fiber, but it serves the following useful purposes:

1. It adds mechanical strength to the fiber and protects the fiber from absorbing surface contaminants with which it may come in contact.
2. The cladding is capable of reducing the scattering loss of light resulting from dielectric discontinuities at the core surface.

In low and medium-loss fibers, the core material is generally glass which is usually surrounded by a glass or plastic cladding. Higher-loss plastic fibers with plastic coating are also available. In addition, most fibers are encapsulated in an elastic abrasion resistant plastic materials. These materials add further strength to the fiber and mechanically isolate to buffer the fiber from small geometrical irregularities of the adjacent surfaces.

1.6 NUMERICAL APERTURE: STEP INDEX FIBER

Refraction of an electromagnetic wave at the surface of separation of two media is determined by Snell's law and is illustrated in Fig. 1.6(*a*). A ray of light which is an electromagnetic wave is passing from a medium of higher refractive index n_1 into a medium of lower refractive index n_2.

For $0 < \theta_1 < \theta_c$ and $0 < \theta_2 < \pi/2$

$$\frac{n_2}{n_1} = \frac{\sin \theta_1}{\sin \theta_2} \tag{1.6.1}$$

where θ_1 and θ_2 are the angles of incidence and refraction respectively.

When the angle of refraction θ_2 is equal to $\pi/2$ (or 90°) as shown in Fig. 1.6(*b*), the angle of incidence $\theta_1 = \theta_c$ and the angle θ_c is known as the "critical angle". Thus

$$n_1 \sin \theta_c = n_2 \sin \theta_2 = n_2 \sin (\pi/2) = n_2 \quad (1.6.2)$$

or

$$\sin \theta_c = (n_2/n_1) \quad (1.6.3)$$

when the angle of incidence θ_1 is greater than the critical angle θ_c, the "total internal reflection" of the light takes place with no loss at the boundary as shown in Fig. 1.6(*c*). Let us consider a cylindrical glass fiber consisting of an inner core of refractive index n_1 and the outer cladding of refractive index n_2 where n_1 is greater than n_2. The end face of the fiber is cut at right angle to the fiber axis. Fig. 1.7 illustrates that a ray of light enters into end face from air outside with refractive index n_a. The ray will propagate unattenuated along the fiber by means of multiple internal reflections provided the angle of incidence at the core-cladding interface, θ_1 is greater than the critical angle θ_c. This requires that the angle of obliqueness to the fiber axis $\phi = (\pi/2 - \theta_1)$ be less than $\phi_m = (\pi/2 - \theta_c)$ and the angle of incidence α of the incoming ray into the end face of the fiber will be less than a certain angle α_m. In order to calculate α_m and ϕ_m we assume that the refractive index n_a of air is equal to 1 and apply Snell's law (Fig. 1.7).

$$n_a \sin \alpha_m = n_1 \sin \phi_m \quad (1.6.4)$$

putting $n_a = 1$ and $\phi_m = (\pi/2 - \theta_c)$ we can write Eq. (1.6.4) as

$$\sin \alpha_m = n_1 \sin (\pi/2 - \theta_c) = n_1 \cos \theta_c$$

We know from Eq. (1.6.3) that

$$\sin \theta_c = n_2/n_1$$

and

$$\cos \theta_c = (1 - \sin^2 \theta_c)^{1/2} = \left[1 - \frac{n_2^2}{n_1^2}\right]^{\frac{1}{2}} \quad (1.6.5)$$

Substituting the value of cos θ_c from Eq. (1.6.5) we get

$$\sin \phi_m = \cos \theta_c = \left[1 - \frac{n_2^2}{n_1^2}\right]^{\frac{1}{2}} \quad (1.6.6)$$

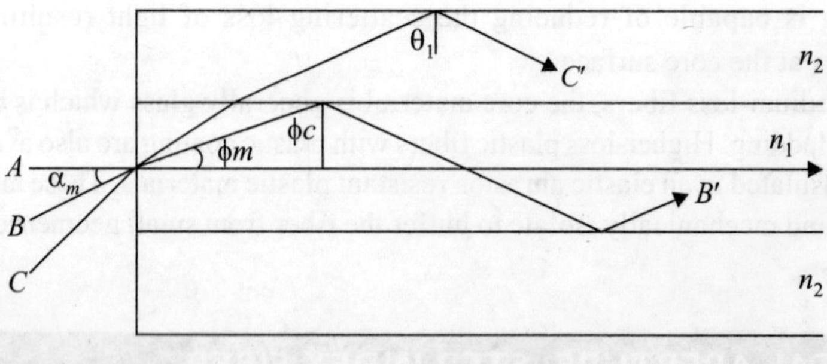

Fig. 1.7 Propagation of a light wave in an optical fiber *AA'* is an axial ray *BB'* is the critical ray for total internal reflection at $n_1 - n_2$ interface; CC' is too obliquely for reflection at the interface, and passes into the cladding by refraction

$$n_a \sin \alpha_m = n_1 \sin \phi_m = n_1 \cos \phi_c = \frac{n_1 \left[n_1^2 - n_2^2\right]^{\frac{1}{2}}}{n_1} = \left[n_1^2 - n_2^2\right]^{\frac{1}{2}}$$

$$n_a \sin \alpha_m = [(n_1 + n_2)(n_1 - n_2)]^{1/2} \quad (1.6.7)$$

If we denote $n_1 - n_2 = \Delta n,\ n_1 + n_2 = 2n$ and $n_a = 1$

We get, $$\sin \alpha_m = (2n\Delta n)^{1/2} \quad (1.6.8)$$

The greater the value of α_m, the greater is the portion of the incident light on the end face can be collected by the fiber and the propagation takes place by total internal reflection. By analogy, with the term used to define the light gathering power of microscope objectives ($n_a \sin \alpha_m$) is known as the numerical aperture (NA) of the fiber. Thus, putting $n_a = 1$.

Numerical aperture $$\text{NA} = \sin \alpha_m = (2n\ \Delta n)^{1/2} \quad (1.6.9)$$

and the acceptance angle,

$$\alpha_m = \sin^{-1} (2n\Delta n)^{1/2} = \sin^{-1} (\text{NA}) \quad (1.6.10)$$

1.7 LIGHT COLLECTED BY A FIBER FROM DIFFUSE SOURCE

We shall now show that when a light is emitted from a diffuse source situated on the fiber axis near the end face, only a fraction $(\text{NA})^2$ of the total amount of light emitted can be collected by the fiber and propagated along the fiber. Let us consider a small diffuse light source such as an isotropic (Lambertian) radiator as shown in Fig. 1.8(*a*). The radiated light power per unit solid angle in a direction at an angle θ normal to the surface is given by

$$I_\theta = I_0 \cos \theta \quad (1.7.1)$$

where I_0 and I_θ represent the power radiated per unit solid angle in the direction of the axis of the fiber and at an angle θ normal to the surface respectively.

The power radiated in a small solid angle $d\Omega$ in the direction θ with the perpendicular to the surface is (Fig. 1.8*b*) $I(\theta)\, d\Omega = I_0 \cos \theta\, d\Omega$. The elementary annular ring, whose radius substends and angle θ and width substends an angle $d\theta$ at the source itself, substends a solid angle $d\Omega = 2\pi \sin \theta\, d\Omega$.

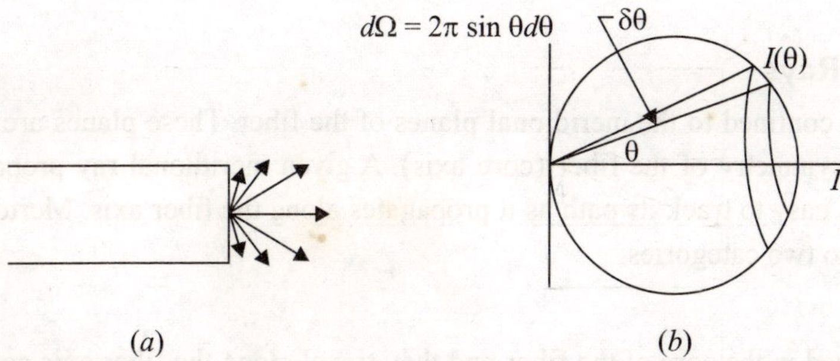

Fig. 1.8 (*a*) A diffuse light source (*b*) the solid angle between θ and $\theta + d\theta$ is $d\Omega = 2\pi \sin \theta d\theta$

The total power Φ_0 emitted by the source is obtained by integrating over all forward direction

$$\Phi_0 = \int_o^{\pi/2} I_o \cos \theta\, d\Omega = \int_o^{\pi/2} I_o \cos \theta\, 2\pi \sin \theta\, d\theta$$

$$= +2\pi I_0 \int_o^{\pi/2} \cos \theta \sin \theta\, d\theta = 2\pi I_o \left[\frac{-\cos^2 \theta}{2}\right]_o^{\pi/2}$$

$$= -2\pi I_o \left[o - \frac{1}{2} \right] = \pi I_0 \tag{1.7.2}$$

But the power that can be collected by the adjacent fiber whose core diameter is greater than the diameter of the source is given by Φ where

$$\Phi = \int_o^{\alpha_m} I_o \cos\theta\,(2\pi)\sin\theta\,d\theta$$

$$\Phi = \int_o^{\alpha_m} (I_o \cos\theta)\,2\pi\sin\theta\,d\theta$$

$$= -\pi I_o[\cos^2\theta]_o^{\alpha_m}$$

$$= \pi I_o[\cos^2\alpha_m - 1] = \pi I_o \sin^2\alpha_m \tag{1.7.3}$$

From Eq (1.7.3) and Eq (1.7.2)

Thus,

$$\frac{\Phi}{\Phi_o} = \sin^2\alpha_m = (\text{NA})^2 = 2n\Delta n \tag{1.7.4}$$

Thus, in order to collect as much light as possible it is necessary to make n and Δn large and the best we can do is to use a glass with high refractive index as core without any cladding.

1.8 RAY OPTICS REPRESENTATION

Two types of rays can propagate through an optical fiber. These are:

(*a*) Meridional rays and

(*b*) Skew rays.

(a) Meridional Rays

Meridional rays are confined to the meridional planes of the fiber. These planes are the planes which contain the axis of symmetry of the fiber (core axis). A given meridional ray propagates in a single plane and hence it is easy to track its path as it propagates along the fiber axis. Meridional rays can be further classified into two categories:

1. *Bound Rays*

These rays are trapped in the core of the fiber and they travel along the fiber axis according to Snell's laws of reflection and refraction.

2. *Unbound Rays*

These rays are refracted out of the fiber core according to the Snell's laws of refraction and cannot be trapped in the core of the fiber.

(b) Skew Rays

The Skew rays can propagate without passing through the core of the fiber. They are not confined to a single plane, but they follow a helical path along the fiber as shown in Fig. 1.9(*a*). It is quite difficult to track these rays as they travel along the fiber, since they do not lie in a single plane. It is quite difficult to visualize the path of the skew rays in a two-dimensional picture. Figure 1.9(*b*) shows the helical path through the fiber by the skew rays and gives a change of direction of an angle 2γ at each reflection where γ is the angle between the projection of the ray in two dimensions and the radius of the fiber core at the point of reflection. The point of emergence of the skew rays from the fiber in air depends on the number of reflections they undergo rather than the input condition of the optical fiber. When the light input to the fiber is not uniform, the skew rays will exhibit a smoothing effects on the distribution of light transmitted. As a result, the light output due to the skew rays will be quite uniform.

Another advantageous feature of the transmission of the skew rays is that their numerical aperture of the fiber will be greater than that for the meridional rays.

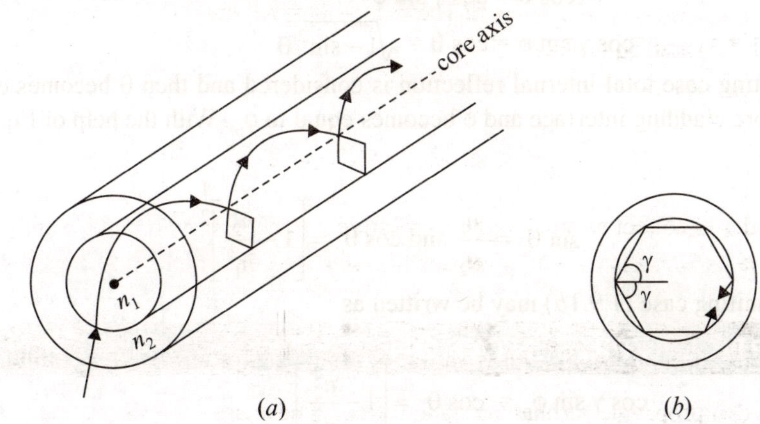

Fig. 1.9 (*a*) Helical path followed by a skew ray in an optical fiber (*b*) Cross-sectional view of the fiber

1.9 ACCEPTANCE ANGLE OF THE SKEW RAYS

To calculate the acceptance angle for a skew ray, it is necessary to determine the direction of the ray in two perpendicular planes as shown in Fig. 1.10. The figure shows that the skew ray is incident on the fiber core at point *A* at an angle α to the end face of the fiber and the ray is refracted at angle γ at air-core interface before going to point *B* on the same plane.

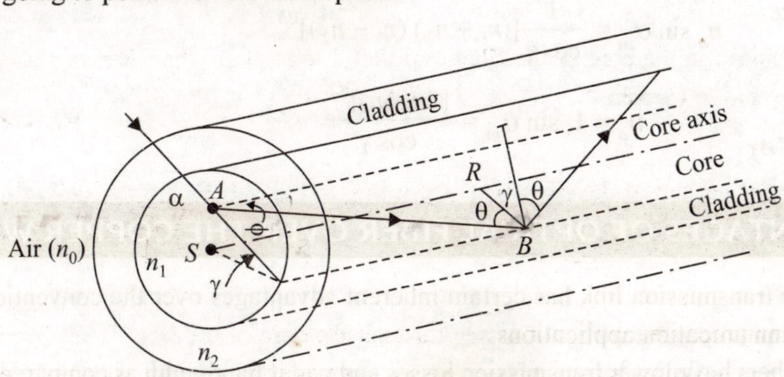

Fig. 1.10 The ray path within the fibre core for skew ray incident at an angle as to the normal at air-core interface.

When we consider the ray between A and B it is necessary to resolve the ray path AB to the core radius at the point B. When the incident and reflected rays are in the same plane, the angle of incidence is equal to the angle of reflection. However, if the two perpendicular planes through the ray path that AB traverses are considered, then γ is the angle between the core radius and the projection of the ray on the plane BRS normal to the core axis, an ϕ is the angle between the ray and the line AT drawn parallel to the core axis. Thus, to resolve the ray path AB relative to the radius BR in these two perpendicular planes, one requires the multiplication of $\cos\gamma$ and $\sin\phi$.

At the reflection point at B, the plane containing incident ray and the normal at the point of reflection makes an angle γ with the plane containing the reflected ray and the normal at the point of reflection at B, the angle Φ may be given by

$$\cos\theta = \cos\gamma \sin\phi \tag{1.9.1a}$$

or

$$\cos\gamma \sin\phi = \cos\theta = \sqrt{1-\sin^2\theta} \tag{1.9.1b}$$

In the limiting case total internal reflection is considered and then θ becomes equal to critical angle θ_c for the core cladding interface and ϕ becomes equal to ϕ_m. With the help of Eq. (1.6.3) we can write

$$\sin\theta_c = \frac{n_1}{n_2} \text{ and } \cos\theta_c = \left[1-\frac{n_2^2}{n_1^2}\right]^{\frac{1}{2}}$$

Thus, in limiting case (1.9.1b) may be written as

$$\cos\gamma \sin\phi_m = \cos\theta_c = \left[1-\frac{n_2^2}{n_1^2}\right]^{\frac{1}{2}} \tag{1.9.2}$$

Apply Snell's law at the point

$$n_a \sin\alpha_m = n_1 \sin\phi_m \tag{1.9.3}$$

where n_a is the refractive index of air in which the fibre is situated and in the limiting case α becomes α_m. Combining Eq. (1.9.2) and Eq. (1.9.3), we get

$$n_a \sin\alpha_m = n_1 \sin\phi_m = \frac{n_1}{\cos\gamma}\left[1-\frac{n_2^2}{n_1^2}\right]^{\frac{1}{2}} \tag{1.9.4}$$

$$n_a \sin\alpha_m = \frac{1}{\cos\gamma}[(n_1+n_2)(n_1-n_2)]^{1/2} \tag{1.9.5}$$

when

$$n_a = 1, \sin\alpha_m = \frac{[2n(\Delta n)]^{1/2}}{\cos\gamma} \tag{1.9.6}$$

1.10 ADVANTAGES OF OPTICAL FIBER OVER THE COPPER WIRE SYSTEM

The optical fiber transmission link has certain inherent advantages over the conventional copper wire system in telecommunication applications:

1. Optical fibers have lower transmission losses and wider bandwidth as compared to copper wires. This means that with optical system more data can be sent over the long distance by decreasing the number of wires and reducing the number of repeaters.

2. The light weight and the small hair thin sized dimension of the fibers offer a distinct advantage over heavy bulky cables in crowded underground city ducts. It is also important in an aircraft where small light weight cables are advantageous.
3. An especially important advantageous features of optical fibers relates to their dielectric nature. This provides optical immunity to the electromagnetic interferences, such as inductive peak up from the signal carrying wires and lightning, and freedom from the electromagnetic pulse effects (EMP effects).
4. Silica is the principal material of which optical fibers are made. This material is abundant and inexpensive, since the main source of silica is sand.

SOLVED EXAMPLES

Example 1.1: *A silica glass optical fiber has a core refractive index of 1.500 and the cladding refractive index of 1.450. Calculate:*

(a) the critical angle for the core-cladding interface.

(b) the acceptance angle α_m in air for the fiber and corresponding angle of obliqueness ϕ_m.

(c) the numerical aperture (NA) of the fiber and

(d) the percentage of light collected by the fiber with respect to the incident light (It is given that the core diameter of the fiber is greater than the diameter of the light source).

Solution:

(*a*) According to Eq. (1.6.3), the critical angle θ_c is given by $\sin \theta_c = n_2/n_1$. Substituting the value of $n_1 = 1.500$ and $n_2 = 1.450$, we get

$$\sin \theta_c = \frac{1.450}{1.500} = 0.966$$

or $$\theta_c = \sin^{-1} 0.966 = 75°$$

Critical angle of obliqueness $\phi_m = 90 - \theta_c = 90 - 75° = 15°$ (Fig. 1.7). The critical angle for the core-cladding interface is 75° and corresponding angle of obliqueness ϕ_m is 15°.

(*b*) $n_a \sin \alpha_m = n_1 \sin \phi_m = n_a \sin (90 - \theta_c) = n_1 \cos \theta_c$.

$$\frac{\sin \alpha_m}{\sin \phi_m} = \frac{n_1}{n_a}$$

where $$n_1 = 1.5 \text{ and } n_a \text{ of air} = 1.$$

$$\sin \alpha_m = n_1/n_a \sin \phi_m = 1.5 \sin \phi_m = 1.5 \sin 15° = 1.5 \times 0.26 = 0.39$$

or $$\alpha_m = \sin^{-1} 0.39$$

$$a_m = 23° \text{ (acceptance angle)}$$

The acceptance angle α_m of the fiber is 23°.

(*c*) Numerical aperture, $\text{NA} = [(n_1 + n_1)(n_1 - n_2)]^{1/2}$

$$= [(1.500 + 1.450)(1.500 - 1.450)]^{1/2}$$

$$= [2.95 \times 0.050]^{172} = (0.1475)^{172} = 0.3841$$

The numerical aperture of the fiber is 0.3841.

(*d*) Percentage of light collected $= (\text{NA})^2 \times 100 = 0.1475 = 14.7\%$

14.7 per cent of incident light is collected by the fiber.

Example 1.2: *The numerical aperture of an optical fiber is 0.3. Calculate*

(a) the acceptance angle for the meridional rays and

(b) also that for the skew rays which change direction by 90° at each reflection (assume the refractive index n_a of air is 1).

Solution:

(*a*) When the refractive index of air, $n_a = 1$, the acceptance angle of the meridional rays is given by

Acceptance angle, $\alpha_m = \sin^{-1}(\text{NA}) = \sin^{-1}(0.3) = 17°30'$

The acceptance angle for the meridional rays is 17°30′.

(*b*) The skew rays change direction by 90° at each reflection, there $2\gamma = 90°$ and $\gamma = 45°$

Hence, the acceptance angle for the skew rays is given by

$$\alpha_{sm} = \frac{\sin^{-1}(\text{NA})}{\cos\gamma} = \frac{\sin^{-1}(0.3)}{\cos 45°} = \frac{17°\,30'}{0.707} = 24°$$

It is found that the acceptance angle for the skew rays is 24° which is greater than that for the meridional rays.

Example 1.3: *The refractive index of the core of W-step index fiber is 1.46 and the relative refractive index difference between the core and the cladding of the fiber is 2%. Estimate (a) the numerical aperture (NA), (b) the solid acceptance angle in air for the fiber whose core index is 1.46 and (c) the critical angle at the core-cladding interface within the fiber.*

Solution:

(*a*) $\Delta = \dfrac{n_1 - n_2}{n_1} = 0.02$ and $n_1 = 1.46$

Thus $n_1 - n_2 = 0.02 \times 1.46$ and $n_2 = 1.46 - 0.02 \times 1.46 = 1 - 0.0292$

or $n_2 = 1.43$

$\text{NA} = [(n_1 + n_2)(n_1 - n_2)]^{1/2} = (2.89 \times 0.03)^{1/2} = (0.0867)^{1/2} = 0.29$

The numerical aperture of the fiber is 0.29.

(*b*) For small angle, the solid acceptance angle in air, ϕ, is given by

$$\phi = \pi(\text{NA})^2 = \pi \times 0.0867 = 0.272 \text{ radians.}$$

(*c*) $\sin\theta_c = \dfrac{n_2}{n_1} = \dfrac{1.43}{1.46}$ or $\theta_c = \sin^{-1}\dfrac{1.43}{1.46} = \sin^{-1} 0.98 = 78°$

The critical angle at the core cladding interface is 78°.

REVIEW QUESTIONS

1.1 Describe the structures of different types of optical fibers with ray paths. What is the approximate diameter of an optical fiber in each case?

1.2 Describe a simple block diagram of an optical fiber transmission link and explain the function of each element in the link.

1.3 What do you mean by the "attenuation property" of a fiber? Why does the attenuation of a signal take place in an optical fiber?

1.4 What are the typical wavelengths used in fiber communications in earlier days and also in recent time?

1.5 How was it possible to manufacture the fibers of low loss in the 1.1 to 1.70 μm wavelength region?

1.6 What are the light sources commonly used in fiber communications? Compare them.

1.7 What is the most important factor which made the above sources suitable for the communication system?

1.8 What is the material used for the fabrication of sources having wavelength in the range of 0.8 to 0.9 μm?

1.9 When an optical signal passes through a fiber, it is attenuated and distorted progressively. What are the factors responsible for attenuation and distortion?

1.10 Discuss the curve on attenuation versus wavelength at different times, what are the wavelengths where the attenuation is minimum in early 80?

1.11 What are the devices used for detecting optical signal after it comes out of the fiber? Compare their characteristics.

1.12 What are the main materials used for the fabrication of the detector in the 1.1 to 1.6 μm wavelength region?

1.13 What is an optical fiber repeating system? Describe it briefly.

1.14 What are the advantages of cladding on a fiber?

1.15 What do you mean by single mode and multimode fibers?

1.16 What do you mean by "numerical aperture" and the "acceptance angle" of a fiber? Derive expression for them.

1.17 Determine with the help of a diagram the ratio between the power accepted by a fiber and the optical power impinging on it and give the expression for the ratio.

1.18 Describe briefly with suitable diagrams the difference between the paths of a meridional rays and the skew rays in a step-index fiber. Derive an expression for the acceptance angle for meridional ray. Also derive the expression (acceptance angle) for a skew ray which changes direction by an angle 2Y at each reflection in a step index fiber.

1.19 An optical fiber in air has a numerical aperture of 0.4. Calculate the acceptance angle for (1) meridional rays and also for a skew ray which changes direction by 100° at each reflection (Refractive index of air $n_a = 1$).

1.20 An optical fiber has a numerical aperture of 0.20 and the cladding refractive index of 1.48. Calculate the acceptance angle in water which has a refractive index of 1.33. Calculate the critical angle at the core cladding interface.

1.21 The refractive index difference between the core and cladding of an optical fiber is 5 per cent. Calculate the numerical aperture and solid angle in air for the fiber when the core refractive index is 1.50. Calculate the critical angle at the core-cladding interface.

1.22 Compare the optical fiber over copper wire communication system.

1.23 The core refractive index of a silica optical fiber is 1.5 and the cladding index is 1.47 (*a*) Determine critical angle at core-cladding interface (*b*) Numerical aperture for the fiber and (*c*) acceptance angle in air for the fiber. [78.5°, 0.30 and 17.4°]

1.24 The typical relative refractive index difference for an optical fiber is 1%. Determine the numerical aperture and solid acceptance angle in air for the fiber when the core index is 1.46 and calculate the critical angle at the core-cladding interface. [0.21, 0.13 rad 81.9°]

CHAPTER 2

Electromagnetic Wave Propagation Through Optical Fiber

CHAPTER FOCUS

- *Introduction: Theory of circular waveguide*
- *Phase and group velocity*
- *Maxwell's field equations*
- *Field components in optical waveguide*
- *Wave equations for E_z and H_z*
- *Different modes of propagation of electromagnetic waves*
- *Solution of wave equations for a step index fiber in the core ($r < a$)*
- *Electric and magnetic field components in the core region ($r < a$)*
- *Determination of E_ϕ and H_ϕ in terms of E_z and H_z*
- *Electric and magnetic field components in the cladding ($r > a$)*
- *Determination of E_ϕ and H_ϕ in terms of E_z and H_z in the cladding ($r > a$)*
- *Range of values of phase constant β for guided mode*
- *Modal equations from boundary conditions*
- *Lower order modes in step index fiber*
- *Table for roots of Bassel function $J_n(x)$*
- *V-parameters and number of modes*
- *Design of a singlemode fiber*
- *Relationship between V and number of modes in multimode fiber*
- *Cut off condition of fiber modes*
- *Monomode and multimode step index fibers*
- *Power flow in step index fibers*
- *Weakly guiding mode solutions*
- *Solved examples and review questions*

2.1 INTRODUCTION: THEORY OF CIRCULAR WAVEGUIDE

In order to have a clear understanding of the optical power propagation mechanism in a fiber, it is necessary to solve Maxwell's field equations subject to the cylindrical boundary conditions of the fiber. Figure 2.1 shows a planer dielectric slab waveguide which consists of a dielectric slab of refractive index n_1 sandwiched between two dielectric slab materials each having a refractive index n_2 where n_1 is greater than n_2. This structure represents a simplest form of an optical waveguide. The outer dielectric medium is known as cladding. To understand the wave propagation in an optical fiber, we shall examine qualitatively the appearance of modal electric field in the slab in this slab waveguide. Fig. 2.2 illustrates the electric field patterns of the electromagnetic wave for several lower order modes. These are obtained by solving Maxwell's field equations using appropriate boundary conditions of the slab waveguide. The order of the mode is equal to the electric field maxima across the guide. The number of the modes is equal to the number of electric field maxima across the guide. The order of the mode is related to the angle that the corresponding mode makes with the waveguide (or the axis of the fiber). The steeper the angle, the higher is the order of the mode. The figure shows that the electric fields of the guided modes are not entirely confined within the dielectric slab and are not totally reduced to zero at the core-cladding interface but they extend partially into the cladding. The fields vary sinosoidally at the guiding region of refractive index n_1 but decay exponentially outside this region. For the lower order mode such as TE_1, the field is restricted to the centre of the dielectric slab (or the axis of the fiber) having a small penetration in the cladding region. On the other hand, for the higher order modes (TE_2 and TE_3) the fields are distributed toward the edge of the slab waveguide and penetrate into the cladding region.

In addition to the bound and refracted modes, the third category of the modes are called "leaky modes" which is partially confined in the core region and gets attenuated by continuously radiating optical energy out of the core as they propagate along the fiber. A mode remains bound in the core as long as it satisfies the condition that the phase constant β lies within

$$\frac{2\pi}{\lambda} n_1 \geq \beta \geq \frac{2\pi}{\lambda} n_2 \tag{2.1.1}$$

where n_1 and n_2 are the refractive indices of the core and cladding respectively and λ is the wavelength.

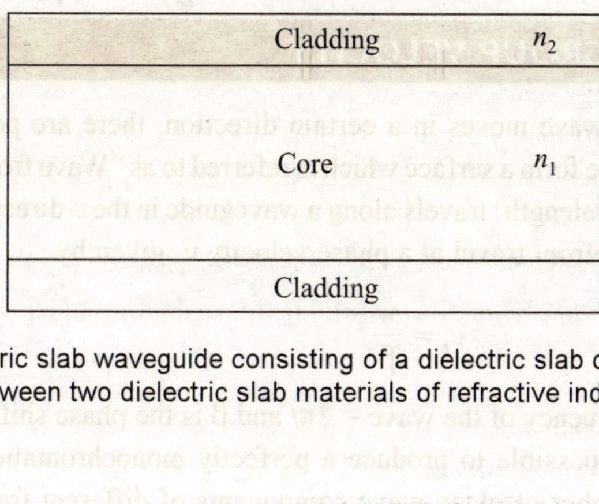

Fig. 2.1 A planer dielectric slab waveguide consisting of a dielectric slab of refractive index n_1 sandwiched between two dielectric slab materials of refractive index n_2 ($n_1 > n_2$).

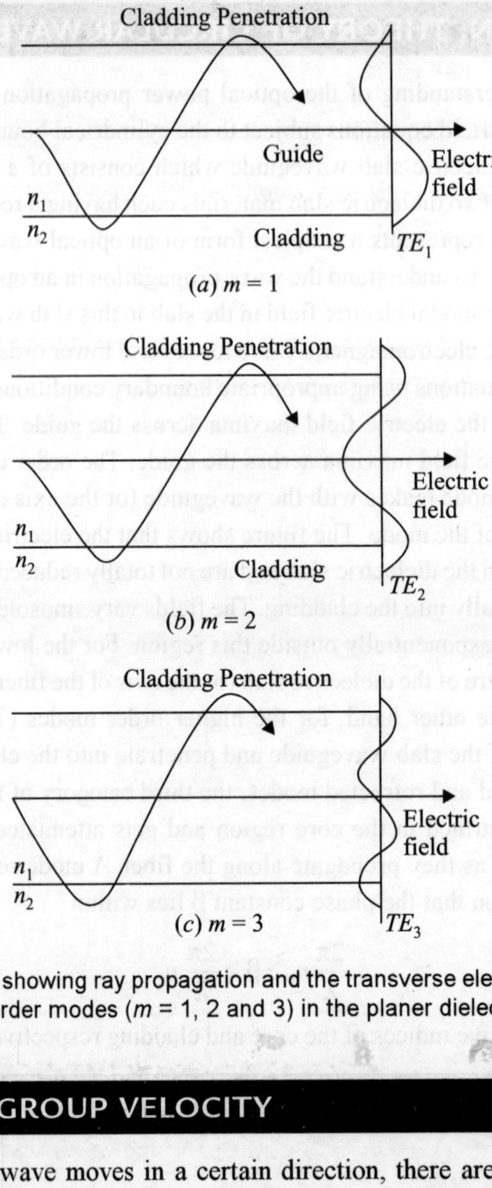

Fig. 2.2 Physical model showing ray propagation and the transverse electric field (TE) patterns of three lower order modes (m = 1, 2 and 3) in the planer dielectric slab waveguide

2.2 PHASE AND GROUP VELOCITY

When an electromagnetic wave moves in a certain direction, there are points of constant phase and these points of constant phase form a surface which is referred to as "Wave front". When a monochromatic wave (wave with single wavelength) travels along a waveguide in the z-direction, these points of constant phase constituting the wavefront travel at a phase velocity v_p given by

$$v_p = \frac{w}{\beta} \tag{2.2.1}$$

where w is the angular frequency of the wave = $2\pi f$ and β is the phase shift constant.

In practice, it is impossible to produce a perfectly monochromatic wave and light is usually composed of a sum of number of plane wave components of different frequencies. When a group of waves with closely similar frequencies travel, their resultant forms a packet of waves. Figure 2.3 illustrates a wavepacket resulting from the combination of two waves of slightly different wavelengths propagating together. This wavepacket does not travel at the phase velocity of the individual waves, but is found to propagate at a group velocity v_g given by

$$v_g = \left(\frac{\delta w}{\delta \beta}\right) \tag{2.2.2}$$

when the propagation of a wave takes place in an infinitely extended medium of refractive index n_1, the phase shift constant β may be written as

$$\beta = \frac{2\pi}{\lambda} n_1 = \frac{2\pi f}{f\lambda} n_1 = \frac{w}{c} n_1 \tag{2.2.3}$$

where c and λ are the velocity and wavelength of the wave in free space respectively f is the frequency of the wave. On combining Eq. (2.2.1) and Eq. (2.2.3) yields

$$v_p = \frac{wc}{wn_1} = \frac{c}{n_1} \tag{2.2.4}$$

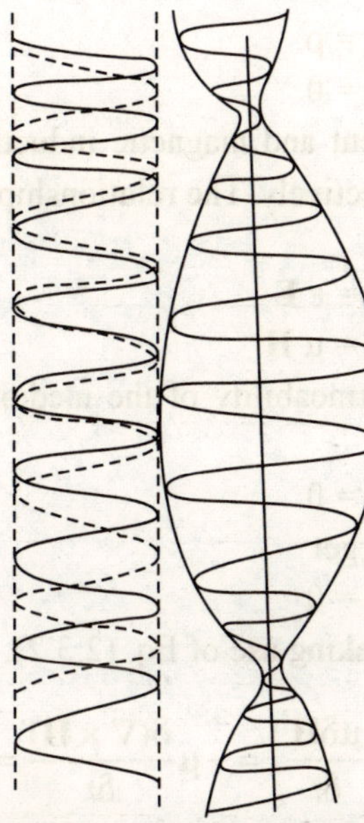

Fig. 2.3 The formation of a wavepackage or group of waves from the combination of two waves with nearly equal frequencies. The envelope of the wavepacket or group of waves travel with a group velocity v_g

Similarly, using Eq. (2.2.4) wherein the limit $(\delta w/\delta\beta)$ becomes $(dw/d\beta)$, the group velocity v_g is given by

$$v_g = \frac{d\lambda}{d\beta}\cdot\frac{dw}{d\lambda} = -\left[\frac{d}{d\lambda}\left(\frac{2\pi}{\lambda}n_1\right)\right]^{-1}\left(\frac{w}{\lambda}\right) = \left(\frac{w}{2\pi\lambda}\right)\left(\frac{1}{\lambda}\cdot\frac{dn_1}{d\lambda} - \frac{n_1}{\lambda^2}\right)^{-1}$$

or $$v_g = \frac{c}{n_1 - \lambda\dfrac{dn_1}{d\lambda}} = \left(\frac{c}{n_g}\right) \tag{2.2.5}$$

The parameters n_g is known as the group refractive index of the medium.

2.3 MAXWELL'S FIELD EQUATIONS

The analysis of the propagation of light wave through an optical fiber waveguide can be made by solving Maxwell's field equations with appropriate boundary conditions. The fiber waveguide consists of non-dissipative cylindrical core of refractive index n_1 and of radius 'a' embedded in a non-dissipative dielectric cladding medium of refractive index n_2 and of radius b, where n_1 is greater than n_2. The core medium is assumed to be a linear, isotropic dielectric medium having no free charges (charge density $\rho = 0$) and no currents (current density vector $J = 0$).

Maxwell's field equations are of the form

$$\nabla \times \mathbf{E} = -\frac{\delta \mathbf{B}}{\delta t} = -\mu \frac{\delta \mathbf{H}}{\delta t} \tag{2.3.1}$$

$$\nabla \times \mathbf{H} = \mathbf{J} + \frac{\delta \mathbf{D}}{\delta t} = \mathbf{J} + \varepsilon \frac{\delta \mathbf{E}}{\delta t} = \varepsilon \frac{\delta \mathbf{E}}{\delta t} \tag{2.3.2}$$

$$\nabla \cdot \mathbf{D} = \rho \tag{2.3.3}$$

$$\nabla \cdot \mathbf{B} = 0 \tag{2.3.4}$$

where D and B are electric displacement and magnetic induction vectors respectively, E and H are electric and magnetic field vectors respectively. The relationship between E and D, and that between H and B are given by

$$\mathbf{D} = \varepsilon\, \mathbf{E} \tag{2.3.5}$$

$$\mathbf{B} = \mu\, \mathbf{H} \tag{2.3.6}$$

where ε and μ are permittivity and permeability of the medium respectively. From Eq. (2.3.3) and Eq. (2.3.5), we get

$$\nabla \cdot \mathbf{E} = 0 \tag{2.3.7}$$

From Eq. (2.3.4) and (2.3.6), we get

$$\nabla \cdot \mathbf{H} = 0 \tag{2.3.8}$$

Taking curl of Eq. (2.3.1) and making use of Eq. (2.3.2), we get

$$\nabla \times (\nabla \times \mathbf{E}) = \nabla \times \left(\frac{\mu \delta \mathbf{H}}{\delta t} \right) = -\mu \frac{\delta (\nabla \times \mathbf{H})}{\delta t} = -\mu \varepsilon \frac{\delta^2 \mathbf{E}}{\delta t^2} \tag{2.3.9}$$

on using the following vector identity and making use of Eq. (2.3.7) yield

$$\nabla \times (\nabla \times \mathbf{E}) = \nabla (\nabla \cdot \mathbf{E}) - \nabla^2 \mathbf{E} = \mu \varepsilon \frac{\delta^2 \mathbf{E}}{\delta t^2} \tag{2.3.10}$$

or

$$\nabla^2 \mathbf{E} = \mu \varepsilon \frac{\delta^2 \mathbf{E}}{\delta t^2} \tag{2.3.11}$$

Similarly, taking curl of Eq. (2.3.2) and making use of Eq. (2.3.1) and Eq. (2.3.6), we get

$$\nabla^2 \mathrm{H} = \mu \varepsilon \frac{\delta^2 \mathbf{E}}{\delta t^2} \tag{2.3.12}$$

Eq. (2.3.11) and Eq. (2.3.12) are the two standard wave equations.

2.4 FIELD COMPONENTS IN OPTICAL WAVEGUIDE

Consider that an electromagnetic wave is propagating along a cylindrical fiber waveguide as shown in Fig. 2.4. In order to make analysis of the propagation of wave in cylindrical fiber waveguide, it is natural to use cylindrical coordinate system (r, ϕ and z) and it is assumed that the z-axis is lying along the axis of the fiber waveguide. When the electromagnetic wave propagates along the direction of z-axis, the electric and magnetic field vectors have components (E_r, E_ϕ and E_z) and (H_r, H_ϕ and H_z) respectively. They have the functional dependence of the form:

$$\mathbf{E} = \mathbf{E}_o(r, \phi)\, e^{j(wt - \gamma z)} = E_o(r, \phi) e^{j(wt - \beta z)} \tag{2.4.1a}$$

$$\mathbf{H} = \mathbf{H}_o(r, \phi)\, e^{j(wt - \gamma z)} = H_o(r, \phi) e^{j(wt - \beta z)} \tag{2.4.1b}$$

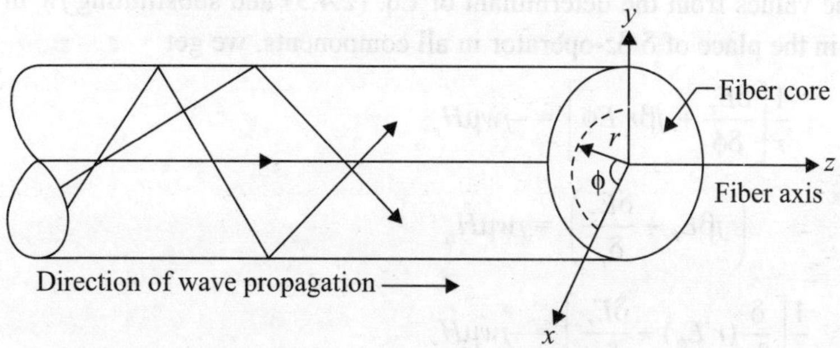

Fig. 2.4 Cylindrical coordinate system (r, ϕ, z) used for making analysis of electromagnetic wave propagating in an optical fiber

where γ is the propagation constant and is given by

$$\gamma = \alpha + j\beta$$

where α and β are the attenuation and phase constants respectively. In a non-dissipative medium (such as the fiber) the attenuation constant α is zero and the angular frequency $w = 2\pi f$ where f is the frequency and $j = \sqrt{-1}$.

From Eq. (2.4.1a) and Eq. (2.4.1b) we can write

$$\frac{\partial \mathbf{E}}{\partial t} = jwE_o(r, \phi)\, e^{j(wt - \beta z)} = jw\mathbf{E} \tag{2.4.2a}$$

$$\frac{\delta \mathbf{H}}{\delta t} = jwE_o(r, \phi)\, e^{j(wt - \beta z)} = jw\mathbf{H} \tag{2.4.2b}$$

$$\frac{\delta \mathbf{E}}{\delta z} = -\, j\beta E_o(r, \phi)\, e^{j(wt - \beta z)} = -\, j\beta\mathbf{E} \tag{2.4.2c}$$

$$\frac{\delta \mathbf{H}}{\delta z} = -\, j\beta H_o(r, \phi)\, e^{j(wt - \beta z)} = -\, j\beta\mathbf{H} \tag{2.4.2d}$$

The above relationships are correct for all six components of the electric and magnetic field vectors. We shall replace $\delta/\delta t$ and $\delta/\delta z$ operators by jw and $-j\beta$ respectively. All components are harmonic in time and coordinate 'z'. The parameter β is the 'z'-component of the propagation vector and the value of β will be determined from the boundry conditions in the electromagnetic fields at the core-cladding interface of the fiber. On transforming Maxwell's first equation (Eq. 2.3.1) in polar coordinate system (r, ϕ and z), we get

$$\begin{bmatrix} \frac{1}{r}\mathbf{a_r} & \mathbf{a}_\phi & \frac{1}{r}\mathbf{a_z} \\ \frac{\delta}{\delta r} & \frac{\delta}{\delta \phi} & \frac{\delta}{\delta z} \\ E_r & rE\phi & E_z \end{bmatrix} = \left[\mu\frac{\delta H_r}{\delta t}\mathbf{a_r} + \mu\frac{\delta H_\phi}{\delta t}\mathbf{a}_\phi + \mu\frac{\delta H_z}{\delta t}\mathbf{a_z}\right] \tag{2.4.3}$$

where a_r a_ϕ and a_z are the unit vectors of E and H in r, ϕ and N-directions respectively.

Writing the values from the determinant of Eq. (2.4.3) and substituting jw in the place of $\delta/\delta t$ operator and $-j\beta$ in the place of $\delta/\delta z$-operator in all components, we get

$$\frac{1}{r}\left[\frac{\delta E_z}{\delta \phi} + j\beta r\, E\phi\right] = -jw\mu H_r \tag{2.4.3a}$$

$$\left(j\beta E_r + \frac{\delta E_z}{\delta_r}\right) = jw\mu H_\phi \tag{2.4.3b}$$

$$\frac{1}{r}\left[\frac{\delta}{\delta r}(r\,E_\phi) - \frac{\delta E_r}{\delta \phi}\right] = -jw\mu H_z \tag{2.4.3c}$$

Adopting the similar procedure with Eq. (2.3.2), we get

$$\frac{1}{r}\left[\frac{\delta H_z}{\delta \phi} + j\beta r H_\phi\right] = jw\varepsilon E_r \tag{2.4.4a}$$

$$\left(j\beta H_r + \frac{\delta H_z}{\delta r}\right) = -jw\varepsilon E\phi \tag{2.4.4b}$$

$$\frac{1}{r}\left[\frac{\delta}{\delta r}(rH_\phi) - \frac{\delta H_r}{\delta}\right] = jw\varepsilon E_z \tag{2.4.4c}$$

Solving Eq. (2.4.3b) and Eq. (2.4.4a), we get

$$E_r = \frac{-j}{w^2\mu\varepsilon - \beta^2}\left[\beta\frac{\delta E_z}{\delta r} + \frac{\mu w}{r}\frac{\delta H_z}{\delta \phi}\right] \tag{2.4.5a}$$

$$H_\phi = \frac{-j}{w^2\mu\varepsilon - \beta^2}\left[\frac{\beta}{r}\frac{\delta H_z}{\delta \phi} + w\varepsilon\frac{\delta E_z}{\delta r}\right] \tag{2.4.5b}$$

Similarly, solving Eq. (2.4.3a) and Eq. (2.4.4b), we get

$$E_\phi = \frac{-j}{w^2\mu\varepsilon - \beta^2}\left[\frac{\beta}{r}\mu\frac{\delta E_z}{\delta \phi} - \mu w\frac{\delta H_z}{\delta r}\right] \tag{2.4.6a}$$

$$H_r = \frac{-j}{w^2\mu\varepsilon - \beta^2}\left[\beta\frac{\delta H_z}{\delta r} - \frac{w\varepsilon}{r}\frac{\delta E_z}{\delta \phi}\right] \tag{2.4.6b}$$

2.5 WAVE EQUATIONS FOR E_z AND H_z

The z-component of wave equation for E_z is represented by component of Eq. (2.3.11)

$$\frac{\delta^2 E_z}{\delta x^2} + \frac{\delta^2 E_z}{\delta y^2} + \frac{\delta^2 E_z}{\delta z^2} = \mu\varepsilon \frac{\delta^2 E_z}{\delta t^2} \tag{2.5.1}$$

Replacing $\left(\frac{\delta^2}{\delta z^2}\right)$ operator by $-\beta^2$ and $\left(\frac{\partial^2}{\partial t^2}\right)$ operator by $-w^2$ in Eq. (2.5.1), we get

$$\frac{\delta^2 E_z}{\delta x^2} + \frac{\delta^2 E_z}{\delta y^2} + (w^2\mu\varepsilon - \beta^2)\, E_z = 0 \tag{2.5.2a}$$

or

$$\frac{\delta^2 E_z}{\delta x^2} + \frac{\delta^2 E_z}{\delta y^2} + k_c^2 E_z = 0 \tag{2.5.2b}$$

where $k_c^2 = w^2\,\mu\varepsilon - \beta^2$.

Eq. (2.5.2b) can be expressed in cylindrical coordinate system as

$$\frac{\delta^2 E_z}{\delta r^2} + \frac{1}{r}\frac{\delta E_z}{\delta r} + \frac{1}{r^2}\frac{\delta^2 E_z}{\delta \phi^2} + k_c^2\, E_z = 0 \tag{2.5.3}$$

Similarly, taking the z-component of the wave equation (2.3.12) and replacing $\left(\frac{\partial^2}{\partial t^2}\right)$ and $\left(\frac{\delta^2}{\partial z^2}\right)$ operators by $-w^2$ and $-\beta^2$ respectively and transforming the result in the cylindrical coordinates, we get

$$\frac{\delta^2 H_z}{\delta r^2} + \frac{1}{r}\frac{\delta H_z}{\delta r} + \frac{1}{r^2}\frac{\delta^2 H_z}{\delta \phi^2} + k_c^2\, H_z = 0 \tag{2.5.4}$$

where $k_c^2 = w^2\,\mu\varepsilon - \beta^2 = K^2 - \beta^2$ where $K^2 = w^2\,\mu\varepsilon$.

2.6 DIFFERENT MODES OF PROPAGATION OF ELECTROMAGNETIC WAVES

The electromagnetic waves propagate in different modes. These are:

1. Transverse electric magnetic (TEM) mode
2. Transverse electric (TE) mode
3. Transverse magnetic (TM) mode
4. Hybrid EH mode
5. Hybrid HE mode.

TEM Mode

When an electromagnetic wave propagates through free space or through a transmission line which may be of parallel wire or coaxial type, the propagation takes place by transverse electric magnetic (TEM) mode. Under this condition, the electric and magnetic fields associated with the wave lie in the plane perpendicular to the direction of propagation and the magnetic and electric fields are at right angle to each other (Fig. 2.5*a*).

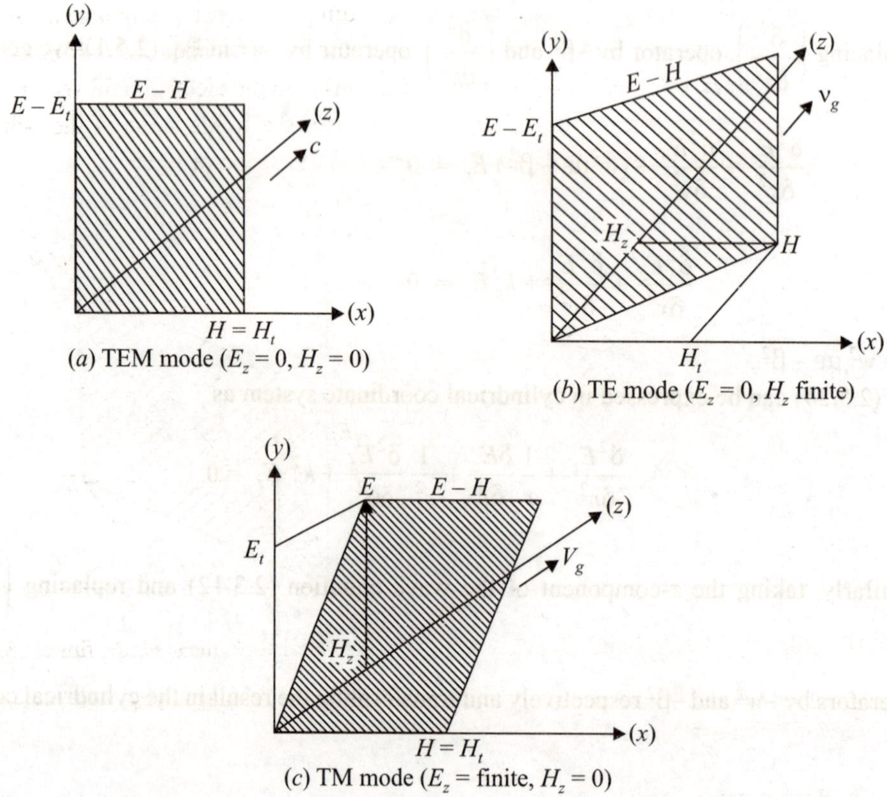

Fig. 2.5 Different conventional modes of propagation of electromagnetic wave: (*a*) TEM mode (*b*) TE mode and (*c*) TM mode.

TE Mode

When an electromagnetic wave is confined within a metallic waveguide which may be either rectangular or cylindrical, it propagates either in TE or TM mode. In the TE mode the electric field vector lies entirely in transverse plane at right angle to the direction of propagation (which is *z*-direction in this case) and the magnetic field vector has a component in the direction of propagation *z*-direction). Under this condition, $E_z = 0$ but H_z is finite. The propagation takes place with the group velocity v_g, while the *E*-*H* plane moves at an angle normal to itself with the speed of light (Fig. 2.5*b*).

TM Mode

In the transverse magnetic (TM) mode the magnetic vector lies entirely in transverse plane (in a plane right angle to 'z'-direction in this example), while electric field components have both transverse component E_t and longitudinal component E_z so that $H_z = 0$ and E_z is finite. The propagation takes place with the group velocity v_g, while *E-H* wavefront travels at an angle normal to itself in the *Y-Z* plane with the speed of light (Fig. 2.5*c*).

Hybrid Modes: (EH and HE Modes)

In hybrid modes (EH or HE mode) both electric field vector *E* and magnetic vector *H* have components E_z and H_z respectively along the *z*-axis so the E_z and H_z are both finite. If the transverse component E_t of the *E* vector is larger than (more prominent) the corresponding transverse component H_t of the *H* vector and as such, the *E* field contributes to the transverse field, the mode is called the *E-H* mode as shown in Fig. 2.6(*a*). If the vector H_t is larger than (more predominant) than the electric field vector E_t. The mode is called the HE mode as shown in Fig. 2.6(*b*). In both cases the fields move in the *z*-direction with group velocity v_g. Both *E-H* plane moves normal to itself in the speed of light.

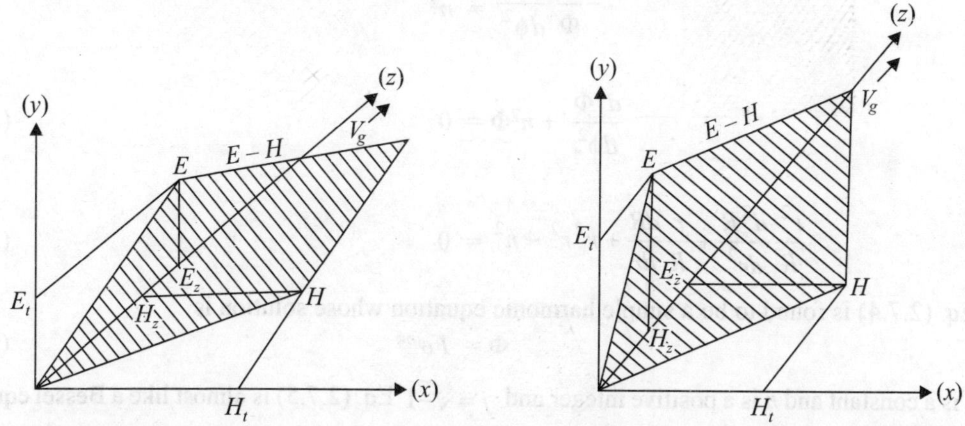

(*a*) HE hybrid mode E_zH_z finite, $H_t > E_t$ (*b*) HE hybrid mode (E_z,H_z finite) ($E_t > H_t$)

Fig. 2.6 Different hybrid modes of propagation of e.m waves in an optical fiber: (*a*) EH mode and (*b*) HE mode.

Meridional Rays and Skew Rays

The propagation of meridional rays takes place only in the TE or TM mode which is completely guided.

The skew rays propagate entirely in the hybrid EH or HE mode and some of these modes contribute to the losses through leakage and radiation.

2.7 SOLUTION OF WAVE EQUATIONS IN A STEP INDEX FIBER

The solution of wave equations can be used to study the propagation of the guided modes in a step index fiber. The separation of variable method is the standard procedure in solving the above equations for E_z and H_z.

We assume that $E_z = R(r)\,\Phi(\phi)$, where $\Phi(\phi)$ is the function of ϕ only and $R(r)$ is the function of *r* only. The time *t* and *z*-dependent factor given by $e^{j(wt-\beta z)}$ is omitted for the time being. Thus,

$$E_z = R(r)\,\Phi(\phi) \tag{2.7.1}$$

On substitution of the value of E_z from Eq. (2.7.1) in Eq. (2.5.3) and dividing of R/r^2 yield

$$\left[\frac{r^2}{R}\frac{d^2R}{dr^2}+\frac{r}{R}\frac{dR}{dr}+k_c^2\,r^2\right]+\left[\frac{1}{\Phi}\frac{d^2\Phi}{d\phi^2}\right]=0 \tag{2.7.2}$$

$$\left[\frac{r^2}{R}\frac{d^2R}{dr^2}+\frac{r}{R}\frac{dR}{dr}+k_c^2 r^2\right]=-\frac{1}{\Phi}\frac{d^2\Phi}{d\phi^2}=n^2 \tag{2.7.3}$$

where n is a constant integer, because the left-hand terms in the parenthesis depend on r only, whereas the right hand term depends on ϕ only but they are equal. Thus both sides are equal to a constant n^2.

Eq. (2.7.3) gives two following equations

$$-\frac{1}{\Phi}\frac{d^2\Phi}{d\phi^2} = n^2$$

$$\frac{d^2\Phi}{d\phi^2} + n^2\Phi = 0 \tag{2.7.4}$$

and

$$\frac{r^2}{R}\frac{d^2R}{dr^2}+\frac{r}{R}\frac{dR}{dr}+k_c^2r^2-n^2 = 0 \tag{2.7.5}$$

Eq. (2.7.4) is found to be a simple harmonic equation whose solution is

$$\Phi = Fe^{jn\phi} \tag{2.7.6}$$

where F is a constant and n is a positive integer and $j=\sqrt{-1}$ Eq. (2.7.5) is almost like a Bessel equation and it would be exactly like one if k_c is replaced by unity. A simple change in variables will transform the equation in the appropriate form.

Let $$R(r) = R_1(k_c r) \tag{2.7.7}$$

on substitution of this in Eq. (2.7.5) leads to

$$\frac{d^2R_1}{d(k_c r)}+\frac{1}{k_c r}\frac{dR_1}{d(k_c r)}+\left(1-\frac{n^2}{(k_c r)^2}\right)R_1 = 0 \tag{2.7.8}$$

This is preciesely a Bessel equation of order n with $k_c r$ as the argument. The solution of this standard equation is

$$R_1(k_c r) = R(r) = PJ_n(k_c r) + QK_n(k_c r) \tag{2.7.9a}$$

where P and Q are the arbitrary constants, $J_n(k_c r)$ and $K_n(k_c r)$ are the Bessel functions of first kind and the modified Bessel function of second kind of order n repsectively and

$$k_c^2 = w^2\mu\varepsilon - \beta^2 = \frac{w^2}{c^2} - \beta^2 = \left(\frac{2\pi}{\lambda}\right)^2 n_1^2 - \beta^2 \tag{2.7.9b}$$

The complete solution of Eq. (2.7.8) for E_z may be written as

$$E_z = R\,\Phi = [PJ_n(k_c r) + Q\,K_n(k_c r)]\,Fe^{jn\phi} \tag{2.7.10}$$

we shall now examine the nature of Bessel function $J_n(x)$ and the modified Bessel function $K_n(x)$. Figure 2.7 shows the variation of $J_n(x)$ with x for the lower values of $n = 0, 1, 2$. Figure 2.8 shows the variation of $K_n(x)$ with x for $n = 0, 1, 2$. It is observed that K_n tends to infinite as x tends to zero.

For the configuration of a step index fiber we consider a homogeneous core of refractive index n_1 and of radius 'a' which is surrounded by an infinite cladding of lower refractive index n_2. The reason for assuming thick cladding is that the guided modes in the core have an exponential decay field outside the core which must have an insignificant value at the outer boundary of the cladding. We have already discussed the field patterns in a slab waveguide of Fig. 2.1 and for lower order modes in Fig. 2.2.

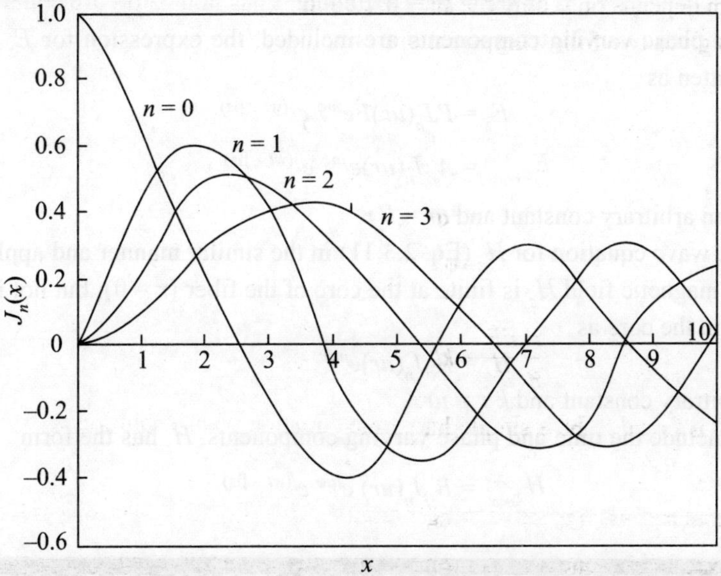

Fig. 2.7 Variation of the Bessel function $J_n(x)$ with x for the first three orders (n = 0, 1, 2, 3, 4).

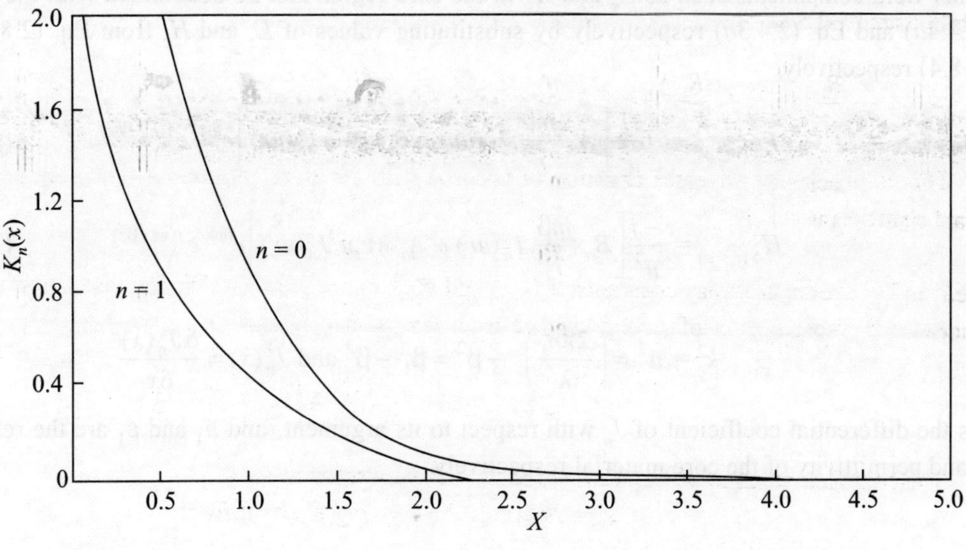

Fig. 2.8 Variation of the Bessel function of second kind $K_n(x)$ with x for n = 0, 1.

2.8 ELECTRIC AND MAGNETIC FIELD COMPONENTS IN CORE REGION ($r < a$)

The modified Bessel functions of second kind $K_n(k_c r)$ are all infinite at the origin where $r = 0$ as apparent from Fig. 2.8. The value of the arbitrary constant Q in Eq. (2.7.10) must be equal to zero if the field E_z is to remain finite at the origin ($r = 0$). Thus, the electric field at the core region where $r < a$ is represented by the Bassel function of first kind and of order n. For this function, we use the commn designation $J_n(ur)$ where

$$k^2c = u^2 = w^2\mu\varepsilon - \beta^2 \tag{2.8.1}$$

when the time and phase varying components are included, the expression for E_z in the core region ($r < a$) may be written as

$$E_z = PJ_n(ur)Fe^{in\phi}\, e^{j(wt - \beta z)}$$

or

$$E_{z(r<a)} = A_o J_n(ur)e^{jn\phi}\, e^{j(wt - \beta z)} \tag{2.8.2}$$

where $A_o = PF =$ an arbitrary constant and $ur = k_c r$.

Solving the wave equation for H_z (Eq. 2.3.11) in the similar manner and applying the boundary condition that the magnetic field H_z is finite at the core of the fiber ($r = 0$), but not infinite, we get the expression for H_z in the core as

$$H_z = B_o J_n(ur)e^{jn\phi} \tag{2.8.3}$$

where B_o is an arbitrary constant and $k_c r = ur$.

When we include the time and phase varying components, H_z has the form

$$H_{z_{r<a}} = B_o J_n(ur)\, e^{jn\phi}\, e^{j(wt - \beta z)} \tag{2.8.4}$$

2.9 DETERMINATION OF E_ϕ AND H_ϕ IN TERMS OF E_z AND H_z

The other field components such as E_ϕ and H_ϕ in the core region can be determined with the help of Eq. (2.4.4*a*) and Eq. (2.4.3*a*) respectively by substituting values of E_z and H_z from Eq. (2.8.2) and Eq. (2.8.4) respectively.

$$E_{\phi\,(r<a)} = -\frac{j}{u^2}\left[A_o \frac{jn\beta}{r} J_n(ur) - B_o\, w\mu u\, J_n'(ur)\right] e^{in\phi}\, e^{j(wt-\beta z)} \tag{2.9.1}$$

$$H_{\phi\,(r<a)} = \frac{-j}{u^2}\left[B_o \frac{jn\beta}{r} J_n(ur) + A_o\, w\varepsilon_1 u\, J_n'(ur)\right] e^{in\phi}\, e^{j(wt-\beta z)} \tag{2.9.2}$$

where $$k_c^2 = u^2 = \left(\frac{2\pi n_1}{\lambda}\right)^2 - \beta^2 = \beta_1^2 - \beta^2 \text{ and } J_n'(x) = \frac{\delta J_n(x)}{\delta x}$$

$J_n'(x)$ is the differential coefficient of J_n with respect to its argument, and n_1 and ε_1 are the refractive index and permittivity of the core material respectively.

2.10 ELECTRIC AND MAGNETIC FIELD COMPONENTS IN CLADDING REGION ($r > a$)

The solution for the field components outside the region of the core can be determined by the modified Bessel function of second kind $K_n(Wr)$, where

$$k_c^2 = W^2 = \beta^2 - \left(\frac{2\pi n_2}{\lambda}\right)^2 = \beta^2 - \beta_2^2 \quad (2.10.1)$$

where n_2 is the refractive index of the cladding region. The reason for assuming $W^2 = \beta^2 - \beta_2^2$ is to make W positive.

The modified Bessel function of the second kind $K_n(Wr)$ for large values of r (*i.e.*, $r >> a$) is represented by

$$K_n(Wr) = \frac{e^{-wr}}{(Wr)^{1/2}} \quad (2.10.2)$$

From Eq. (2.10.2) we infer that $K_n(Wr)$ tends to zero as Wr tends to infinite provided W is a positive real quantity. Thus, the expression for E_z outside the core region (from Eq. 2.7.10)

$$E_{z_{r>a}} = QF\,K_n(Wr)e^{-jn\phi} = C_o K_n(Wr)e^{-jn\phi} \quad (2.10.3)$$

where $QF = C_o =$ an arbitrary constant

When the time and phase varying components are included in the expression for E_z, we get

$$E_{z_{r>a}} = C_o K_n(Wr)e^{-jn\phi}\, e^{j(wt-\beta z)} \quad (2.10.4)$$

Solving Eq. (2.5.4) in the similar manner, applying the boundary condition that the magnetic field H_z does decrease to zero outside the core and including the frequency and phase varying components, we get the expression for H_z outside the core ($r > a$) as

$$H_{z_{r>a}} = D_o K_n(Wr)e^{in\phi}\, e^{j(wt-\beta z)} \quad (2.10.5)$$

2.11 DETERMINATION OF E_ϕ AND H_ϕ IN TERMS OF E_z AND H_z IN THE CLADDING ($r > a$)

The other components such as E_ϕ and H_ϕ in the cladding region (where $r > a$) can be determined with the help of Eq. (2.4.6*a*)

$$E_{\phi_{r>a}} = \frac{j}{W^2}\left[C_o \frac{jn\beta}{r} K_n(Wr) - D_o\, W\mu w K_n'(Wr)\right] e^{jn\phi} e^{j(wt-\beta z)} \quad (2.11.1)$$

and H_ϕ is given by Eq. (2.4.5*b*) and by substituting the values of H_z and E_z from Eq. (2.10.4) and Eq. (2.10.5)

$$H_{\phi_{r>a}} = \frac{j}{W^2}\left[(C_o w\varepsilon_2 WK_n'(Wr) + D_o \frac{jn\beta}{r} K_n(Wr)\right] e^{jn\phi} e^{j(wt-\beta z)} \quad (2.11.2)$$

where $k_c^2 = W^2 = \beta^2 - \beta_2^2 = \beta^2 - \left(\frac{2\pi}{\lambda} n_2\right)^2$. K_n' is the differential coefficient of K_n with respect to its argument so that

$$K_n'(x) = \frac{\delta K_n(x)}{\delta x}$$

ε_1 and ε_2 are the permittivity of the core and cladding materials respectively, and n_2 is the refractive index of cladding. n is a positive integral number.

2.12 RANGE OF VALUES OF PHASE CONSTANT β FOR GUIDED MODE

From the definition of modified Bessel function of second kind $K_n(Wr)$, it is found that

$$K_n(Wr) = \frac{e^{-Wr}}{(Wr)^{1/2}} \tag{2.12.1}$$

tends to zero as Wr tends to infinite. It follow that W must be a real positive quantity. This, in turn, interprets that $\beta \geq K_2$ represents the cut off condition (2.1.1). The cut-off condition is that point at which the mode is no longer bound to the core of the fiber.

A second point can be obtained from the behaviour of $J_n(ur)$ inside the core, u must be real so that $J_n(ur)$ is real. It follows from Eq. (2.9.2) that $K_1 \geq \beta$.

The possible range of β for bound mode is that $n_2 K \leq \beta \leq n_1 K$

or
$$n_2 \frac{2\pi}{\lambda} \leq \beta \leq n_1 \frac{2\pi}{\lambda} \tag{2.12.2}$$

where $K = 2\pi/\lambda$ is the free space phase constant, n_1 and n_2 are the refractive indices of the core and cladding respectively.

2.13 MODAL EQUATIONS FROM BOUNDARY CONDITIONS

The range of values of phase constant β may also be determined from the boundary conditions for the field at the core-cladding interface. The boundary conditions are:

1. the tangential components E_ϕ and E_z of E inside and outside the dielectric interface at $r = a$ must be the same.
2. the tangential components H_ϕ and H_z of H inside and outside the dielectric interface $r = a$ must be the same.

Let us consider first the tangential components of the electric fields inside and the outside the dielectric interface at $r = a$.

For z-components of electric field, we have Eq. (2.8.2) for the inner core-cladding boundary $E_z = E_{z1}$ and Eq. (2.10.3) at the outside of the boundary in the cladding $E_z = E_{z2}$

$$E_{z1} - E_{z2} = A_o J_n(ua) - C_o K_n(Wa) = 0 \tag{2.13.1}$$

Similarly, for the ϕ component of electric field E_ϕ inside and outside the dielectric interface at $r = a$, we have Eq. (2.9.1) and Eq. (2.11.1)

$$E_{\phi 1} - E_{\phi 2} = \frac{j}{u^2}\left[A_o \frac{jn\beta}{a} J_n(ua) - B_o\, W \mu u J_n'(ua) \right]$$

$$- \frac{j}{W^2}\left[C_o \frac{jn\beta}{a} K_n(Wa) - D_o\, W \mu u\, K_n'(Wa) \right] = 0 \qquad (2.13.2)$$

Inside the core $\quad u^2 = K_1^2 - \beta^2 = \left(\frac{2\pi}{\lambda} n_1\right)^2 - \beta^2$

Outside the core $\quad W^2 = \beta^2 - K_2^2 = \beta^2 - \left(\frac{2\pi n_2}{\lambda}\right)^2$

Similarly, the tangential components of magnetic field H_z at the core-cladding interface $r = a$, we can show that z-component from Eq. (2.8.3) and Eq. (2.10.5) as

$$H_{z1} - H_{z2} = B_o J_n(ua) - D_o K_n(Wa) = 0 \qquad (2.13.3)$$

Taking the ϕ component of magnetic fields at the core-cladding interface at $r = a$ from Eq. (2.9.2) and Eq. (2.11.2), we have

$$H_{\phi 1} - H_{\phi 2} = \frac{-j}{u^2}\left[B_o \frac{jn\beta}{a} J_n(ua) + A_o w\, \varepsilon_1\, u\, J_n'(ua) \right]$$

$$- \frac{-j}{W^2}\left[D_o \frac{jn\beta}{a} K_n(Wa) + C_o\, w\, \varepsilon_2\, W\, K_n'(Wa) \right] = 0 \qquad (2.13.4)$$

Eqs. (2.13.1, 2.13.2, 2.13.3 and 2.13.4) are the set of four equations with four unknown coefficients A_o, B_o, C_o and D_o. A solution of these four equations exists only the determinant of these coefficients is zero.

$$\begin{bmatrix} J_n(ua) & 0 & -K_n(Wa) & 0 \\ \frac{\beta n}{au^2} J_n(ua) & \frac{jw\mu}{u} J_n'(ua) & \frac{\beta n K_n(Wa)}{aW^2} & \frac{jw\mu K_n'(Wa)}{W} \\ 0 & J_n(ua) & 0 & -K_n(Wa) \\ -\frac{jw\varepsilon_1}{u} J_n'(ua) & \frac{\beta n}{au^2} J_n(ua) & \frac{-jw\varepsilon_2}{W} K_n'(Wa) & \frac{\beta n}{aw^2} K_n(Wa) \end{bmatrix} = 0 \qquad (2.13.5)$$

The evaluation of the determinant yields the following eigen-value equation for β

$$(I_n + K_n)\,(k_1^2 I_n + k_2^2 K_n) = \left(\frac{\beta n}{a}\right)^2 \left(\frac{1}{u^2} + \frac{1}{W^2}\right)^2 \qquad (2.13.6)$$

where
$$I_n = \frac{J'_n(ua)}{uJ_n(ua)} \tag{2.13.7}$$

and
$$K_n = \frac{K'_n(Wa)}{WK_n(Wa)} \tag{2.13.8}$$

On solving Eq. (2.13.6) for β, it will be found that only the discrete values of β restricted to the range
$$n_2 K \le \beta \le n_1 K \tag{2.13.9}$$
where $K = (2\pi/\lambda)$, will be allowed.

On putting $n_2K = K_2$ and $n_1K = K_1$, Eq. (2.13.9) takes form
$$K_2 \le \beta \le K_1 \tag{2.13.10}$$

Eq. (2.13.6) is a complicated transcendental equation and can be solved by numerical technique. When it is solved, its solution for any particular mode will give all the characteristics of that mode.

2.14 LOWER ORDER MODES IN STEP INDEX FIBER

In the case of a dielectric waveguide, all modes are hybrid ones except for those for which $n = 0$. When $n = 0$, the right hand side of Eq. (2.13.6) vanishes and two different eigen equations result. These are
$$I_o + K_o = 0 \tag{2.14.1}$$

or
$$\frac{J'_o(ua)}{uJ_o(ua)} + \frac{K'_n(Wa)}{WK_o(Wa)} = 0 \tag{2.14.2}$$

Again, $\quad J'_0(x) = -J_1(x)$ and $K'_0(x) = -K_1(x)$

or
$$\frac{J_1(ua)}{uJ_o(ua)} + \frac{K_1(Wa)}{WK_o(Wa)} = 0 \tag{2.14.3}$$

which corresponds to TM_{om} mode ($E_z = 0$) (transverse magnetic mode) and
$$K_1^2 I_o + K_2^2 K_o^2 = 0 \tag{2.14.4}$$

or
$$\frac{K_1^2 J_1(ua)}{uJ_o(ua)} + \frac{K_2^2 K_1(Wa)}{WK_o(Wa)} = 0$$

which corresponds to TE_{om} mode ($H_z = 0$) (transverse electric mode). The electric field patterns in the cross-section of the fiber core for the two $k = 0$ and $m = 1$ modes are shown in Fig. 2.10.

When $n \ne 0$, the situation becomes more complicated and in order solve Eq. (2.13.6) numerical methods are needed. However, when the difference between the core and cladding refractive indices is small [$(n_1 - n_2) \ll n_1$] accurate solution is obtained by Synder[a] and Gloge[b]. The condition $n_1 - n_2 \ll 1$ was referred to by Gloge as giving rise to "Weakly guided modes".

[a] A.W Snyder "Asymptotic expressions for eigenfunctions and eigenvalues of a dielectric or optical waveguide", IEEE Trans. Microwave Theory and Tech. MTT 17 1130–38, 1969.

[b] D. Gloge Weakly guiding Fibers Appl. opt. 10 2252–58, 1971.

2.15 TABLE FOR ROOTS OF BESSEL FUNCTION $J_n(x)$

In order to determine the characteristics of a mode, we shall carefully examine the curve for $J_n(x)$ with x as shown in Fig. 2.7. The Bessel function $J_n(x)$ represents the field amplitude at the core and it is an oscillatory function of x. Because of the oscillatory behaviour of $J_n(x)$, there will be m number of roots of Eq. (2.13.6) for a given value of n and these roots are designated as t_{mn} and the corresponding modes are either TE_{mn} or TM_{mn} or EH_{nm} or HE_{nm}. From Fig. 2.7 the roots are given in Table 2.1.

Table 2.1 For t_{nm}

$t_{01} = 2.405$	$t_{11} = 3.892$	$t_{21} = 6.136$
$t_{02} = 5.520$	$t_{12} = 7.016$	$t_{22} = 8.417$
$t_{03} = 8.654$	$t_{13} = 10.173$	$t_{23} = 11.620$

t_{01}, t_{02} and t_{03} are the first, second and third roots of $J_0(x)$ respectively
t_{11}, t_{12} and t_{13} are the first, second and third roots of $J_1(x)$ respectively
t_{21}, t_{22} and t_{23} are the first, second and third roots of $J_2(x)$ respectively.

2.16 *V*-PARAMETERS AND NUMBER OF MODES

An important parameter associated with the cut off condition is the normalized frequency V which is referred to "V-parameter"

From Eq. (2.8.1) and Eq. (2.10.1), we get

$$u^2a^2 = a^2\left[\left(\frac{2\pi}{\lambda}\right)^2 n_1^2 - \beta^2\right] = a^2[n_1^2 K^2 - \beta^2] = a^2\left[\left(\frac{2\pi}{\lambda}\right)^2 n_1^2 - \beta^2\right] \tag{2.16.1}$$

$$W^2a^2 = a^2\left[\beta^2 - \left(\frac{2\pi}{\lambda}\right)^2 n_2^2\right] = a^2\left[\beta^2 - \left(\frac{2\pi}{\lambda}\right)^2 n_2^2\right] \tag{2.16.2}$$

where $K = 2\pi/\lambda$, n_1 and n_2 are the refractive indices of core and cladding respectively.

By adding Eq. (2.16.1) and Eq. (2.16.2), we get

$$V^2 = a^2[u^2 + W^2] = a^2\left[\left(\frac{2\pi}{\lambda}\right)^2 n_1^2 - \left(\frac{2\pi}{\lambda}\right)^2 n_2^2\right] = a^2(\beta_1^2 - \beta_2^2) \tag{2.16.3}$$

V is an important parameter and associated with the cut off condition. It is also referred to as "V-number" or V-parameter from Eq. (2.16.3)

$$V = \frac{2\pi}{\lambda} a(n_1^2 - n_2^2)^{1/2} = \frac{2\pi a}{\lambda}(NA) \tag{2.16.4}$$

where $(n_1^2 - n_2^2)^{1/2} = NA$ = numerical aperture and a is the radius of the core

Again by putting $\frac{2\pi}{\lambda} = \frac{2\pi f}{f\lambda} = \frac{w}{c}$, we can write from (2.16.4)

$$V = \frac{wa}{c}[n_1^2 - n_2^2]^{1/2} = \frac{w}{w_c} \tag{2.16.5}$$

where $$w_c = \frac{c}{a\left(n_1^2 - n_2^2\right)^{1/2}}$$

Here V is a dimensionless number and determines the number of modes a fiber can support.

Normalized Propagation Function *b*

The number of modes that exists in a fiber waveguide as function of V may be conveniently represented by the normalized propagation function 'b'. The term b is defined by

$$b = 1 - \frac{a^2u^2}{V^2} = \frac{V^2 - a^2u^2}{V^2} = \frac{a^2(u^2 + W^2) - a^2u^2}{V^2} = \frac{a^2W^2}{V^2}$$

From Eq. (2.16.2) and Eq. (2.16.4)

or $$b = \frac{(\beta/K)^2 - n_2^2}{n_1^2 - n_2^2} = \frac{(\beta/K + n_2)\left(\frac{\beta}{K} - n_2\right)}{(n_1 - n_2)(n_1 + n_2)} \tag{2.16.6a}$$

$$b = \frac{(\beta + n_2\ K)\ (\beta - n_2\ K)}{(n_1\ K + n_2\ K)\ (n_1 K - n_2 K)}. \text{ Since } \beta = n_1\ K$$

$$b = \frac{\beta - n_2 K}{n_1 K - n_2 K} = \frac{\frac{\beta}{K} - n_2}{n_1 - n_2} \tag{2.16.6b}$$

Figure 2.9(a) shows a graph of propagation function b as a function of V for a few lower order of modes. The figure shows that each mode can exist only for the values of V that exceeds the limiting value. The mode is cut off when

$$\beta/K = n_2 \tag{2.16.7}$$

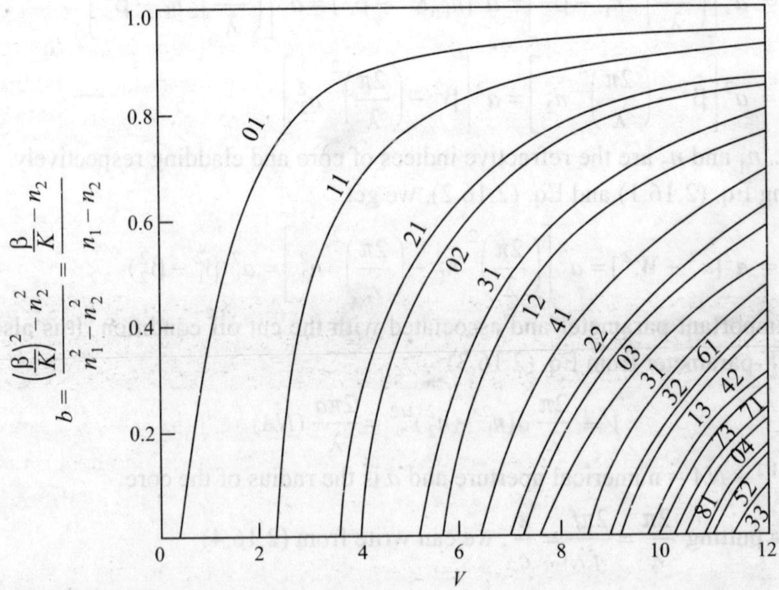

Fig. 2.9 Normalized propagation parameter "b" as a function of V number when $n \neq 1$ the curves numbered nm designate the $HE_{n+1,\,m}$ and $EH_{n-1,\,m}$ modes. For $n = 1$, the curves numbers nm designate HE_{2m} and TE_{om} and TM_{om} modes.

The HE_{11} mode has no cut off and ceases to exist when the core diameter of the fiber is zero. Figure 2.10 shows the electric vector in a fiber core in lowest order modes.

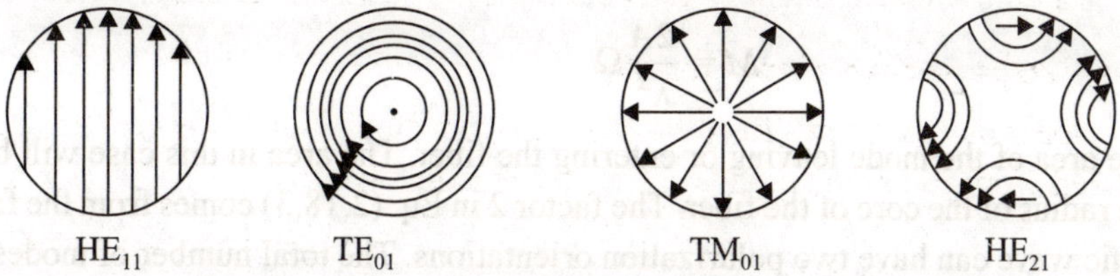

Fig. 2.10 Electric field vectors in the fiber core for a few lower order mode of a Si fiber.

2.17 DESIGN OF A SINGLEMODE FIBER

The basic principle on which a singlemode fiber is designed is to choose the appropriate values of core radius a and the refractive indices n_1 and n_2 of the core and cladding respectively, so

$$V = \frac{2\pi}{\lambda} a[n_1^2 - n_2^2]^{1/2} = t_{01} = 2.405 \qquad (2.17.1)$$

The value of $x = 2.405$ is the first root lowest order Bessel function $J_o(x) = 0$. In this condition all modes except the HE_{11} mode are cut off. The single mode fibers are fabricated by making the dimension of the core diameter to be a few wavelengths (usually 8 to 12) and by having a small index difference between the core and cladding indices. When $V = 2.4$ or less a single mode (HE_{11}) propagation is possible for fairly large variations in the value of the core radius a and the core cladding index difference. However, in the partial design of a singlemode fiber, the core and cladding index difference varies between 0.1 and 0.2 per cent and the core diameter is chosen to be just below the cut off of the first higher mode. The value of V should taken to be less than 2.4. For example, a typical singlemode fiber has a core radius of 3 μm, and the numerical aperture of 0.1 at the wavelength of 0.8 μm.

$$V = \frac{2\pi}{\lambda} a(NA) = \frac{2\pi \times 3.00 \times 10^{-6}}{0.8 \times 10^{-6}} \times 0.1 = 2.356.$$

2.18 RELATIONSHIP BETWEEN *V* AND NUMBER OF MODE IN MULTIMODE FIBER

The parameter V is also related to the number of modes M in a multimode fiber when M is quite large. An approximate relationship can be derived from the ray theory. A ray incident on one end of a step index fiber will be accepted by the fiber if it lies within the angle α_m defined by the numerical aperture.

Numerical aperture, $NA = \sin \alpha_m = (n_1^2 - n_2^2)^{1/2}$ (2.18.1)

In practical case, the value of numerical aperture NA is so small that $\sin \alpha_m \cong \alpha_m$. The solid angle for the fiber is therefore

$$\Omega = \pi \alpha_m^2 = \pi (n_1^2 - n_2^2) \qquad (2.18.2)$$

The wavelength λ of an electromagnetic wave emerging out from a laser diode or waveguide the number of modes per unit solid angle M is given by

$$M = \frac{2A}{\lambda^2}\Omega \tag{2.18.3}$$

where A is the area of the mode leaving or entering the fiber. The area in this case will be equal to πa^2 where a is the radius of the core of the fiber. The factor 2 in Eq. (2.18.3) comes from the fact that a plane electromagnetic wave can have two polarization orientations. The total number of modes M entering in the fiber is obtained from Eq. (2.18.2) and Eq. (2.18.3) as

$$M = \frac{2\pi a^2}{\lambda^2}\pi(n_1^2 - n_2^2) = \frac{V^2}{2} \tag{2.18.4}$$

2.19 CUT OFF CONDITION OF FIBER MODES

A mode is referred to as cut off when it is no longer bound to the core of the fiber and its fields no longer decay outside the core. The cut off conditions for different modes are found by solving Eq. (2.10.4) and it is very complicated. The results are given in Table 2.2 in the lower limit where $w \to 0$ and $\beta = \beta_2$.

Table 2.2

n	*mode*	*Cutoff condition*	*Value of ua*
0	TE_{om}, TM_{om}	$J_{om}(ua) = 0$	$ua = t_{om}$
1	HE_{1m}, EH_{1m}	$J_{1m}(ua) = 0$	$ua = t_{1m}$
2	EH_{nm}	$J_{nm}(ua) = 0$	$ua = t_{nm}$
	HE_{nm}	$\left(\frac{n_1^2}{n_2^2}+1\right) J_{n-1,m}(ua)$	$= \frac{ua}{n-1} J_{nm}(ua)$

Upper Limit of a Mode

At high frequency as $w \to \infty$, $\lambda \to 0$, and it is found that the values of ua approaches the following limiting values so as to make the field vanish as beyond the core-cladding interface.

Table 2.3

n	*mode*	*Cutoff condition*	*Value of ua*
0	TE_{om}, TM_{om}	$J_{1m}(ua) = 0$	$ua = t_{1m}$
	HE_{nm}	$J_{n-1,m}(ua) = 0$	$ua = t_{n-1,m}$
For	EH_{nm}	$J_{n+1,m}(ua) = 0$	$ua = t_{n+1,m}$

In each case u/V approaches zero as w tends to infinite so that $\beta = \beta_1$.

2.20 MONOMODE AND MULTIMODE STEP INDEX FIBERS

It was already mentioned that step index fiber has a core of refractive index n_1 and a cladding of slightly lower refractive index n_2. Mathematically, the refractive index profile of a step index fiber may be expressesd as:

$$\text{Refractive index } n(r) = n_1 \text{ in the core where } r < a \tag{2.20.1}$$

$$n(r) = n_2 \text{ in the cladding where } r > a \tag{2.20.2}$$

where a is the radius of the fiber core and r is the radial distance from the centre of the core of the fiber. Figure 2.11(a) shows a multimode step index fiber of the core diameter of the order of 50 μm or more, which is sufficient to allow the propagation of signals in many modes in the fiber core. This is illustrated in Figure 2.11(a) where there are many ray paths through the fiber core. Figure 2.11(b) shows a singlemode step index fiber which only allows the propagation of only one transverse electromagnetic mode typically HE_{11} and its core diameter must be of the order of 1 to 10 μm. The propagation of signals in singlemode, as illustrated in Fig. 2.11(b), corresponds to a single axial ray path only through the axis of the fiber core. The advantage of a singlemode step index fiber is that it has low value of intermodal dispersion because only one mode is transmitted. But in the case of a multimode fiber considerable amount of dispersion may occur due to different group velocities due to different propagating modes. This restricts the maximum bandwidth attainable with a step index multimode fiber when compared with the singlemode fiber. However, for lower bandwidth applications the multimode fiber has several advantages over the singlemode fibers. These are:

1. Most light emitting diodes cannot be efficiently coupled with the singlemode fiber because of low core diameter.
2. Larger core diameter of a multimode fiber facilitates easier coupling to the optical source.
3. Lower tolerance is required on fiber connector.

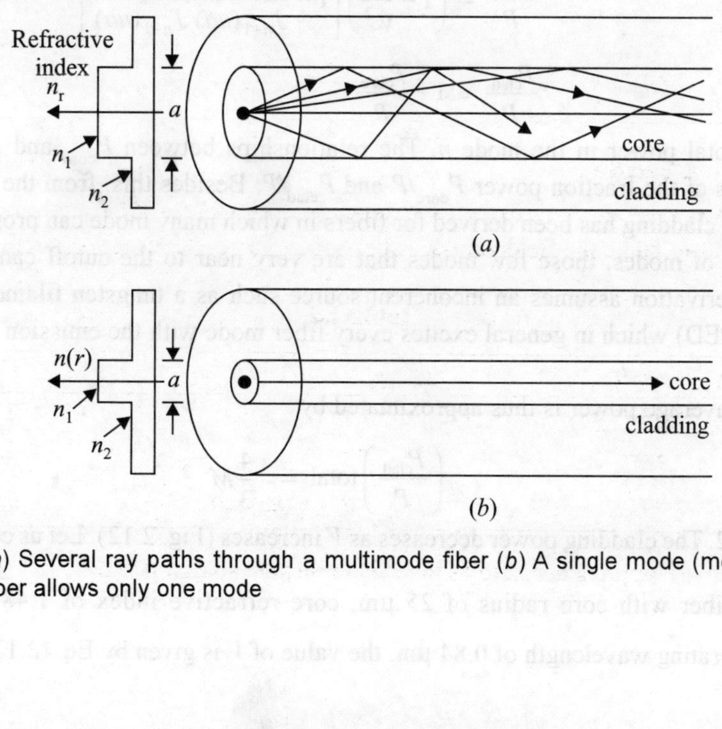

Fig. 2.11 (a) Several ray paths through a multimode fiber (b) A single mode (monomode) fiber allows only one mode

2.21 POWER FLOW IN STEP INDEX FIBERS

As shown in Fig. 2.2 the electric and magnetic fields in a slab waveguide do not reduce to zero at the core-cladding interface. The change takes place from an oscillatory in the core to the exponential decay in the cladding. The electromagnetic energy of the guided mode is carried partially in the core and partially in the cladding. When a mode is further away from its cutoff frequency the energy is more concentrated in its core. As the cutoff frequency is approached, the field penetrates further into the cladding region and a greater percentage of energy travels in the cladding. At the cutoff, the field no longer decays outside the core and the mode becomes a fully radiating mode. The relative amount of power flowing in the core and cladding can be found by integrating the pointing vector S in the direction of the fiber axis (z-direction in this case).

$$\text{Pointing vector } S = 1/2\, R_e\,(E \times H^*)\boldsymbol{a}_z \tag{2.21.1}$$

a_z is the unit vector in the z-direction over the fiber cross-section. Thus the power flow in the core P_{core} and that in the cladding P_{clad} are given by

$$P_{\text{core}} = \frac{1}{2}\int_0^a \int_0^{2\pi} r\,(E_x H_y^* - E_y\, H_x^*)\cdot d\phi\; dr\; \boldsymbol{a}_z \tag{2.21.2}$$

$$P_{\text{clad}} = \frac{1}{2}\int_a^\infty \int_0^{2\pi} r\,(E_x H_y^* - E_y H_x^*)\cdot d\phi\; dr\; \boldsymbol{a}_z \tag{2.21.3}$$

where * denotes the complex conjugate.

When $(n_1 - n_2) << n_1$, the weakly guided mode approximation is done, the relative core and cladding power for a particular mode M are given by

$$\frac{P_{\text{core}}}{P} = \left(1 - \frac{u^2}{V^2}\right)\left[1 - \frac{J_n^2(ua)}{J_{n+1}(ua)\, J_{n-1}(ua)}\right] \tag{2.21.4}$$

and

$$\frac{P_{\text{clad}}}{P} = 1 - \frac{P_{\text{core}}}{P} \tag{2.21.5}$$

where P is the total power in the mode n. The relationships between P_{core} and P_{clad} are shown in Fig. 2.12 in terms of the fraction power P_{core}/P and P_{clad}/P. Besides this, from the cutoff the average total power in the cladding has been derived for fibers in which many mode can propagate. Since there are large number of modes, those few modes that are very near to the cutoff can approximately be neglected. The derivation assumes an incoherent source such as a tungsten filament lamp or a light emitting diode (LED) which in general excites every fiber mode with the emission of same amount of power.

The total average power is thus approximated by

$$\left(\frac{P_{\text{clad}}}{P}\right)\text{total} = \frac{4}{3}M^{-\frac{1}{2}} \tag{2.21.6}$$

M is equal to $V^2/2$. The cladding power decreases as V increases (Fig. 2.12). Let us consider an example of a stepindex fiber with core radius of 25 μm, core refractive index of 1.48 and $\frac{n_1 - n_2}{n_1} = \Delta$ = 0.01. At an operating wavelength of 0.84 μm, the value of V is given by Eq. (2.17.1).

$$V = \frac{2\pi a}{\lambda} n(2\Delta)^{1/2} = \frac{2\pi \times 25 \times 10^{-6}}{0.84 \times 10^{-6}} \times 1.48 \times (2 \times 0.01)^{1/2}$$

$$= \frac{2\pi \times 25}{0.84} \times 1.48\,(2 \times 0.01)^{1/2} = 39$$

and $$M = \frac{V^2}{2} = \left(\frac{39}{2}\right)^2 = 760 \text{ mode}$$

From Eq. (2.21.6), we find

$$\left(\frac{P_{\text{clad}}}{P}\right) = \frac{4}{3} \frac{1}{(760)^{1/2}} = 0.05$$

We calculate that 5 per cent of the power propagate in the cladding. If Δ is decreased to, say, 0.003, then 242 modes propagate in the fiber, then we can calculate to find that 9 per cent of power propagate in the cladding. For the case of a single mode fiber propagating in HE_{11} mode, it is seen in Fig. 2.12 that for $V = 1$, about 70 per cent of power propagates in the cladding, whereas for $V = 2.405$, which is where the TE_{01} mode begins, approximately 84 per cent of the optical power is now within the core.

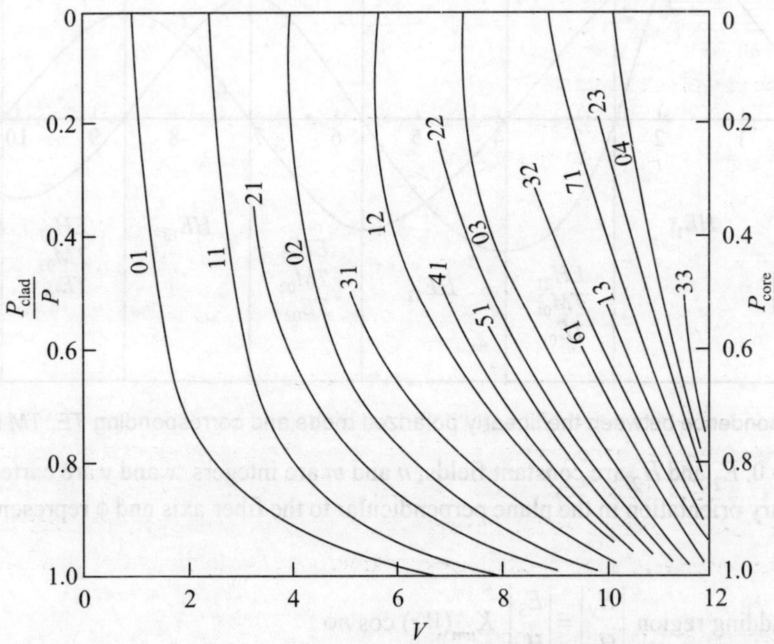

Fig. 2.12 Variation of fractional power P_{core}/P and P_{clad}/P with V number

2.22 WEAKLY GUIDING MODE SOLUTIONS

When the difference between the refractive indices between the core and cladding is very small, a slightly simpler expressions for the mode field patterns can be obtained. In that case $(n_1 - n_2) << n_1$ where

n_1 and n_2 are the refractive index of core and cladding of the fiber. Fibers satisfying this condition are known as "Weakly guiding fibers". Then the electromagnetic field patterns and the propagation constants for the mode pairs $HE_{n+1,m}$ and $EH_{n-1,m}$ are very similar, and those of three modes in each TM_{om}, TE_{om} and HE_{2m} group are also similar. This is illustrated in Fig. 2.13 for $(n, m) = (0, 1)$ and $(2, 1)$. The differences within each pair such as TE_{01} and TM_{01} or group such as HE_{31} and EH_{11} reduces to zero as $\frac{n_1 - n_2}{n_1} \to 0$. This is the weakly guiding limit. Then combining such groups of modes together gives rise to sets of linearly polarized waves in which the transverse electric and magnetic fields are each parallel and mutually orthogonal (at right angle) over the whole fiber cross-section. In view of this fact, the fields can be best expressed in terms of cartesian field components E_x, E_y, H_x, H_y

$$\text{Within the core, } \begin{vmatrix} E_y \\ H_x \end{vmatrix} = \begin{vmatrix} E_2 \\ H_2 \end{vmatrix} J_{n,m}(ur) \cos n\phi \qquad (2.22.1)$$

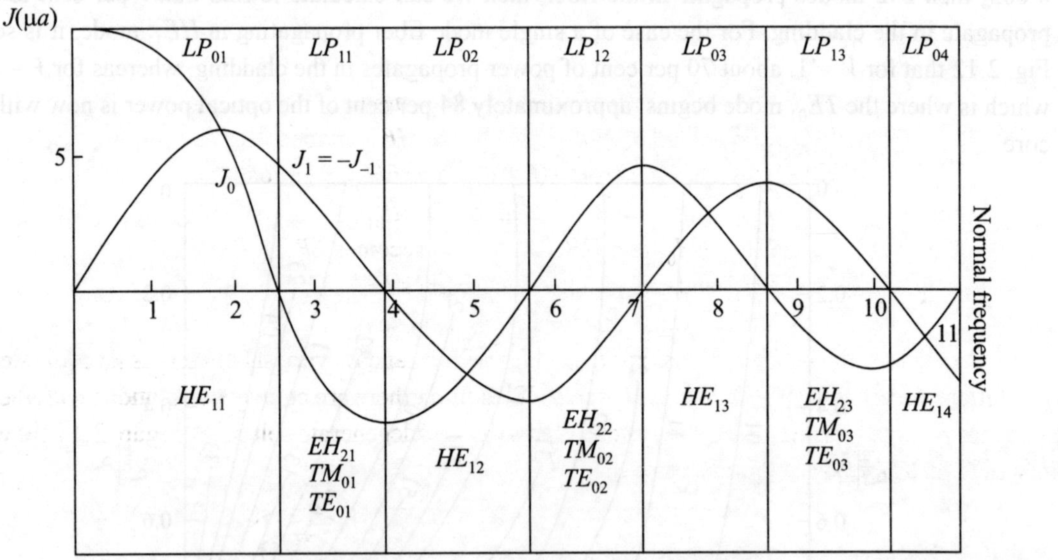

Fig. 2.13 Correspondence between the linearly polarized mode and corresponding *TE*, *TM* modes in a fiber

where $H_x = H_y = 0$. E_2 and H_2 are constant fields, *n* and *m* are integers, *x* and *y* are cartesian coordinate system of arbitrary orientation in the plane perpendicular to the fiber axis and ϕ represent the azimuthal angle.

$$\text{In the cladding region } \begin{vmatrix} E_y \\ H_x \end{vmatrix} = \begin{vmatrix} E_3 \\ H_3 \end{vmatrix} K_{nm}(Wr) \cos n\phi \qquad (2.22.2)$$

Again $E_x = H_y = 0$, E_3 and H_3 are constant fields *n* and *m* are integers.

The (n, m) mode is derived by adding together the $n - 1$ and $n + 1$ solution for E_z and H_z. Thus in the core E_z and H_z are given by

$$E_z = E_o[J_{n-1}(ur) \cos (n-1)\phi] + [J_{n+1}(ur) \cos (n+1)\phi] \qquad (2.22.3)$$

We use the recursion formula of Bessel function as

$$J_{n-1}(ur) + J_{n+1}(ur) = \frac{2n}{ur} J_n(ur) \tag{2.22.4}$$

In the case of H_z

$$H_z = H_o[J_{n-1}(ur) \cos (n-1)\,\phi + J_{n+1}(ur) \cos (n+1)\phi] \tag{2.22.5}$$

and

$$J_{n-1}(ur) + J_{n+1}(ur) = \frac{2n}{ur} J_n(ur) \tag{2.22.6}$$

Instead of deriving the polar components (E_ϕ, E_r, H_ϕ and H_r) of the transverse fields, the Cartesian components (E_y, E_x, H_y and H_x) are obtained by means of transformations $x^2 + y^2 = r^2$ and $y/x = \tan(\phi)$. It is then found that combining two solutions in this way causes the E_x and H_y field components to be almost cancelled out when the change of refractive index at the core-cladding boundary is small. It is also found that the longitudinal field components E_z and H_z are much smaller than the main transverse components of the electromagnetic wave solutions. They are referred to as linearly polarized model with LP_{nm} modes.

In general

each LP_{om} mode is derived from an HE_{1m} mode
each LP_{1m} mode from a TE_{om}, TM_{om} and HE_{2m} mode
each LP_{nm} mode $n \geq 2$ from $HE_{(n+1),\,m}$ and $EH_{(n-1),\,m}$ mode

Thus HE_{11} .... becomes LP_{01} (2.22.7)
TE_{01}, TM_{01} and HE_{21} .... becomes LP_{11}
EH_{11} and HE_{31} .... becomes LP_{21}
HE_{12}............. becomes LP_{02} and so on.

As well as, the set of solutions for E_y and H_x with E_x and H_y very small there is an equivalent set of solutions for E_x and H_y with E_y and H_x small. In addition there are equivalent solutions with the field polarities reversed, making each LP_{nm} mode comprise four degenerate solutions. Figure 2.14 shows the four field distributions of the LP_{11} mode.

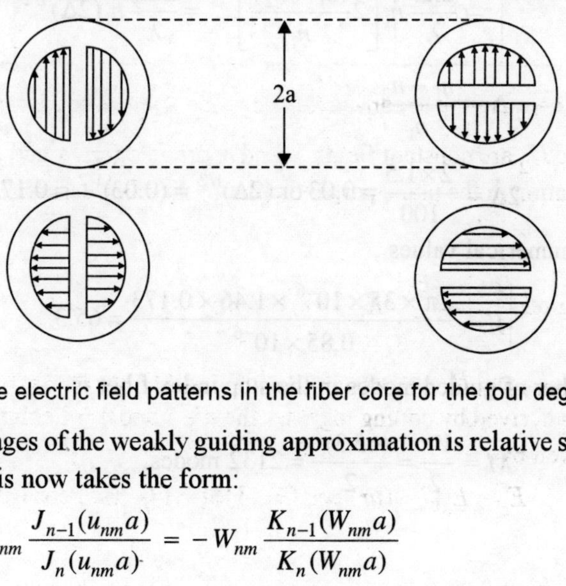

Fig. 2.14 Transverse electric field patterns in the fiber core for the four degenerate LP_{11} modes

One of the advantages of the weakly guiding approximation is relative simplicity of the boundary condition that results. This now takes the form:

$$u_{nm} \frac{J_{n-1}(u_{nm}a)}{J_n(u_{nm}a)} = -W_{nm} \frac{K_{n-1}(W_{nm}a)}{K_n(W_{nm}a)} \tag{2.22.8}$$

In order to determine u_{nm} and W_{nm}, a numerical solution is required and hence β_{nm} for allowed guided mode can be determined.

SOLVED EXAMPLES

Example 2.1: *A multimode fiber has a core diameter of 70 μm and the relative refractive index difference of 1.5 per cent. It operates at the wavelength of 0.85mm. The refractive index of the fiber is 1:46*

Calculate (a) the refractive index of the cladding

(b) the normalized frequency V-number of the fiber, and

(c) the total number of guided modes in the fiber.

Solution:

$$a = \frac{70}{2} = 35\ \mu\text{m},\ \Delta = \frac{n_1 - n_2}{n_1} = 1.5\%,\ n_1 = 1.46$$

$$\lambda = 0.85\ \mu\text{m}.$$

$$(a)\ \Delta = \frac{n_1 - n_2}{n_1} = \frac{1.5}{100} = 0.015$$

or $$n_1 - n_2 = 0.015\, n_1$$

or $$n_2 = n_1 - 0.015 n_1 = 1.46\,(1 - 0.015) = 1.438$$

(*b*) The expression for *V*-number is given by (Eq. 2.16.4)

$$V = \frac{2\pi a}{\lambda}[n_1^2 - n_2^2]^{1/2} = \frac{2\pi a}{\lambda}[(n_1 + n_2)(n_1 - n_2)]^{1/2}$$

$$= \frac{2\pi a}{\lambda}(2n_1)^{1/2}(n_1 - n_2)^{1/2} \quad [\text{Putting } (n_1 + n_2 = 2n_1)]$$

$$= \frac{2\pi a}{\lambda} n_1 \left[2\frac{(n_1 - n_2)}{n_1}\right]^{1/2} = \frac{2\pi a}{\lambda} n_1 (2\Delta)^{1/2}$$

where $$\Delta = \frac{n_1 - n_2}{n_1}$$

$$2\Delta = \frac{2 \times 1.5}{100} = 0.03 \text{ or } (2\Delta)^{1/2} = (0.03)^{1/2} = 0.173$$

Substituting the numerical values

$$V = \frac{2\pi \times 35 \times 10^{-6} \times 1.46 \times 0.173}{0.85 \times 10^{-6}} = 65.3$$

(*c*) The total number of guided modes in the step index fiber is

$$M = \frac{V^2}{2} = \frac{(65.3)^2}{2} = 2132 \text{ modes.}$$

Example 2.2: *The core refractive index of a step index fiber is 1.48 and the relative index difference between the core and cladding is 1.5 per cent. The fiber is operating at an wavelength of 0.85 μm. Calculate*

(a) the core diameter of the step index fiber for single mode operation.

(b) the new maximum core diameter for single mode operation when the difference in refractive index between the core and cladding is 0.15 per cent.

Solution: (*a*) The value of the V-parameter for single mode operation is given by (Eq. 2.17.1)

$$V = t_{01} = 2.405$$

$$2\Delta = \frac{2\times 1.5}{100} = 0.03 \text{ or } (2\Delta)^{1/2} = (0.03)^{1/2} = 0.173$$

$$n_1 = 1.48 \quad \lambda = 0.85 \times 10^{-6} \text{ 'm'}$$

The radius a obtained from Eq. (2.16.4) as

$$a = \frac{V\times\lambda}{2\pi n_1 (2\Delta)^{1/2}} = \frac{2.405\times 0.85\times 10^{-6}}{2\pi\times 1.48\times 0.173} = 1.3\ \mu m$$

The maximum core diameter for single mode operation is 2.6 μm

(*b*) $$2\Delta = 2\times\frac{0.15}{100} = (0.003) \text{ or } (2\Delta)^{1/2} = (0.003)^{1/2} = 0.054$$

Radius 'of' the core, $$a = \frac{V\lambda}{2\pi n_1 (2\Delta)^{1/2}} = \frac{2.405\times 0.85\times 10^{-6}}{2\times 3.14\times 1.48\times 0.054}$$

or $$a = 4\times 10^{-6} \text{ m} = 4\ \mu\text{m}$$

The maximum core diameter for single mode operation is 8μm.

Example 2.3: *The core and cladding refractive indices of a step index fiber are 1.6 and 1.44 respectively. An electromagnetic wave having an wavelength of 0.8μm is propagating through the fiber in guided mode through the core of the fiber. Find out the maximum and minimum values of phase constant β.*

Solution: The mode remains bound to the core when the propagation constant *b* satisfies the following condition

$$\frac{2\pi}{\lambda} n_1 \geq \beta \geq \frac{2\pi}{\lambda} n_2$$

The maximum value of $\beta = \frac{2\pi}{\lambda} n_1 = \frac{2\pi}{0.8} \times 1.6 = 12.56$ rad/μm

The minimum value of $\beta = \frac{2\pi}{\lambda} n_2 = \frac{2\pi}{0.8} \times 1.44 = 11.33$ rad/μm.

Example 2.4: *A single mode step index fiber is operating in the guided mode. The core refractive index and radius are 1.46 and 5 μm respectively. The refractive index difference between the core and cladding is 0.25 per cent. Calculate the cutoff wavelength of the fiber.*

Solution: The cutoff value of V-parameter for the single mode operation is

$$V_c = 2.405 = t_{01},\ a = 5 \times 10^{-6} \text{ m},\ n_1 = 1.6.$$

$$(2\Delta)^{1/2} = \left(2\frac{n_1 - n_2}{n_1}\right)^{1/2} = \left(\frac{2 \times 0.25}{100}\right)^{1/2} = 0.005 \text{ and } (2\Delta)^{1/2}$$

$$= (0.005)^{1/2} = 0.071$$

The cutoff wavelength

$$\lambda_c = \frac{2\pi a}{V_c} n_1 (2\Delta)^{1/2} = \frac{2 \times 3.14 \times 5 \times 10^{-6} \times 1.46 \times 0.071}{2.405}$$

$$\lambda_c = \frac{2 \times 3.14 \times 5 \times 10^{-6} \times 1.46 \times 0.071}{2.405} = 1.48\ \mu\text{m}$$

The cutoff wavelength for the single mode step index fiber is 1.48 μm.

Example 2.5: *A fiber has a core of radius 30 mm. The core and cladding refractive indices are 1.50 and 1.49 respectively. The operating wavelength is 0.85μm. Determine (a) the value of total number of guided mode and (b) the ratio of power flow in the core and cladding.*

Solution: The number of guided modes M is given by

$$M = \frac{1}{2}V^2 = \frac{1}{2}\frac{(2\pi a)^2}{\lambda^2}[n_1^2 - n_2^2] = \frac{2\pi^2 a^2 (n_1^2 - n_2^2)}{\lambda^2}$$

$$M = \frac{2\pi^2 \times (30 \times 10^{-6})^2}{(0.85)^2 \times 10^{-12}}[(1.5 + 1.49)(1.5 - 1.49)]$$

$$= \frac{2\pi^2 \times (30)^2 \times 10^{-12}}{(0.85)^2 \times 10^{-12}}(2.99 \times 0.01) = \frac{2\pi^2 \times 900 \times 0.299}{(0.85)^2}$$

$$= 734 \text{ modes}$$

According to Eq. (2.21.5)

$$\frac{P_{\text{clad}}}{P} = \frac{4}{3}M^{-1/2} = \frac{4}{3}(734)^{1/2} = \frac{40}{795}$$

$$\frac{P_{\text{core}}}{P} = 1 - \frac{P_{\text{clad}}}{P} = 1 - \frac{40}{795} = \frac{755}{795}$$

$$\frac{P_{\text{core}}}{P_{\text{clad}}} = \frac{755}{40} = 18.9$$

REVIEW QUESTIONS

2.1 Write down the Maxwell's field equations for a nondissipative, linear and isotropic dielectric medium with no free charge or current in it. State fundamental laws and theorems from which these equations are obtained. Express the relationship between electric field vector E and displacement vector D and also between magnetic field vector H and magnetic induction vector B.

2.2 Express Maxwell's field equations in Cylindrical coordinate system and find out the values of E_r, E_ϕ, H_r and H_ϕ in terms of E_z and H_z when the electromagnetic wave is propagating in the z-direction.

2.3 Starting from the Maxwell's field equation, derive the wave equation for E_z and H_z for the wave travelling in the z-direction.

2.4 Explain with relevant diagrams the different modes of an electromagnetic wave in an optical fiber. What do you mean by "hybrid mode"? Explain with ray diagram.

2.5 Determine the solution for the wave equation for E_z and H_z in a step index fiber by the solution of partial differential equation method. Determine the value of E_z and H_z in the core and the cladding region. With the help of Maxwell's equations in the cylindrical coordinate system, determine the values of E_r, E_ϕ, H_r and H_ϕ.

2.6 State the range of values of phase constant β for the meridional rays in terms of the refractive indices of the core and cladding of the fiber.

2.7 Write down the modal equations and the boundary conditions at the interface of core and cladding for E_r, E_ϕ, H_r and H_ϕ. Hence determine the values for the meridional rays.

2.8 What do you mean by V-parameter? Give the expression for the same. A single mode step index fiber has a core diameter and numberical aperture *as* of 3μm and 0.1 respectively. Calculate the value of V-parameter when the wavelength of the propagating wave is 0.8 μm.

2.9 Determine the expressions for the power flow through the core and cladding of a step index fiber.

2.10 The core diameter of a multimode step index fiber is 80 μm. The core refractive index is 1.5 and the relative index difference is 1.5%. The fiber is operating at a wavelength of 0.85 μm. Calculate the (*a*) normalized frequency for the fiber and (*b*) the number of guided modes.

2.11 The same fiber of problem 2.10 is operating at the same wavelength of 0.85 μm. Determine the percentages of power through the core and cladding.

CHAPTER 3

Losses and Dispersions in Optical Fiber

CHAPTER FOCUS

- *Introduction: Attenuation or loss*
- *Rayleigh scattering losses*
- *Mie scattering*
- *Non-linear scattering*
- *Stimulated Raman Scattering*
- *Absorption losses*
- *Radiative losses: Bending losses*
- *Constant radius bending*
- *Leaky modes (Skew rays)*
- *Mode-coupling losses*
- *Combined losses in fibres*
- *Critical radius of a bend*
- *Dispersion: Effect Dispersion on pulse transmission*
- *Intermodal or multipath dispersion*
- *Material dispersion*
- *Combined effect of material dispersion and intermodal dispersion*
- *Waveguide dispersion*
- *Effect of source line width*
- *Relationship between V and β,* $\frac{d(Vb)}{dV}$ and $V\frac{d^2(Vb)}{dV^2}$
- *Total dispersion and maximum transmission rate*
- *Maximum bit rate*
- *Total dispersion single mode fiber*
- *Dispersion-adjusted single mode fiber*
- *Solved examples and review questions*

3.1 INTRODUCTION: ATTENUATION OR LOSS

In this chapter we shall discuss the important optical properties of fibers. These are attenuation (loss) and dispersion of an optical signal propagating through fibers.

Attenuation: The attenuation or transmission loss of an optical signal in an optical fiber is one of the most important factors widely used in the study of fiber communications. The maximum transmission distance prior to the signal restoration is determined by attenuation in the channel and the optical fiber communications became attractive when the transmission loss in a fiber became less than in a metallic conductor.

Attenuation Units

Signal atenuation or transmission loss is defined as the ratio of the input transmitted optical power P_{in} into a fiber to the output (received) optical power P_{out} from the fiber. This ratio is a function of the operating wavelength. The symbol α_{dB} is commonly used to express the attenuation in decibels per kilometre.

$$\alpha_{dB} = \frac{1}{L}\left[10 \log_{10}\left(\frac{P_{in}}{P_{out}}\right)\right] \tag{3.1.1}$$

where L is the length in km of the fiber.

An ideal fiber would have no power loss so that $P_{out} = P_{in}$. This is equivalent to 0-dB attenuation. An idal fiber is not found in practice. Suppose for a fiber α_{dB} = 3 dB/km is average power loss, and it means that the optical signal power will decrease by 50 per cent over a path length of one kilometre. The numerical values may be obtained using the relationship.

$$\frac{P_{in}}{P_{out}} = 10^{dB/10} \tag{3.1.2}$$

The basic attenuation mechanisms in a fiber are scattering, absorption and the radiative losses of the optical energy. Absorption is associated with the fiber material and with the structural imperfection in the optical waveguide. Attenuation due to radiative effects originates from both microscopic and macroscopic perturbations of the fiber geometry.

Example 3.1: *The average optical power launched into a 10 km length of fiber is 100 μW and the average output power is 2.5 μW. Calculate (a) the signal attenuation in decibels through the fiber. It is assumed that there are no connectors or splices (b) signal attenuation per km of the fiber, (c) overall signal attenuation for the 11 km optical link using the same fiber with 3 splices each having an attenuation of 0.8 dB and (d) numerical value of the ratio between input and output power.*

Solution:

$$P_{in} = 100 \times 10^{-6}\ \mu W,\ P_{out} = 2.5 \times 10^{-6}\ \mu W, \text{ and } L = 10 \text{ km}$$

With the use of Eq. (3.1.1) the signal attenuation

$$\alpha_{dB} = \frac{1}{L}\left(10 \log_{10} \frac{P_{in}}{P_{out}}\right)$$

or

$$\alpha_{dB} = \frac{1}{10} \times 10 \log_{10} \frac{100 \times 10^{-6}}{2.5 \times 10^{-6}} = \log_{10} 40 = 1.6 \text{ dB per km}$$

Total attenuation = 1.6 × 10 = 16 dB

Total attenuation for 11 km = 1.6 × 11 = 17.6 dB

However, the link has 3 splices each with attenuation of 0.8 dB and the loss due to splices

= 3 × 0.8 = 2.4 dB

Overall attenuation in the link = 17.6 + 2.4 = 20 dB

The value of P_{in}/P_{out} for the 11 km line with splices (using Eq. 3.1.1) is

$$\frac{P_{in}}{P_{out}} = 10^{dB/10} = 10^{20/10} = 10^2 = 100.$$

3.2 RAYLEIGH SCATTERING LOSSES

Rayleigh scattering in glass is the same phenomenon that scatters light from the sun in the atmosphere and explains the formation of a blue sky. Rayleigh type of scattering of light takes place due to the variations in the refractive index in glass.

The glass used in optical fiber is not a crystalline solid but an amorphous one. It is prepared by allowing glass to cool from the molten state at high temperatures until it freezes. When it is still plastic, glass is drawn out in tension into its fiber form. During the forming process following two defects may arise:

1. Glass being amorphous is composed of a randomly connected network of molecules. Such a structure naturally contains regions in which the molecular density is either higher or lower than the average density in the glass.
2. Since the glass is made up of several oxides, such as SiO_2, GeO_2 and P_2O_5, compositional fluctuation may occur.

These two effects give rise to the variations in the refractive index which occur within the glass over distances that are small compared to the wavelength of light. Submicroscopic variations in the glass density and doping impurities are frozen into glass during manufacture and they act as the reflecting and refracting facets to scatter a small portion of light passing through the glass. The careful manufacturing techniques are capable of reducing these anamolies to a minimum but they cannot be totally eliminated.

Structural inhomogeneities and defects are created during the fabrication of fiber and they can also cause the scattering of light out of the fiber. These defects may be in the form of trapped gas bubbles, unreacted starting materials, and crystallized regions in the glass. In general, the preform manufacturing methods have been so improved that they have minimized these extrinsic effects to the point where the scattering arising from them is negligible compared to the intrinsic Rayleigh scattering.

The losses induced due to the scattering effect vary inversely as the fourth power of wavelength (loss λ^{-4}) and the losses decrease dramatically with the increase of wavelength. For wavelengths below about 1 μm it is the dominant loss mechanism in a fiber. These effects are reduced to less than about 0.3 dB/km at a wavelength of 1.3 μm. Fig. 3.1 shows the variation of the intrinsic scattering losses in silica fiber glass with wavelength over the usable portion of the spectrum from 0.7 to 1.6 μm wavelength.

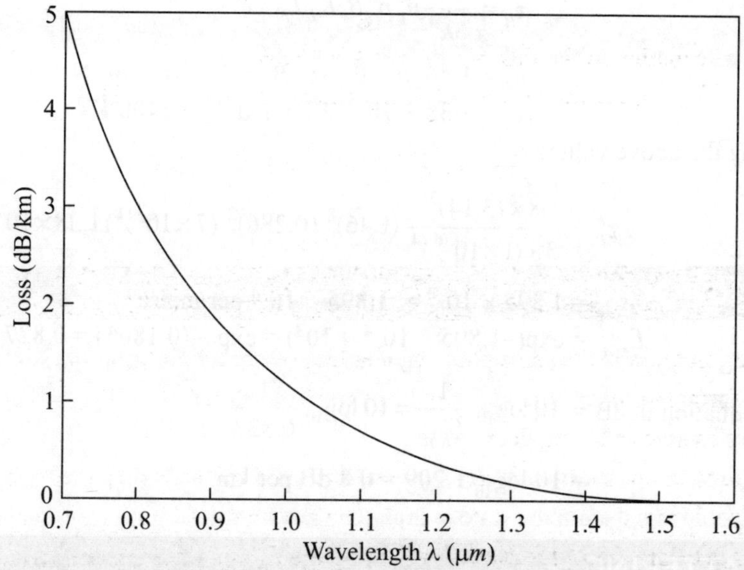

Fig. 3.1 A typical curve for the variation of Rayleigh scattering losses in dB/km as function of wavelength in μm

For a single component glass *q*, the Rayleigh scattering coefficient τ_R is given by[1]

$$\tau_R = \frac{8\pi^3}{3\lambda^4} n_1^8 p^2 B_c K_B T_F \tag{3.2.1}$$

where τ_R is the Rayleigh scattering coefficient, λ is the wavelength of the optical radiation, n_1 is the refractive index of the medium, p is the average photoelastic coefficient, B_c is the isothermal compressibility at a fictive temperature T_F, and K_B is the Boltzmann constant.

The fictive temperature of glass is defined as the temperature at which glass can reach a state of thermal equilibrium and closely related to the anneal temperature. Again the Rayleigh scattering coefficient τ_R is related to the transmission loss factor (transmissivity) of the fiber denoted by $\mathscr{L}$ by the following relationship:

$$\mathscr{L} = \exp(-\tau_R L) \tag{3.2.2}$$

where L is the length of the fiber. It is apparent from Eq. (3.2.2) that the fundamental components of Rayleigh scattering is reduced considerably by operating at the longest wavelength.

Example 3.2: *An estimated value of fictive temperature of silica glass is 1400°K with an isothermal compressibility of* 7×10^{-11} $m^2 N^{-1}$. *The refractive index and photoelastic coefficient for the silica are 1.46 and 0.286 respectively. Calculate the theoretical value of attenuation in dB per km due to the fundamental Rayleigh scattering in the silica at the optical wavelength of 1 μm. Boltzmann constant* $= 1.38 \times 10^{-23}$ JK^{-1}.

[1] R. Olshansky, 'Propagation in glass optical waveguides'. Rev. Mod phys 51 (2) 341–367 (1979)

Solution: According to Eq. (3.2.1) Rayleigh scattering coefficient τ_R is given by,

$$\tau_R = \frac{8\pi^3}{3\lambda^4} n_1^8 p^2 B_c K_B T_F \tag{3.2.3}$$

We have $n_1 = 1.46, p = 0.286, B_c = 7 \times 10^{-11}\ \text{m}^2\ \text{N}^{-1}, \lambda = 1 \times 10^{-6}$ m

$K_B = 1.38 \times 10^{-23}\ \text{JK}^{-1}$ and $T_F = 1400°\text{K}$.

Substituting the above values

$$\tau_R = \frac{8\times(3.14)^3}{3\times(1\times10^{-6})^4}(1.46)^8\ (0.286)^2\ (7\times10^{-11})1.38\times10^{-23}\times1400$$

$$= 1.895 \times 10^{-4} = 1.895 \times 10^{-4} \text{ per metre.}$$

$$\mathcal{L}_{KM} = \exp(-1.895 \times 10^{-4} \times 10^3) = \exp - (0.1805) = 0.827$$

$$\text{Attenuation in dB} = 10 \log_{10} \frac{1}{\mathcal{L}_{KM}} = 10 \log_{10} \frac{1}{0.827}$$

$$= 10 \log_{10} 1.209 = 0.8 \text{ dB per km.}$$

3.3 MIE SCATTERING

Linear scattering may occur at inhomogeneities which are comparable to the size of the guided wavelength. When the size of scattering inhomogeneities is greater than $\lambda/10$, the scattered intensity has an angular dependence and can be quite large. The scattering occurring due to such inhomogeneities is mainly in the forward direction and this type of scattering known as "Mie Scattering". Depending in the fiber material, design and manufacture, Mie scattering can cause considerable power loss. The inhomogeneities can be minimized by

1. Reducing imperfection during glass manufacturing process
2. Carefully controlled extrusion and coating on the fiber, and
3. Increasing the fiber guidance by increasing relative refractive index between core and cladding.

3.4 NON-LINEAR SCATTERING

When a considerable amount of power is passed through a fiber, a high value of electric field leads to the presence of non-linear scattering. Such scattering causes a significant power loss to be scattered in forward, backward or side way directions depending on the nature of interaction. A frequency shift of scattered light is associated with the non-linear scattering. The important non-linear scattering are:

1. Brillouin Scattering and
2. Raman Scattering.

The power density is defined as the power per unit area and at normal power density the non-linear interactions are negligible but at a higher power density a significant amount of energy may be removed by light wave. Such power may be encountered with more powerful laser diode through a small fiber core.

Stimulated Brillouin Scattering

Brillouin scattering may be regarded as the modulation of light through thermal molecular vibration within the fiber. The incident photons of light undergoes non-linear interaction to produce vibrational energy or "phonons" in the glass as well as the scattered light or "photons". The scattered light is found to be frequency modulated by the thermal energy, and both upward and downward frequency shifts are observed. The amount of the frequency shift and the strength of scattering vary as the function of the scattering angle maximum occurring in the backward direction and the minimum or zero being observed in the forward direction. Thus Brillouin scattering occurs mainly in the backward direction which directs the power to the source (*i.e.* from receiver) and the power in the receiver is reduced.

The optical power level at which Brillouin scattering becomes significant in a single mode fiber is given by an empirical formula. The threshold power level P_B is given by[2]

$$P_B = 4.4 \times 10^{-3}\, d^2 \lambda^2 \alpha_{dB}\, \Delta\tau \text{ Watts} \tag{3.4.1}$$

where d and λ are the core diameter and operating wavelength of the fiber respectively and both measured in micrometers, α_{dB} is the fiber attenuation in dB/km. $\Delta\tau$ is the source bandwidth (*i.e.* injection laser) in giga-Hertz. The above equation enables us to determine the threshold power which must be launched in a single mode fiber before stimulated Brillouin scattering occurs.

3.5 STIMULATED RAMAN SCATTERING

The non-linear interaction in Raman scattering produces a high frequency phonon and a scattered photon, whereas low frequency phonons are produced in Brillouin scattering. In Raman scattering the scattered light is predominately in the forward direction and thus the power is not reduced in the receiver.

The threshold power level for the significant Raman scattering to occur is given by

$$P_R = 5.9 \times 10^{-2}\, d^2 \lambda\, \alpha_{dB} \text{ Watts} \tag{3.5.1}$$

where d is the diameter of the fiber core in µm, λ is the wavelength emitted by the source in µm, α_{dB} is the fiber loss in dB/km and P_R is the threshold optical power.

Example 3.3: *When a long single mode fiber is operating at a wavelength of 1.3 µm, the attenuation is 0.5 dB/km. The core diameter of the fiber is 6 µm and the laser source bandwidth is 600 MHz. Calculate the threshold optical powers for the stimulated Brillouin and Raman scattering.*

Solution: The threshold optical power level for stimulated Brillouin scattering in the monomode fiber is given by Eq. (3.4.1)

Values are $d = 6$ µm, $\lambda = 1.3$ µm, $\alpha_{dB} = 0.5$ dB/km and $\Delta\tau = 0.6$ GHz.

$$\begin{aligned} P_B &= 4.4 \times 10^{-3} \times d^2\lambda^2 \times \alpha_{dB}\, \Delta\tau \text{ watts} \\ &= 4.4 \times 10^{-3} \times 6^2 \times 1.3^2 \times 0.5 \times 0.6 \text{ W} \\ &= 80.3 \text{ mW} \end{aligned}$$

The threshold optical power level for stimulated Raman Scattering is given by Eq. (3.5.1)

$$\begin{aligned} P_R &= 5.9 \times 10^{-2} \times d^2\lambda\, \alpha_{dB} \\ &= 5.9 \times 10^{-2} \times 6^2 \times 1.3 \times 0.5 \\ &= 1.38 \text{ Watt.} \end{aligned}$$

In this example, the Brillouin threshold occurs at an optical power level of 80 mW, while the Raman threshold occur at 1.38 watts which is seventeen times larger.

---

[2] RH Stoler, Nonlinearity in Fiber Transmission, *Proc. JEEE,* 68 (10) pp. 1232-1236 (1980).

Nonlinear Scattering and Frequency Shift

The frequency shift associated with the inelastic scattering can be quite small (less than 1 cm^{-1}) in Brillouin scattering where the frequency shifts occurs due to acoustic phonon. In the case of Raman scattering large frequency shifts greater than 100 cm^{-1} are observed and in this case shifts occur due to the molecular vibration or optical frequency phonon. The important feature of inelastic scattering processes is that they not only result in a frequency shift but for a sufficiently high intensity of the incident optical power they exhibit an optical gain at the shifted frequency. The incident optical frequency is known as the pump frequency (w_p) that gives two shifted frequency components known as stokes (w_s) and antistoke's (w_a) components of the radiation.

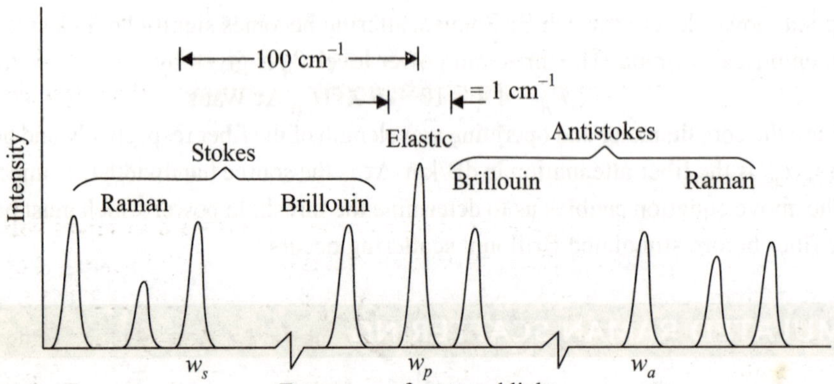

Fig. 3.2 Spectrum of scattered light showing inelastic scattering process (not drawn to the scale). The intensities of anti-stokes Raman lines are far less than those of stokes Raman lines

For a typical fiber the power of the pump frequency around 1 Watt in hundred metres of fiber results in a Raman gain of about a factor of 2. By contrast the peak Brillouin gain is much greater than the Raman gain but the Brillouin frequency shift and gain-bandwidth are much smaller. Moreover Brillouin gain only exist for the propagation of light in opposite direction to that of pump light, while Raman amplification will occur for light propagation in either direction. Figure 3.2 shows the spectrum of scattered light illustrating inelastic processes. The intensity of the anti-stoke Raman lines is far less than that of stokes Raman lines.

3.6 ABSORPTION LOSSES

There are three mechanisms that contribute to the absorption losses. These are: (1) Ultraviolet absorption, (2) Infrared absorption and (3) Ion resonance absorption.

Ultraviolet Absorption

Ultraviolet absorption takes place, because for fused silica valance electrons absorb light and can be ionized to conduction electrons with a centre wavelength of about 0.14 μm. This wavelength corresponds to the energy level of about 8.5 eV. This ionization gives rise to an energy loss in the light fields and contributes to the transmission loss in the fiber. This absorption does not only occur at this fixed wavelength but takes place over a broad band of frequencies up to the visible portion of the spectrum. This absorption loss decreases with the increase of the wavelength. The UV absorption 'tail' falls off causing a negligible amount of loss in the wavelength band from 1.2 to 1.3 μm. The ultraviolet edge of the electron absorption band in both crystalline and amorphous materials follows the following empirical relationship:

$$\alpha_{uv} = Ce^{E/E_0} \tag{3.6.1}$$

which is known as Urbach's rules. Here C and E_0 are empirical constants. E is the photon energy. α_{uv} = attenuation constant in the ultraviolet region.

In order to modify the refractive index of a fiber, glass is purposely doped with induced impurities, such as germanium dioxide. The refractive index of glass increases with the amount of dopant. The presence of the impurity causes some increase in the UV absorption tail, because of an upward shift in the wavelength of the UV absorption peak. The typical value of loss at 1.2 μm wavelength for most fibers does not exceed 0.1 dB/km. Figure 3.3 shows the absorption tail for pure silica glass and germanium doped silica glass.

Infra Red Absorption

Infra red absorption occurs because the photons of light energy are absorbed by atoms within the glass molecules and converted to random mechanical random vibrations typical of heating. For silica glass the IR absorption exhibits a main spectral peak at 8 μm with the minor peaks at 3.3, 3.8 and 4.4 μm. Again the peaks are broad falling off in the visible portion of the spectrum and the typical values of losses at the wavelength of 1.5 μm are less than 0.5 dB/km. Figure 3.3 shows typical infrared absorption for silica glass in the usable range.

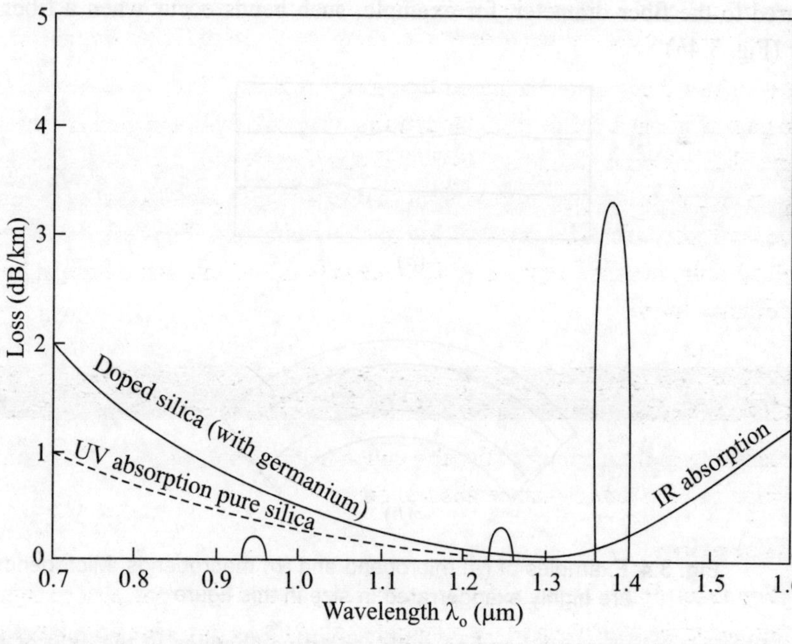

Fig. 3.3 Absorption loss effect in pure glass silica fibers: Graphs for the variation of loss in dB/km with wavelength in μm. OH^- ions in silica glass exhibits absorption peaks at 1.39 μm, 1.25 μm and 0.95 μm

Ion-Resonance Absorption

Minute quantities of water molecules may be trapped in the glass at the time of manufacture and these contribute OH-ions to the materials. The OH-ions exhibit absorption peaks at 0.95, 1.25 and 1.39 μm within the visible spectrum and the main absorption peak occurs at the wavelength of 1.39 μm. When it is desired that these peaks are not to spread and merge to predominate the loss spectrum of the fiber the water content of the glass has to be kept below 0.01 part per million parts. Figure 3.3 shows the three peaks with a typical value for a low-loss fiber.

The presence of some other metals and impurities may give rise to undesirable losses within the usable portion of the spectrum. The presence of some impurities like iron, copper and chromium is to be avoided. In order to reduce the presence of these impurities to acceptable levels good refining technique for purifying raw materials must be followed. Zone refining technique which is usually used for preparing pure silicon chips for integrated circuits is usually employed.

3.7 RADIATIVE LOSSES; BENDING LOSSES

Radiative losses occur whenever an optical fiber undergoes a bend of finite radius of curvature. Fibers can be subjected to two types of bends:

1. *Microbendings* are the random microscopic bends of the fiber axis that can arise when the fibers are incorporated into cables (Fig. 3.4*a*).
2. *Constant Radius Bending or Macrobendings* These are the bends having radii that are large compared to the fiber diameter, for example, such bends occur when a fiber cable turns a corner (Fig. 3.4*b*).

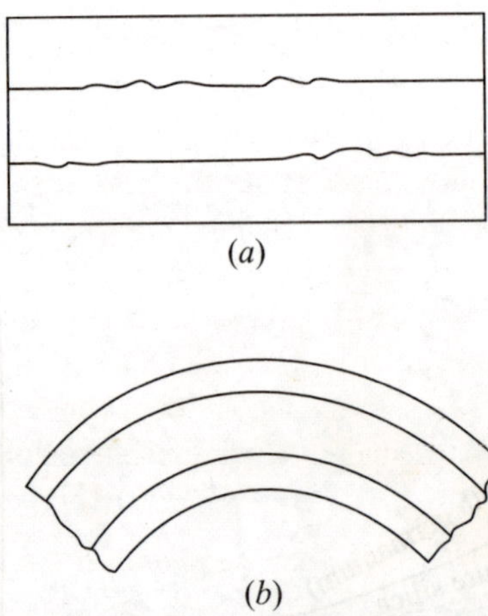

Fig. 3.4 Examples of (*a*) microbend and (*b*) macrobends. Microbends are highly exaggerrated in size in this figure

Microbending is a microscopic bending with repetitive changes in the axis of the core of the fiber (Fig. 3.4) and it takes place due to the slightly different contraction rate between the core and the cladding materials. Microbending occurs due to nonuniform lateral pressure created during the cabling of the fiber. When the fibers are subsequently wound on a multilayer cable or wound on spool for transportation of the fiber from one place to other, then microbending can also be introduced into the fiber. Losses due to microbending takes place, because the small bends act as the scattering facets and these facets cause mode coupling to occur. Energy from the guided modes is cross coupled to the leaky mode and is lost through the cladding. Microbendings are randomly distributed over the length of the fiber and the losses due to them will be uniformly distributed so that the total amount of loss for a fiber can be determined. Careful precautions in manufacturing and handling of fibers will be able to reduce these losses. One method of minimizing the microbending losses is done by extruding a compressible jacket over the fiber. When external forces are applied to this configuration, the jacket will be deformed, but the fiber will tend to stay relatively straight as shown in Fig. 3.5.

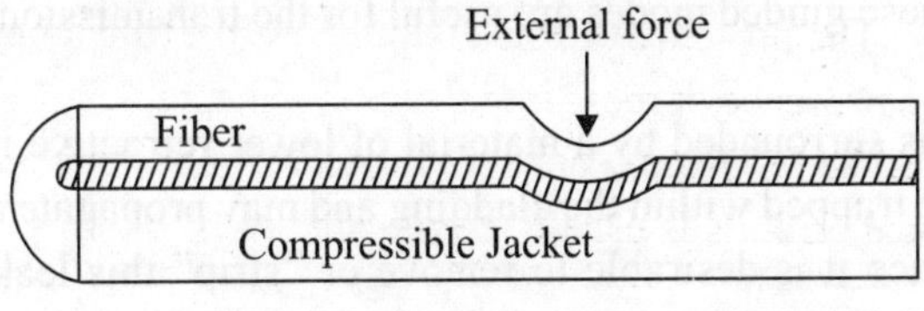

Fig. 3.5 A compressible jacket extruded over the fiber minimizes microbending resulting from external forces

3.8 CONSTANT RADIUS BENDING

When the optical fibers are installed in cable ducts or on a pole for a transmission line, it is necessary to introduce bends in the fiber to join corners. Sometime these bends may be quite sharp. For example, at a pole hanger point where the cable has a sag between two poles or at right angle turns in building conduits, the sag is quite sharp. These large radius bends will also introduce some loss of light energy in the fiber.

Figure 3.6 illustrates the manner in which a bend introduces a loss in a fiber. In the figure a ray of light which is fully guided, reaches the reflection point A at an angle ϕ_A. If the fiber were not bent, it arrives at B_1 as the next reflection point. But with the bend in the fiber, the ray encounters the fiber wall early at the outside edge of the bend at point B_2 at an incidence angle ϕ_B. The angle ϕ_B is smaller than angle ϕ_A in the straight section. When angle ϕ_B is smaller than the critical angle, the ray will manage to escape from the fiber core and some of the energy in the corresponding mode will be lost. If a bare fiber is bent sharply, the light escaping from the core can be observed as a bright light in the form of a line along the outer edge of the fiber bend.

A sharp bend is responsible for more modes to occur and light to escape at the bend. When a fiber is installed, it is necessary to ensure that no sharp bends are going to occur.

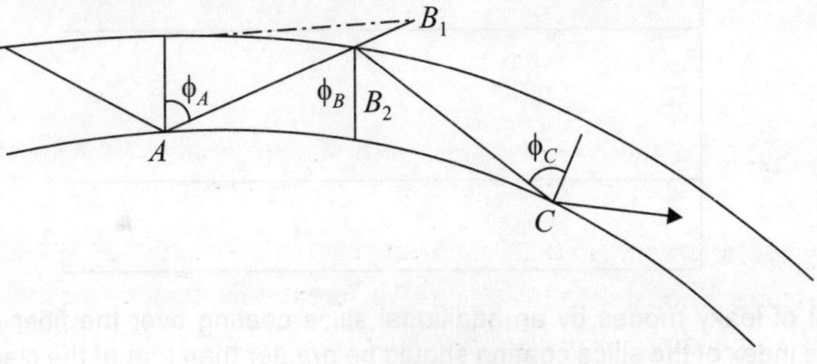

Fig. 3.6 Loss due to ray propagation in a bent fiber

3.9 LEAKY MODES (SKEW RAY)

At each reflection of a ray propagating in helical type of path (skew mode) the angle of incidence θ_1 at the core-cladding surface is more than the critical angle θ_C the mode will propagate through the fiber. However, at some higher order modes, the angle of incidence θ_1 may be less than the critical angle θ_C and a part of the propagating ray will escape from the core by refraction and the remainder of the ray will be partially reflected. Successive partial reflections will cause the leaky skew ray to decrease rapidly in intensity as it propagates, and thus, at some distance along the cable only the completely guided modes will exist. These guided modes are useful for the transmission of information through the fiber.

When the cladding is surrounded by a material of lower refractive index, the partial refraction from the skew mode may be trapped within the cladding and may propagate along the length of the fiber up to the receiver. In practice it is desirable to remove or "strip" this leaky mode from the core and cladding as rapidly as possible so that the signal dispersion will be reduced. This is accomplished by surrounding the thin cladding layer on the fiber core by a third layer of pure silica. The refractive index of pure silica should be greater than that of the cladding, but less than that of the core (Fig. 3.7).

This silica coating on the cladding provides mechanical strength to the fiber and acts to remove the partially refracted ray from the leaky mode and possesses rays from the cut off modes by total refraction as shown in Fig. 3.7. In the figures m_1 represents the path of the ray in one of the completely guided meridional modes which is completely reflected by the core-cladding interface, and m_2 represents a skew ray which is partially reflected. Some of the ray energy is refracted through the cladding and then through the silica coating to be finally absorbed in an opaque jacket. m_3 represents a cut off meridional mode which is completely refracted out of the core and dissipated in the opaque jacket.

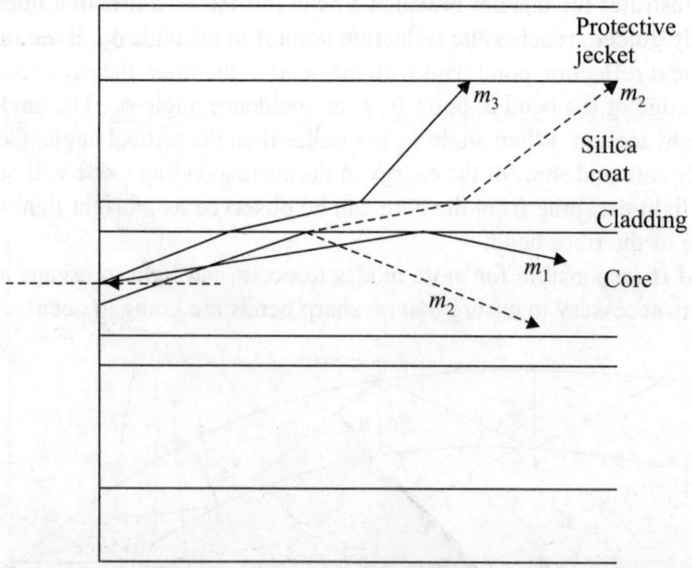

Fig. 3.7 Removal of leaky modes by an additional silica coating over the fiber cladding. The refractive index of the silica coating should be greater than that of the cladding but less than that of the core $n_1 > n_3 > n_2$

The leaky modes launched in the transmitting end of the fiber usually contain only a small percentage of the total guided power and these are rapidly reduced near the transmitter. When the fiber is complete and unspliced a fixed loss will occur in the fiber only once. However, when splices are introduced in the cable, the mode propagating in the first section may not be properly coupled to the next section and the result is that the power is transferred to the leaky mode at the splice. This power is lost just after the splice by the "stripping effect". Thus, each splice will contribute to the loss.

3.10 MODE-COUPLING LOSSES

Power may be launched successfully into a propagating mode but it may be coupled into a leaky or radiating mode at some point further down the fiber. This type of improper coupling can occur for several reasons.

Small imperfections in the core glass or in the core-cladding interface may occur due to the small variations in the core diameter, cross sectional shape or bubbles in the glass and these are responsible for causing the energy to be coupled into one of the leaky modes. Losses from these sources will be uniformly distributed along the length of the fiber.

Another source of mode coupling is the imperfectly formed splices or imperfectly aligned connectors. In a typical case, this loss may be about 0.5 dB per connector coupling or about 0.2 dB per splice. These are discrete losses which can be reduced by decreasing the number of splices or connectors in a given fiber line.

3.11 COMBINED LOSSES IN FIBERS

Now we can summarize the four types of losses inherent to a fiber. These are (1) Rayleigh scattering losses, (2) material absorption losses (3) leaky mode losses and (4) losses due to mode coupling because of scattering.

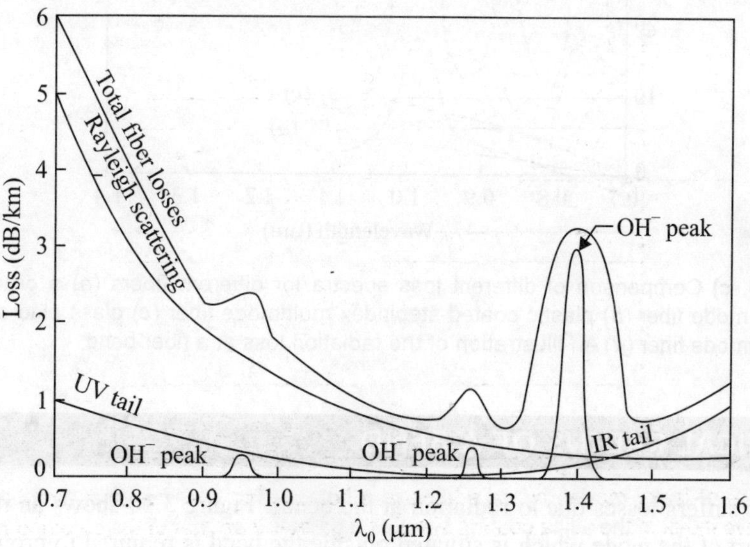

Fig. 3.8 Total loss spectrum of a typical optical fiber

During manufacturing process this must be reduced to a minimum. Of course, losses due to Rayleigh scattering and material absorption are predominant and every step should be taken during manufacture to reduce these losses. Figure 3.8 shows the losses in a typical multimode fiber as a function of wavelength so that the losses due to Rayleigh scattering and material absorption can be compared. This fiber has relatively low losses compared to some other fiber which may have a loss as high as 20 dB/km.

All fibers have also spectrum curve of this general shape but the actual loss values and peaking wavelengths may vary depending on the type of fiber. Figure 3.9 (*a*) (*b*) (*c*) shows actual experimental curves for three commercial fibers drawn on the same coordinates as Fig. 3.3. These fibers are used for multimode transmission in the 0.8 μm spectrum gap, with the information specified for the range 0.5 to 1.1 μm instead of 0.7 to 1.6 μm wavelength band. All of the curves have a "window" in the spectrum of losses at about 0.8 μm which coincides with the spectral output of many LED sources especially available for fiber communications. Within the window, losses at their lowest, typically between 5 and 10 dB/km. Other spectrum windows occur at 1.2 μm and at 1.3 μm in the typical spectra for stepindex glass-clad fibers and since this lies within the gap between the ultraviolet and infrared tails, the lowest possible losses for any fibers can be obtained within this window. Typical losses are in the order of 0.5 to 2 dB/km that can be obtained in carefully designed fibers. These fibers are used for long distance monomode transmission system with widely spaced repeaters.

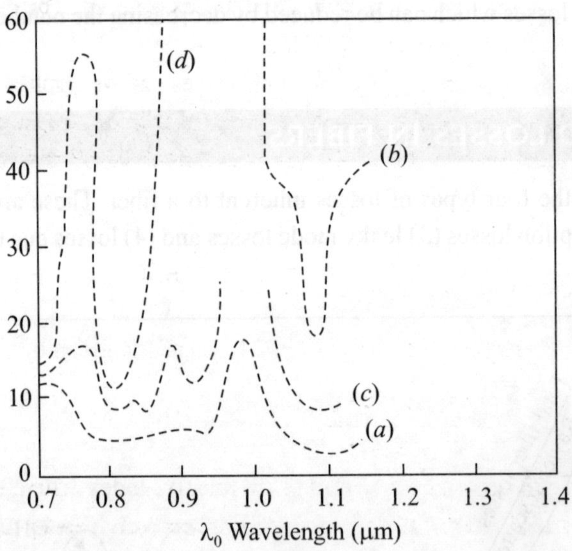

Fig. 3.9 (*a*, *b*, *c*) Comparison of different loss spectra for different fibers (*a*) a gradedindex multimode fiber (*b*) plastic coated stepindex multimode fiber (*c*) glass clad stepindex multimode fiber (*d*) An illustration of the radiation loss at a fiber bend.

3.12 CRITICAL RADIUS OF A BEND

The optical fiber suffers losses due to radiation at the bends. Figure 3.10 shows an illustration of this situation. The part of the mode which is situated outside the bend is required to propagate faster than that on the inside so that the wave front perpendicular to the direction of propagation is maintained. Thus, the part of the mode in the cladding is required to travel faster than the velocity of light in that medium. Since this is not possible the energy associated with this part of the mode is lost through radiation. This loss can be represented by a radiation attenuation coefficient α_r. The relationship between the radius of curvature of the bend and α_r is given by[3]

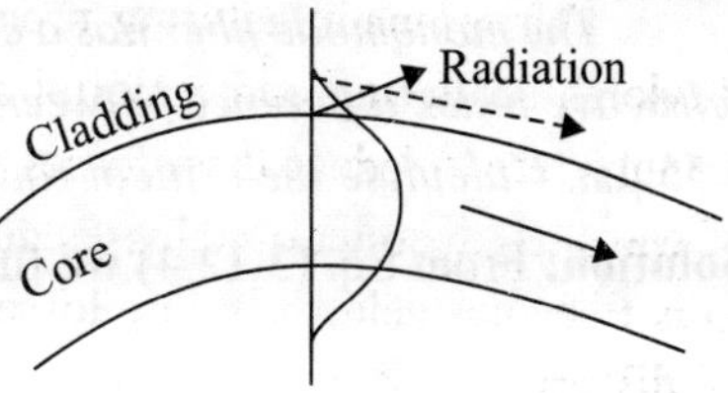

Fig. 3.10

$$\alpha_r = C_1 \exp(-C_2 R) \quad (3.12.1)$$

where R is the radius of curvature of the fiber bend, and C_1 and C_2 are the constants which are independent of R. Moreover, large bending losses tend to occur in multimode fiber at a critical radius of curvature R_c which may be calculated from the following relationship[4]

$$R_c = \frac{3n_1^2 \lambda}{4\pi\left(n_1^2 - n_2^2\right)^{3/2}} \quad (3.12.2)$$

where n_1 and n_2 are the refractive indices of the core and cladding of the fiber respectively.

It may be noted from the above expression that the potential microbending may be minimized by

(*a*) Designing fibers with large relative refractive index difference between the core and cladding.

(*b*) Operating at the shortest possible wavelength.

The above criteria for minimizing the bending losses can be applied to singlemode fiber. One theory based on single quasi-guided mode gives an expression from which the critical radius of curvature for a singlemode fiber R_{cs} can be determined. The expression for the critical radius of curvature is given by[5]

$$R_{cs} = \frac{20\,\lambda}{(n_1 - n_2)^{3/2}}\left[2.748 - 0.996\frac{\lambda}{\lambda_c}\right]^{-3} \quad (3.12.3)$$

where λ_c is the cut off wavelength of the single mode fiber and is given by

$$\lambda_c = \frac{2\pi a n_1 (2\Delta n)^{1/2}}{2.405} \quad (3.12.4)$$

Hence for a specific singlemode fiber with fixed relative index difference and cutoff wavelength, the critical wavelength of the radiated light becomes progressively shorter as the radius of the bend is reduced.

Example 3.4: *Two step index fibers one of which is a mono-mode type and the other one is of multimode type have the following parameters:*

---

[3] M.M. Ramsay and G.A. Hochham, Propagation in Optical Fiber Waveguide.

[4] H.F. Wolf, Optical waveguide in H.F. Wolf edition Handbook Fiber Optics, Theory and Applications, pp. 43–152, Granada 1979.

[5] W.A. Gambling, H. Matsumura and G.M. Ragdale, Curvature and Microbending Losses in Single Mode Optical Fiber, Opt Quantum Electronics, pp. 43–59, 1979.

The monomode fiber has a core diameter of 8 μm and core refractive index of 1.500. The relative refractive index difference between the core and cladding is 0.3 per cent. The operating wavelength is 1.55 μm. Calculate the critical radius of curvature at which bending losses occur in the fiber.

Solution: From Eq. (3.12.4) we find that the cut off wavelength λ_c of a monomode fiber is given by

$$\lambda_c = \frac{2\pi\, an_1\,(2\Delta n)^{1/2}}{2.405}$$

It is given: $a = 4 \times 10^{-6}$, $n_1 = 1.500$ and $\Delta = n_1 - n_2 = \frac{0.3}{100} = 0.003$

so that $2\Delta = 0.006$

$$\lambda_c = \frac{2\pi \times 4 \times 10^{-6} \times 1.500 \times (0.006)^{1/2}}{2.405} = \frac{12\pi \times (0.0776)}{2.405} = 1.214\ \mu\text{m}$$

From Eq. (3.12.3) the critical radius R_{cs} is given as

$$R_{cs} = \frac{20\lambda}{(n_1 - n_2)^{3/2}}\left[2.748 - 0.996\frac{\lambda}{\lambda_c}\right]^{-3}$$

$$= \frac{20 \times 1.55 \times 10^{-6}}{(0.003)^{3/2}}\left[2.748 - \frac{0.996 \times 1.55 \times 10^{-6}}{1.214 \times 10^{-6}}\right]^{-3} = 34\ \text{mm}.$$

Example 3.5: *The multimode fiber has a core refractive index of 1.500 and the relative refractive index difference is 3 per cent and the operating wavelength is 0.80 μm. Calculate the critical radius of curvature at which large bending losses occur.*

Solution: The relative refractive index difference is given as

$$\Delta = \frac{n_1^2 - n_2^2}{2\,n_1^2} = \frac{n_1 - n_2}{n_1} = 0.03$$

$$n_2^2 = n_1^2 - n\Delta\, n_1^2 = (1.500)^2 - 2(0.03) \times (1.5)^2$$
$$= 2.225 - 0.1350 - 2.115$$

From Eq. (3.12.3)

$$R_c = \frac{3n_1^2\,\lambda}{4\pi\left(n_1^2 - n_2^2\right)^{3/2}} = \frac{3 \times 2.225 \times 0.80 \times 10^{-6}}{4\pi(2.250 - 2.115)^{3/2}} = 9\ \mu\text{m}$$

The critical radius of curvature of the multimode fiber is 9 μm.

3.13 DISPERSION: EFFECT OF DISPERSION ON PULSE TRANSMISSION

It may be expected that a light pulse with a given width and amplitude injected into one end of a fiber should arrive at its other end with the shape and width unchanged, and only its amplitude is reduced by losses. If the losses are extremely large, the pulse amplitude at the receiving end will be too small to be detected and a repeater has to be included to boost up the signal level before its entering into next section.

Several dispersion effects are encountered by the pulse of light propagating through a fiber and these act to spread out the pulse in the time domain changing the shape of the pulse. As a result, the pulse may be got merged in the previous or in the succeeding pulses becoming indistinguishable at the receiver as illustrated in Fig. 3.11. This effect is known as inter symbol interference (ISI). The pulses may be separated by increasing the time interval between pulses, but thereby the maximum bit rate will be reduced.

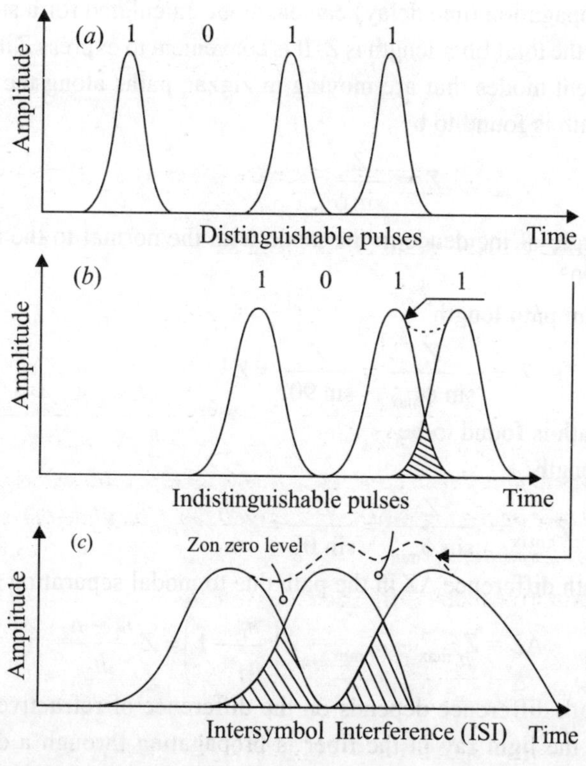

Fig. 3.11 An illustration of pulse broadening using a digital bit pattern 1011. The broadening of the light pulses occurs as they are transmitted through the fiber producing intersymbol interference (ISI) (*a*) fiber input (*b*) the output pulse of the fiber at distance L_1 (*c*) fiber output at a distance L_2 where $L_2 > L_1$.

There are three kinds of dispersion due to three separate mechanisms existing in the fiber. These are (1) Intermodal Dispersion or Multipath dispersion, (2) Material or chromatic dispersion and (3) waveguide dispersion.

3.14 INTERMODAL OR MULTIPATH DISPERSION

The effective group velocity of light wave is different for each mode that a waveguide will support, even though the phase velocity in different mode ray may be identical. This happens because total path followed by a ray, in each of the mode is zigzag in nature and the path in a particular mode has different total length than each of the other mode rays.

When a pulse of light is transmitted in a fiber, it will propagate over several different paths corresponding to the excited modes and come out of the far end at slightly different times. The received pulse at the far end is the summation of these mode pulses each delayed by different amount of time. The shortest time delay will be suffered by the plane wave propagating straight along the axis of the fiber core: the longest time delay will be suffered by the wave propagating with the highest order supported mode where the angle of incidence is slightly greater than the critical angle θ_c. The difference in the arrival time Δt (propagation time delay) can easily be calculated for a stepindex fiber.

Let us assume that the total fiber length is Z. It is convenient to express Z in kilometers. Figure 3.12 shows two rays of different modes that are moving in zigzag paths along the length of the fiber. The length Z_t of the zigzag path is found to be

$$Z_t = \frac{Z}{\sin\phi} \tag{3.14.1}$$

Where ϕ is the angle of incidence in the mode with the normal to the fiber wall. In the lowest order mode, $\phi = \phi_{max} = 90°$

Thus, the minimum path length

$$Z_t = \frac{Z}{\sin\phi_{max}} = \frac{Z}{\sin 90°} = Z \tag{3.14.2}$$

and at ϕ_{min} the longest path is found to be

Maximum path length,

$$Z_{t\,max} = \frac{Z}{\sin\phi_{min}} = \frac{Z}{\sin\theta_c} = Z\frac{n_1}{n_2} \tag{3.14.3}$$

The maximum path difference ΔZ in the path due to modal separation is found to be

$$\Delta Z = Z_{t\,max} - Z_{t\,min} = Z\left(\frac{n_1}{n_2} - 1\right) = Z\frac{n_1 - n_2}{n_2} \tag{3.14.4}$$

The maximum path difference depends on the difference of refractive index between the core and the cladding. Since the light ray in the fiber is propagating through a dielectric medium with a dielectric constant ε_r which is greater than unity, the ray travels more slowly than that they would move in free space. It is assumed that the relative permeability μ_r of the dielectric medium is unity. The phase velocity in the dielectric along the ray path is

Phase velocity,

$$v_p = \frac{1}{\sqrt{\mu\varepsilon}} = \frac{1}{\sqrt{\mu_o\varepsilon_o}}\frac{1}{\sqrt{\mu_r\,\varepsilon_r}} = \frac{c}{\sqrt{\varepsilon_r}}\ (\mu_r = 1) \tag{3.14.5}$$

Since $c = \frac{1}{\sqrt{\mu_o\varepsilon_o}}$ velocity of light in free space $= 3 \times 10^8$ m/s

Thus, the dielectric constant ε_r is

$$\sqrt{\varepsilon_r} = \frac{c}{\sqrt{v_p}} = n = \text{refractive index of the medium} \tag{3.14.6a}$$

Thus, we find that the relationship between the dielectric refractive index n_1 is

$$\varepsilon_r = n_1^2 \tag{3.14.6b}$$

The phase velocity in the glass core of the fiber with refractive index n_1 is given by

$$\text{Phase velocity, } v_{p(\text{glass})} = \frac{c}{n_1} \tag{3.14.7}$$

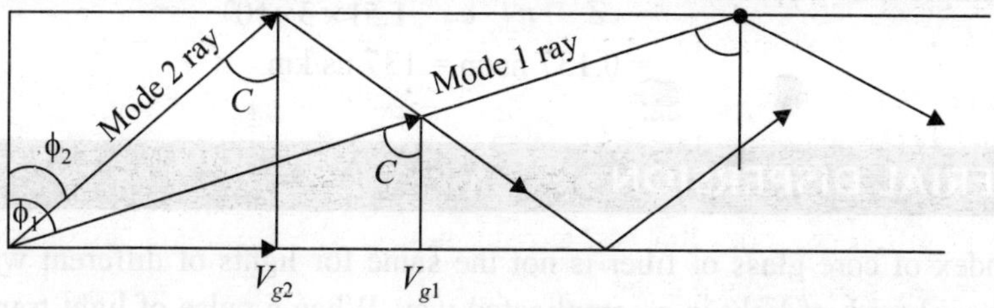

Fig. 3.12 Two rays of different modes are moving in a zigzag path along a fiber

Finally, the maximum difference in the time delay Δt between the rays of highest and lowest order modes is obtained by dividing the difference in the path length by the phase velocity. Denoting this by Δt and is given by

$$\Delta t = \frac{\Delta Z}{v_p} = \frac{n_1 Z}{c}\left[\frac{n_1 - n_2}{n_2}\right] = \frac{Z}{c}\frac{n_1 (n_1 - n_2)}{n_2} \tag{3.14.8}$$

Denoting the difference between the refractive indices at the core and cladding $(n_1 - n_2)$ by Δn, the maximum time difference becomes

$$\Delta t = \frac{Z\, n_1}{c\, n_2}\Delta n \tag{3.14.9}$$

or $$\frac{\Delta t}{Z} = \frac{n_1}{n_2 c}\Delta n$$

The term $\Delta t/Z$ is known as the "multiple time difference" of the fiber.

For Example: For an unclad glass fiber, Core refractive index $n_1 = 1.5$, Cladding refractive index $n_2 = 1.0$ and Velocity of light 'c' = 3×10^8 m/sec, we find that $\Delta n = n_1 - n_2 = 1.5 - 1 = 0.5$

$$\frac{\Delta t}{Z} = \frac{1.5 \times 0.5}{1 \times 3 \times 10^8} = 2.5 \times 10^{-9} = 2.5 \text{ ns/m} = 2.5\ \mu\text{s/km}$$

$(\Delta t/Z)$ is usually expressed in units of nanosecond per kilometer (or ns/km). This type of dispersion does not occur in single mode fibers and does not depend on the wavelength of the propagating wave. The intermodal dispersion is the characteristic of the fiber itself. The effect of cladding with glass having a slightly lower refractive index than that of the core is to reduce the intermodal dispersion considerably.

Example 3.6: *The core of an optical fiber is made of glass of refractive index 1.55 and is clad with another glass of refractive index 1.51. The launching of light is done from air of refractive index 1.0. Determine (1) numerical aperture, (2) acceptance angle and (3) multiple time dispersion.*

Solution:

$\Delta n = 1.55 - 1.51 = 0.04$ and $2n = 1.55 + 1.51 = 3.06$

Numerical aperture, $(NA) = (2n\,\Delta n)^{1/2} = (3.06 \times 0.04)^{1/2} = 0.352$

Acceptance angle $\alpha_m = \sin^{-1}(NA) = \sin^{-1} 0.352 = 20.6°$.

Using Eq. (3.14.9)

$$\frac{\Delta t}{Z} = \frac{n_1}{n_2}\frac{\Delta n}{c}$$

$$\text{"Multiple time dispersion"} = \frac{\Delta t}{Z} = \frac{n_1}{n_2}\frac{\Delta n}{c} = \frac{1.55 \times 0.04}{1.51 \times 3 \times 10^8}$$

$$= 0.137 \text{ ns/m} = 137 \text{ ns/km.}$$

3.15 MATERIAL DISPERSION

The refractive index of core glass of fiber is not the same for lights of different wavelengths but it depends on the wavelength of light in a complicated way. When a pulse of light transmitted contains components of several wavelengths centered about a centre wavelength λ_o as shown in Fig. 3.13. The pulse component having shorter wavelengths will experience more delay than those of larger wavelengths. As a result there is an effective time dispersion Δt of the pulse at the receiving end of the fiber.

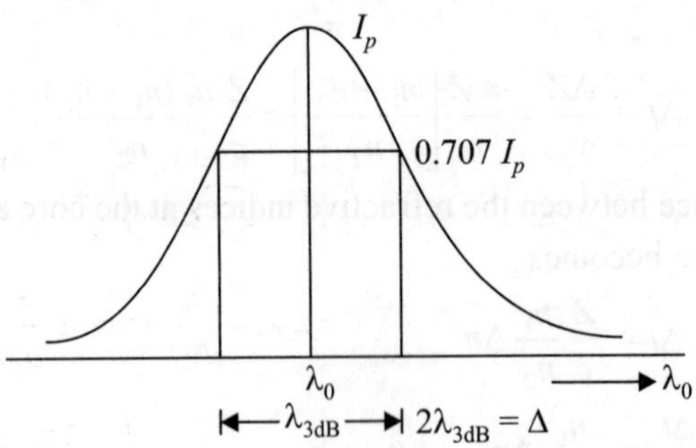

Fig. 3.13 Defining the spectral width of the light sources: Central wavelength λ_0 : $2\lambda_{3dB} = \Delta$

Relationship between the Material Dispersion and Wavelength

When a disturbance or signal is superimposed into a wave, the wavepacket does not propagate at the phase velocity v_p of the wave, where

$$v_p = \frac{w}{\beta} \tag{3.15.1}$$

but at the group velocity v_g which is given by

$$v_g = \frac{dw}{d\beta} = \frac{1}{\frac{d\beta}{dw}} \tag{3.15.2}$$

where β is the phase constant and ω is the angular velocity of the wave. In a non-dispersive medium the phase velocity v_p is independent of the wave frequency so that the group velocity and the phase velocity are the same. But in a dispersive medium where by definition the phase velocity is the function of frequency, v_g and v_p will be different.

$$v_g = \frac{1}{\frac{d\beta}{dw}} = \frac{1}{\frac{1}{v_p} - \frac{w}{v_p^2}\left(\frac{dv_p}{dw}\right)} = \frac{v_p}{1 - \frac{w}{v_p}\left(\frac{dv_p}{dw}\right)} \tag{3.15.3}$$

This is important, since it is the signal velocity with which we are concerned in communications. When a light pulse travels through a dispersive medium with a speed v_g, the pulse is obviously attenuated and to some extent distorted due to dispersion during travel through the medium, we define the group refractive index N such that

Group index, $$N = \frac{c}{v_g} \tag{3.15.4a}$$

and in a dispersive medium N will be different from the ordinary refractive index or phase index n. The velocity of propagation of light pulses in the ray model of Fig. 3.12 is c/n so that the travel time difference Δt between the axial ray pulses and those travelling by the most oblique path should be given by

$$\frac{\Delta t}{z} = \frac{N_1 \Delta n}{n_2 c} \tag{3.15.4b}$$

where N_1 is the group refractive index of the core. We are required to express v_g and N in terms of phase index n and λ.

Substituting the value of v_g from (3.15.2)

$$N = \frac{c}{v_g} = c\frac{d\beta}{dw} = \frac{c}{v_p}\left[1 - \frac{w}{v_p}\frac{d\,v_p}{dw}\right] = n - \frac{c\,w}{v_p^2}\frac{d(c/n)}{v_p dw}$$

or $$N = n + \frac{cw}{v_p^2}\frac{c}{n^2}\frac{dn}{dw} = n + w\frac{dn}{dw} = n + w\frac{dn}{d\lambda}\frac{d\lambda}{dw} \tag{3.15.5}$$

$$w = \frac{2\pi c}{\lambda} \text{ and } \frac{dw}{d\lambda} = \frac{d}{d\lambda}(2\pi f) = \frac{d}{d\lambda}\left(\frac{2\pi c}{\lambda}\right) = -\frac{1}{\lambda^2}2\pi c$$

in Eq. (3.15.4)

$$N = n - \frac{2\pi}{\lambda}\frac{dn}{d\lambda}\left(\frac{-\lambda^2}{\lambda}\right) = n - \lambda\frac{dn}{d\lambda} \tag{3.15.6}$$

Therefore, the group velocity $$v_g = \frac{c}{N} = \frac{c}{n - \lambda(dn/d\lambda)} \tag{3.15.7}$$

An impulse of light then travels through a distance Z in the medium in time Δt and with the help of Eq. (3.15.4b) Δt is given by

$$\Delta t = \frac{Z}{v_g} = \frac{NZ}{c} = \left(n - \lambda\frac{dn}{d\lambda}\right)\frac{Z}{c} \tag{3.15.8}$$

If the light contains a range of free space wavelength $\Delta\lambda$ about the centre wavelength λ_o in Fig. 3.12 and if the medium is dispersive, the impulse spreads out as it propagates and arrives over a range of time Δt which is given by (using Eq. 3.15.8)

$$\Delta_{t\lambda} = \frac{d\Delta t}{d\lambda}\Delta\lambda = \frac{Z}{c}\left[\frac{dn}{d\lambda} - \frac{dn}{d\lambda} - \lambda\frac{d^2n}{d\lambda^2}\right]\Delta\lambda = \frac{-Z}{c}\lambda\frac{d^2n}{d\lambda^2}\Delta\lambda \tag{3.15.9}$$

It is customary to define the spectral width $\Delta\lambda$ of an optical source as the range of wavelengths over which the spectral power exceeds 50 per cent of the peak spectral power. It is convenient to define a relative spectral width γ which is defined by (Fig. 3.13)

$$\gamma = \left|\frac{\Delta\lambda}{\lambda}\right| = \left|\frac{\Delta w}{w}\right| \tag{3.15.10}$$

Thus, an impulse of light after propagating for a distance Z through a dispersive medium will have broadened into a pulse of half power width Γ where Γ is

$$\Delta t_\lambda = \Gamma = \frac{Z}{c}\gamma\left[\lambda^2 \frac{d^2n}{d\lambda^2}\right] = \frac{Z}{c}\frac{\Delta\lambda}{\lambda}\left(\lambda^2 \frac{d^2n}{d\lambda^2}\right) \tag{3.15.11}$$

Equation (3.15.11) is obtained by comparing Eq. (3.15.9) and Eq. (3.15.10) That is

$$\frac{\Gamma}{Z} = \frac{\gamma}{c}|Y_m| \tag{3.15.12}$$

where $$Y_m = \lambda^2 \frac{d^2n}{d\lambda^2}$$

Y_m represents a material dispersion coefficient. If we write $\Delta f = \frac{1}{4\Gamma}$ as the approximate signal bandwidth of the fiber, this gives $\Delta f\, Z = c/(4\gamma\,|Y_m|)$. Figures 3.14 and 3.15 show the variation of Y_m and $(\lambda/c)\,(d^2n/d\lambda^2)$ with wavelength repectively for bulk sample of pure and doped silica.

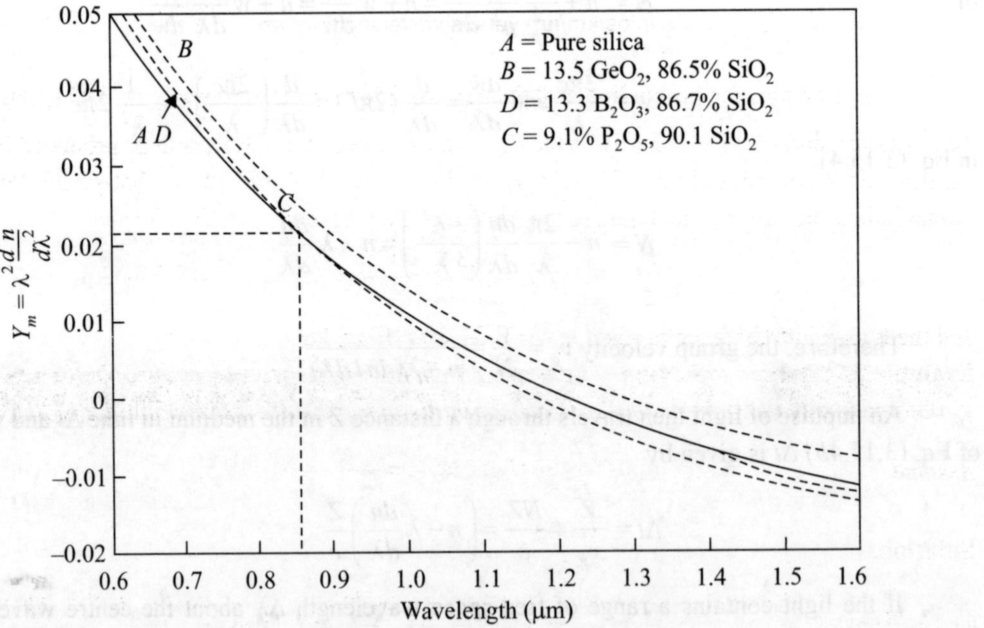

Fig. 3.14 A plot of material dispersion perameter $Y_m = \lambda^2\left(\frac{d^2n}{d\lambda^2}\right)$ in (ps/km/nm) versus wavelength λ in μm. Y_m = 0.0221 when λ = 0.85 μm

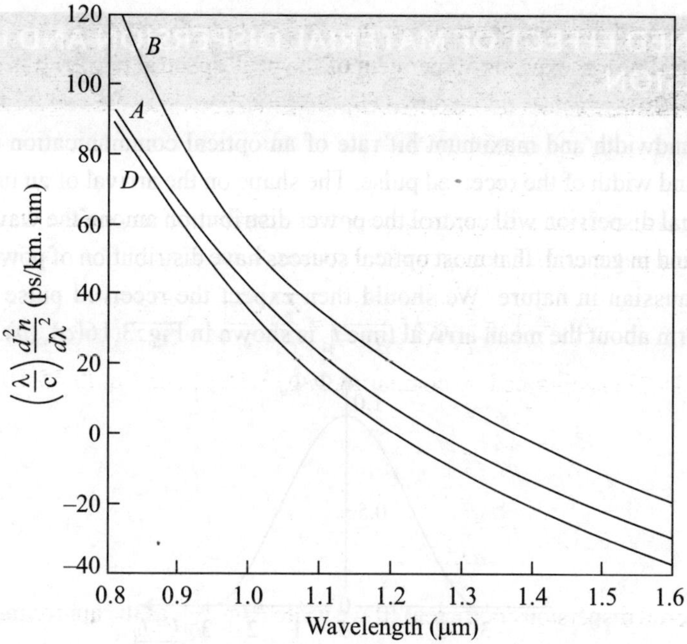

Fig. 3.15 A plot between dispersion perameter $Y_m = \left(\frac{\lambda}{c}\right)\left(\frac{d^2n}{d\lambda^2}\right)$ in ps/km/nm versus wavelength in μm

For example, for pure silica at 0.85 μm we find from the graph $Y_m = 0.021$, thus $\frac{\Gamma}{Z} = 7.2 \times 10^{-11} \gamma\ (s/m)$, and $(\Delta f)\ Z = 3.5 \times 10^9/\gamma [m/s]$, a typical range of a wavelength Δ*f* produced by GaAs light emitting diode source is about 0.30 nm at central wavelength of 0.85 μm so that γ = 30/850 = 0.035 and in bulk silica the rate at which pulses spread out is then (using Eq. 3.15.12)

$$\frac{\Gamma}{Z} = \frac{\gamma}{c}|Y_m| = \frac{0.35 \times 0.021}{3 \times 10^8} = 2.5 \times 10^{-12} = 2.5 \text{ ns/km}$$

and the bandwidth distance product is $(\Delta f)\ Z = 100$ MHz. km.

Example 3.7: *A laser source emits light at 0.85 μm having a spectrum width of 0.003 μm. The value of γ_m obtained from the graph is 0.021. Find the value of bandwidth distance product.*

Assume $\Delta f \approx \frac{1}{4\Gamma}$ *so that* $(\Delta f)\ Z = \frac{c}{\tau(\gamma_m)}$.

Solution: $$\gamma = \frac{0.003}{0.85} = 0.0035$$

Thus, $$\frac{\tau}{Z} = \frac{\gamma}{c}|\gamma_m| = \frac{0.0035 \times 0.021}{3 \times 10^8} = 2.5 \text{ ns/km}$$

Bandwidth-distance product

$$(\Delta f)Z = \frac{c}{4\gamma|\gamma_m|} = \frac{3 \times 10^8}{4 \times 0.021 \times 0.0035}$$

$$= 1 \text{ GHz. km}$$

3.16 COMBINED EFFECT OF MATERIAL DISPERSION AND INTERMODAL DISPERSION

In estimating the bandwidth and maximum bit rate of an optical communication channel we need to consider the shape and width of the received pulse. The shape on the arrival of an impulse that has been broadened by material dispersion will control the power distribution among the wavelengths making up the pulse. We shall find in general, that most optical sources have distribution of power with wavelengths that are roughly Gaussian in nature. We should then expect the received pulse shape to take on a similar Gaussian form about the mean arrival time t_o is shown in Fig. 3.16(*a*), (*b*). Now let us assume

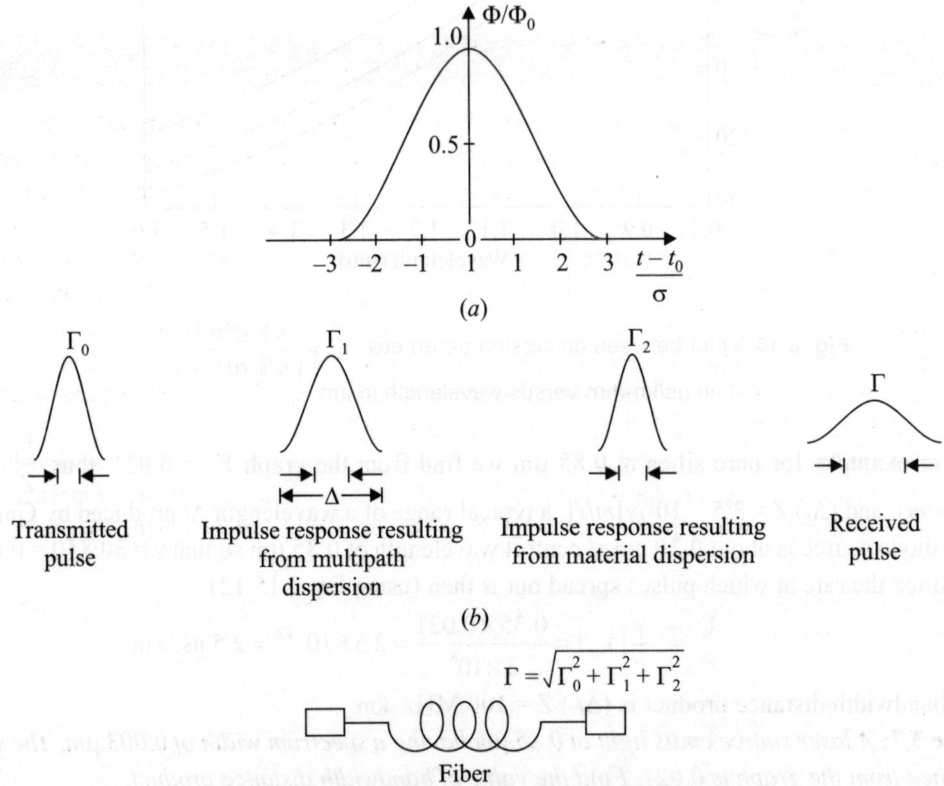

Fig. 3.16 Dispersive effect in combination

(*a*) The theoretical pulse shape $\mu(t) = \exp(t - t_0)/2\sigma^2$

Halfwidth = 2.355 σ

Real pulse will fall to zero infinite time

(*b*) Nevertheless the mechanisms causing the pulse broadening can be treated in practice that they are going to be independent of pulse broadening mechanisms.

$$\Gamma = \sqrt{\Gamma_0^2 + \Gamma_1^2 + \Gamma_2^2}$$

that the impulse is broadened by both material and multi path dispersions, and that the two mechanisms are uncorrelated and they independently lead to approximately Gaussian pulses having their half height pulse width Γ_1 and Γ_2 respectively. The two mechanisms will then combine to produce pulse which remains roughly Gaussian in shape and which will have a half width Γ which is given by

$$\Gamma = \sqrt{\Gamma_1^2 + \Gamma_2^2} \tag{3.16.1}$$

When the transmitted pulse is not an impulse to start with but is also a Gaussian with a full width at the half height value of Γ_0, then we may extend this argument as illustrated in Fig. 3.15 to suggest that Γ becomes

$$\Gamma = \sqrt{\Gamma_0^2 + \Gamma_1^2 + \Gamma_2^2} \tag{3.16.2}$$

where Γ_0 is the initial pulse width, Γ_1 is the half height pulse width that would result from multipath broadening alone and Γ_2 represents the broadening due to material dispersion alone.

3.17 WAVEGUIDE DISPERSION

When a fiber could be so operated that the multipath and material dispersions are all eliminated as should be the case for a single mode fiber operation near $\lambda = 1.3$ μm, then a third dispersion mechanism will predominate except in the case of ideally monochromatic light. Unfortunately ideal monochromatic sources are not available so that the light transmitted consists of components of several wavelengths near the central wavelength λ_o within an envelope like that in Fig. 3.13. The result is that the constant phase wave front separation as illustrated in Fig. 3.13 will be slightly different for different wavelengths and the corresponding angle of incidence will also be slightly different. This will cause a corresponding shift in the group velocity. The group velocity varies in complicated way with the wavelength and the net result is the broadening of the received pulse. This effect is only the result of the guiding characteristics of the fiber and hence it is known as the "waveguide dispersion".

For a step index fiber the dispersion coefficient resulting from waveguide dispersion mechanism has a peak value of about 0.6 ps/nm. The value is an ideal lower limit on the dispersion at the wavelength where the material dispersion disappears. A partial cancellation does take place between the material dispersion and waveguide dispersion which has a zero cross point as shown in Fig. 3.17 slightly higher in wavelength. By tuning the light source and keeping the bandwidth low it is possible to obtain practical value of loss less than 6.6 ps/nm.km.

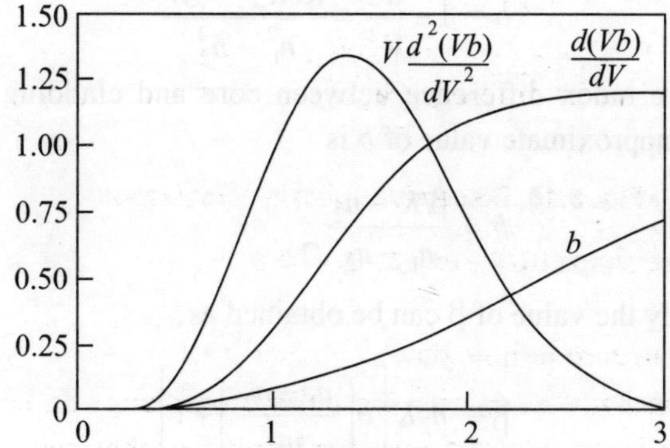

Fig. 3.17 Graph between b, $d(Vb)/dV$ and $V[d^2(Vb)/dV^2]$ versus V for the lowest order mode. When $V\left(\frac{d^2(Vb)}{dV^2}\right) = 0.480$, $V = 2.14$

Dispersion Units

Modal dispersion is independent of the line width of the source and usually expressed independent of line width of this source. It is specified for a fiber by the characteristic pulse spread per kilometre length in the units of ns km^{-1}. For multimode fiber dominated by modal dispersion on the specification can be given alternatively as an analog bandwidth—distance product in units of GHz.km.

Both material and waveguide dispersions depend on the source linewidth. To remove the dependence the specification for these quantities can also be given as ns.km^{-1}.nm^{-1}.

3.18 WAVEGUIDE DISPERSION

When an optical fiber operates at a wavelength around 1.23 μm the material dispersion is low but the waveguide dispersion becomes important near this wavelength. Generally this type of dispersion is negligible in multimode fibers but in a singlemode fiber operating at the wavelength around 1.23 μm, it is not negligible except in the case of monochromatic light. Unfortunately ideal monochromatic sources are not available so that the light transmitted consists of components of several wavelengths near the central wavelength λ_0 in Fig. 3.13. The result is that the constant phase wave front separation will be slightly different for different wavelength and the corresponding angle of incidence will also be slightly different. This will cause a corresponding shift in the group velocity. The group velocity varies in complicated way with the wavelength and the net result is the broadening of the received pulse. The effect is only the result of guiding characteristics of the fiber and hence it is known as "waveguide dispersion". The amount of delay due to waveguide dispersion is given by

$$\Gamma_{wg} = \frac{Z}{c}\frac{d\beta}{dK} \tag{3.18.1}$$

where Z is the length of the fiber, c is the velocity of ligh in a vacuum β is the phase constant for a guided mode within fiber core of refractive index n_1 and K is the propagation for the mode in a vacuum. In Chapter 2 we have defined the normalized propagation constnat b as (using 2.16.6*a*)

$$b = 1 - \frac{u^2}{V^2} = \frac{(\beta/K)^2 - n_2^2}{n_1^2 - n_2^2} \tag{3.18.2}$$

When the refractive index difference between core and cladding is small we can assume $\beta/K + n_2 \approx n_1 + n_2$ so that approximate value of b is

$$b \approx \frac{\beta/K - n_2}{n_1 - n_2} \tag{3.18.3}$$

Rearranging properly the value of β can be obtained as

$$\beta = n_2 K\left[b\left(\frac{n_1 - n_2}{n_2}\right) + 1\right]$$

$$\approx n_2 K(b\Delta + 1) \tag{3.18.4}$$

where $\quad \dfrac{n_1 - n_2}{n_2} = \Delta$ and $K = \dfrac{2\pi}{\lambda}$

From Eq. (3.18.2) we find that b is the function of V and K. On substitution of Eq. (3.18.4) in Eq. (3.18.1), we get

$$\Gamma_{wg} = \frac{Z}{c}\left(n_2 + n_2\Delta\frac{d(Kb)}{dK}\right) \quad (3.18.5)$$

We already know that

$$V = \frac{2\pi}{\lambda}\, a\, n_1\sqrt{2\Delta}$$

$$= \frac{2\pi}{\lambda}\, a\, n_2\sqrt{2\Delta}.\ \text{Since } n_1 - n_2 << n_1$$

$$V = K\, a\, n_2\sqrt{2\Delta} \quad (3.18.6)$$

We can now write from Eq. (3.15.6) and Eq. (3.15.6). The amount of delay Γ due to waveguide dispersion

$$\Gamma_{wg} = \frac{Z}{c}\left(n_2 + n_2\Delta\frac{d(Vb)}{dV}\right) \quad (3.18.7)$$

Neglecting the constant term in delay, the wavelength dependent term in time delay

$$\Gamma_{wg} = \frac{Z}{c} n_2\Delta\frac{d(Vb)}{dV}$$

and $$\Delta\Gamma_{wg} = \frac{d\Gamma_{wg}}{dV} = \frac{Z}{c} n_2\,\Delta\frac{d^2(Vb)}{dV^2} \quad (3.18.8)$$

Figure 3.18 show a plot of $\frac{d^2(Vb)}{dV^2}$ versus V for the lowest order mode HE_{11} mode. In a single mode fiber the modal dispersion is not present and in the wavelength region near 1.27 μm the wave guide dispersion is comparable with material dispersion. We like to discuss the waveguide dispersion in that region of the spectrum.

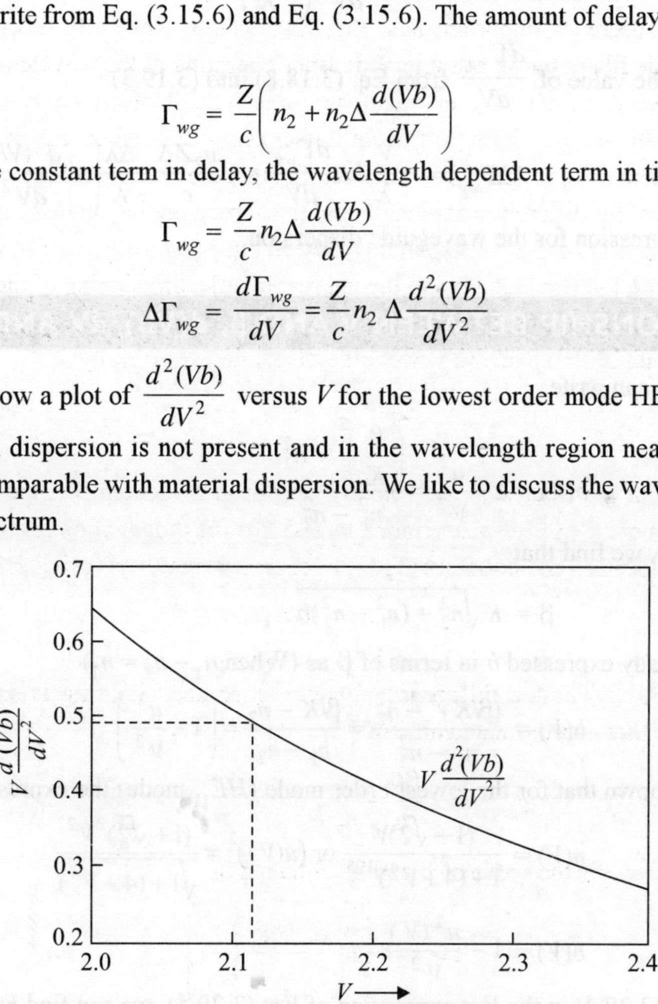

Fig. 3.18 A graph between $V\frac{d^2(Vb)}{dV^2}$ and V for the first order fiber mode in the case of a practical singlemode fiber. When $V = 2.14$, $V\frac{d^2(Vb)}{dV^2} = 0.480$.

3.19 EFFECT OF SOURCE LINE WIDTH

The difference in the propagation time $\Delta\Gamma_{wg}$ due to finite line width $\Delta\lambda$ of the signal of wavelength λ is

$$\Delta\Gamma_{wg} = \Delta\lambda \frac{d\Gamma_{wg}}{d\lambda} = \Delta\lambda \frac{d\Gamma_{wg}}{dV} \cdot \frac{dV}{d\lambda} \tag{3.19.1}$$

We have

$$V = \frac{2\pi a}{\lambda} n_1 \sqrt{2\Delta}, \text{ and } \frac{dV}{d\lambda} = -\frac{2\pi a n}{\lambda^2}\sqrt{2\Delta} = -\frac{V}{\lambda} \tag{3.19.2}$$

From Eq. (3.19.1) and Eq. (3.19.2), we get

$$(\Delta\Gamma_{wg}) = \Delta\lambda \frac{d\Gamma_{wg}}{dV}\left(-\frac{V}{\lambda}\right) \tag{3.19.3}$$

Substituting the value of $\frac{d\Gamma_{wg}}{dV}$ from Eq. (3.18.8) into (3.19.3)

$$(\Delta\Gamma_{wg}) = \frac{V}{\lambda}\Delta\lambda \frac{d\Gamma_{wg}}{dV} \approx \frac{n_2 Z\Delta}{c} \cdot \frac{\Delta\lambda}{\lambda}\left(V\frac{d^2(Vb)}{dV^2}\right) \tag{3.19.4}$$

This is an expression for the waveguide dispersion.

3.20 RELATIONSHIP BETWEEN V AND β, d(Vb)/dV AND V d²(Vb)/dV²

From Eq. 3.18.2, we can write

$$b(V) = 1 - \frac{u^2}{V^2} = \frac{\left(\frac{\beta}{K}\right)^2 - n_2^2}{n_1^2 - n_2^2} \tag{3.20.1}$$

Solving for β, we find that

$$\beta = K\sqrt{n_2^2 + (n_1^2 - n_2^2)b} \tag{3.20.2}$$

We have already expressed b in terms of β as (When $n_1 - n_2 \approx n_1$)

$$b(V) = \frac{(\beta/K)^2 - n_2^2}{n_1^2 - n_2^2} \approx \frac{\beta/K - n_2}{n_1 - n_2} = \left(1 - \frac{u^2}{V^2}\right) \tag{3.20.3}$$

Gloge[6] has shown that for the lowest order mode (HE_{11} mode) the expression for $u(V)$ is

$$u(V) = \frac{(1+\sqrt{2})V}{1+(4+V^4)^{0.25}} \text{ or } (u(V))^2 = \frac{(1+\sqrt{2})^2 V^2}{\sqrt{1+(4+V^4)}} \tag{3.20.4}$$

$$b(V) = 1 - \frac{u^2(V)}{V^2} \tag{3.20.5}$$

Substituting (3.20.4) in the last expression of Eq. (3.20.5), we can find $b(V)$

$$b(V) = 1 \frac{(1+\sqrt{2})^2}{\sqrt{1+(4+V^4)}} \tag{3.20.6}$$

[6] Pl. see the reference on page 32

Fig. 3.16 shows a plot $b(V)$ with V. From this expression we can obtain plots of $d(Vb)/dV$ and $Vd^2(Vb)/dV^2$ as a function V for a single mode fiber the value V varies from 2 to 2.405 as shown in Fig. 3.17. The value $Vd^2(Vb)/dV^2$ decreases monotomically from 0.64 down to 0.25. Hence the wave guide dispersion has a small negative value.

Example 3.8: *Calulate the waveguide dispersion in unit of ps. km^{-1} for a 9/125 single mode fiber with n_1 = 1.48, D = 0.22% operating at 1.3 mm, core diameter = 9 mm cladding diameter 125 mm.*

Solution: Data given:

$$\lambda = 1.3 \times 10^{-6}, a = 4.5 \times 10^{-6}; \Delta = 0.0022\%; n_1 = 1.48$$

We calculate the value of V from Eq. (3.19.2)

$$V = \frac{2\pi a}{\lambda} n_1 \sqrt{2\Delta} = \frac{2\pi \times 4.5 \times 10^{-6} \times 1.48}{1.3 \times 10^{-6}} \sqrt{2 \times 0.002}$$

$$= 2.14$$

$$n_2 = n_1 - \Delta n_1 = 1.48(1 - 0.002) = 1.477$$

In the case of a single mode fiber the value V lies in the range $2.0 < V < 2.405$ for HE_{11} mode.

From Fig. 3.16, we find that $V \dfrac{d^2(Vb)}{dV^2} = 0.480$ at $V = 2.14$ so that we get Eq. (3.19.1)

$$\frac{\Delta\Gamma_{wg}}{Z\Delta\lambda} = \frac{(-n_2\Delta)}{c}\frac{1}{\lambda} V \frac{d^2(Vb)}{dV^2} = \frac{-1.477 \times 0.0022}{3 \times 10^8} \cdot \frac{1(0.48)}{1.3 \times 10^{-6}}$$

$$= -400 \times 10^{-6} = -4 \text{ ps km}^{-1}.\text{nm}^{-1}$$

3.21 TOTAL DISPERSION AND MAXIMUM TRANSMISSION RATE (PULSES MAY BE GAUSSIAN)

Each of three dispersions gives rise to the pulse broadening during the transmission through the fiber. Each contributes more or less depending on the type of fiber and its operating conditions. In the case of a multimode step index fiber the intermodal (or multipath) dispersion will be predominant. While the waveguide dispersion will be predominant when a single mode fiber is operated at material dispersion null wavelength of about 1.3 μm. The total dispersion at any wavelength is equal to the root mean square combination of these three types of dispersion or

$$\Delta t_{\text{total}} = \sqrt{(\Delta t_{im})^2 + (\Delta t_{md})^2 + (\Delta t_{wgd})^2} \tag{3.21.1}$$

Of course, each of these values of dispersion in Fig. 3.15 is a theoretical value and because of departure from the ideal practical fibers the actual values will be somewhat higher. However, the result is satisfactory in determining an estimate of the behaviour of the fiber.

In order to find out how the dispersion affects the maximum bit rate or information rate we examine Fig. 3.13. The figure illustrates the manner in which pulses arriving at the receiver with different amount of delay and form a dispersed or broadened pulse. Pulse A represents a pulse which experiences an average delay. Pulse B is one which experiences the minimum, and pulse C is one which experiences the maximum delay. The resultant received pulse is obtained by adding all pulse components between the limits B and C. Figure 3.11 gives an illustration using a digital bit pattern 1011 of the broadening of light pulses as they are transmitted along a fiber Fig. 3.11(a) shows pulse at fiber input. Figures 3.11(b) and 3.11(c) pulses at distance L_1 and L_2 where $L_2 > L_1$.

3.22 MAXIMUM BIT RATE

The pulse width Γ_0 denotes the width of the transmitted pulse, while Γ represents the width of the received pulse. Γ is greater than Γ_0 by total dispersion time Δ_{total}. Thus,

$$\Gamma = \Gamma_0 + \Delta_{total} \tag{3.22.1}$$

If it is desired to detect the next pulse without any ambiguity, then it must not be transmitted until a period Γ after the beginning of the next pulse and it will allow the time for the spreading to subside. The bit rate corresponding to this lead situation will be

Bit rate, $$B = 1/\Gamma \tag{3.22.2}$$

In actual case the dispersion factors have a tendency to become larger than the ideal in practice, and the pulse tail tend to spread beyond the average pulse limit. It is essential to allow a safety factor of about 5 times on the received pulse width Γ. Thereby the desired error rates in transmission will be minimum. In order to obtain the minimum bit rate, the width of the transmitted pulse can be made considerably smaller than the total dispersion Δ_{total} encountered. Thus, if $\Gamma_0 << \Delta_{total}$, the equation for the bit rate becomes

$$B_{max} = \frac{1}{5(\Delta t_{total} + \Gamma_0)} \approx \frac{1}{5\,\Delta t_{total}} \tag{3.22.3}$$

Provided $\Delta t_{total} >> \Gamma_0$

Example 3.9: *A single mode fiber operating at the wavelength of 1.3 μm is found to have a total material dispersion of 2.81 ns, and a total waveguide dispersion of 0.495 ns. Determine the received pulse width and approximate bit rate of the fiber if the transmitted pulse has a width of 0.5 ns. Intermodal dispersion does not occur in a single mode fiber.*

Solution: From Eq. (3.21.1)

Total dispersion $$\Delta t_{total} = \sqrt{(\Delta t_{imd})^2 + (\Delta t_{md})^2 + (\Delta t_{wgd})^2}$$

$$= \sqrt{0^2 + 2.81^2 + (0.495)^2} = 2.85 \text{ ns}$$

Width of the received pulse $$\Gamma = \Gamma_0 + \Delta t_{total}$$

$$= 0.5 + 2.85 = 3.35 \text{ ns}$$

Approximate bit rate $$B_{max} = \frac{1}{5 \times 3.35 \times 10^{-9}} \approx 60 \text{ MHz}.$$

3.23 TOTAL DISPERSION: SINGLE MODE FIBER

In order to minimise the total dispersion of a signal through a single mode fiber, it is essential to operate it at a wavelength longer than 1.37 μm. At that wavelength the small value of material dispersion cancels the small waveguide dispersion. Thus, the total dispersion is reduced to zero as shown in Fig. 3.19. The zero dispersion point occurs near 1300 nm wavelength. At this wavelength there is also a small attenuation of the signal through the fiber, this helps the operation of a data link at this wavelength. Thus a low dispersion at this wavelength gave an impetus for the development of a family of sources and receivers which will operate at this wavelength.

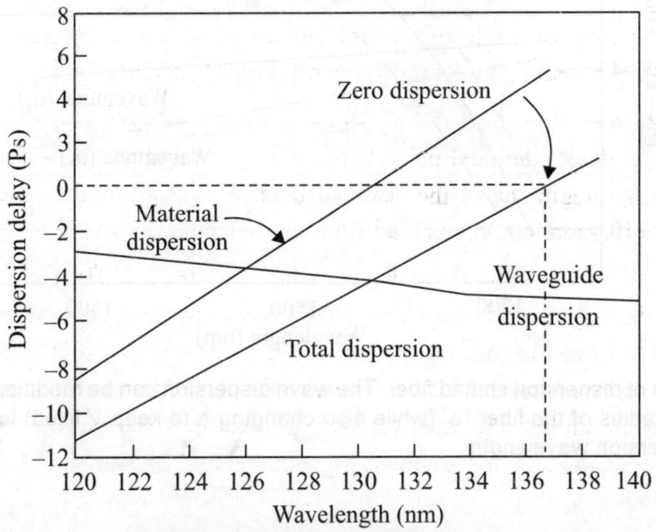

Fig. 3.19 Addition of fiber material dispersion and waveguide dispersion of a single mode fiber to achieve zero dispersion at an wavelength of 1.37 μm.

The amount of waveguide dispersion has been found to depend on the amount of doping level as well as the core radius "*a*" and the refractive index gradient Δ between the core and cladding. As for the triangular doping profile (Fig. 3.19) zero dispersion is attained at other wavelengths between 1300 nm and 1700 nm. Thus it was possible to develop fibers that have a minimum attenuation and zero dispersion at the wavelength of 1500 nm. This process is called the "dispersion shifting" and the fiber is referred to as the "dispersion shifted fiber."

3.24 DISPERSION-ADJUSTED SINGLE-MODE FIBER

We have observed that the lowest attenuation occurs at 1500 nm wavelength, while the lowest total dispersion occurs at the wavelength 1300 nm. Since these two wavelengths are very near to each other, it has become an aim to combine these two features so that the optimum performance in terms of distortion (which controls the bandwidth) and loss occur at the same wavelength. Out of these two parameters the dispersion is most adjustable and is the one which is manipulated. These manipulations will lead to two adjustments; one is to shift the zero dispersion wavelength to a higher value so that dispersion is flattened in the vicinity of 1550 nm (Fig. 3.20) and the other is to minimise or flatten dispersion over a range of wavelength value.

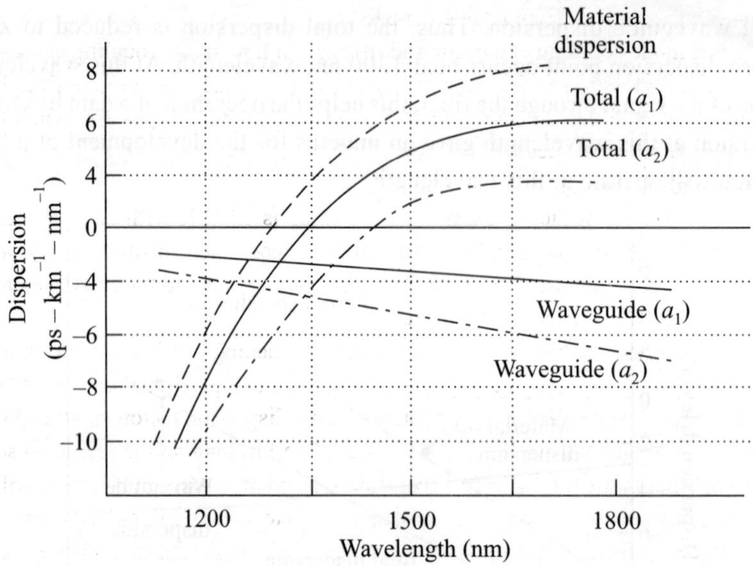

Fig. 3.20 Examples of dispersion shifted fiber. The wave dispersion can be modified by changing the core radius of the fiber "*a*" (while also changing Δ to keep *V* fixed) to increase the zero dispersion wavelength.

REVIEW QUESTIONS

3.1 Define attenuation or losses in an optical signal through a fiber. Determine an expression for the attenuation in dB per km in an optical fiber. The average power lunched in a 15 km length of an optical fiber is 200 μW and the average output power available is 4 μW. Estimate the signal attenuation or loss in dB through the fiber without any connectors splices (*b*) signal attenuation per km. (*c*) the attenuation by 10 km length of fiber with three splices each with an average attenuation of 0.5 dB (*d*) what is the numerical ratio between the input and output for the 10 km length of fiber with 3 splices each with attenuation of 0.5 dB.

3.2 What are the basic attenuation mechanisms? On what factor these mechanisms depend?

3.3 Explain in brief Rayleigh scattering mechanisms. What are two important factors on which the scattering of glass depends? Write down the equation for the scattering component Γ_r and attenuation in dB per km due to fundamental Rayleigh scattering in silica.

3.4 What are the three important mechanisms that are responsible for absorption losses in signal through an optical fiber. Explain in brief the curve for wavelength versus attenuation for different ranges of the signal.

3.5 Explain the bending losses.

What is micro bending? How can it be reduced? What is the method of minimizing micro bending. Illustrate with diagram how the micro bending is avoided by a compressible jacket.

3.6 Two step index fibers one of which is of monomode type and the other one is of multimode type have the following values of parameters.

The multimode fiber has a core refractive indeed of 1.520 and relative index difference between the core and cladding of 5 per cent. The operating wavelength is 0.80 μm. Calculate the critical radius of curvature at which the bending losses will occur in the fiber.

3.7 Explain how energy is lost from a "leaky mode".
Explain how a mode coupling can occur and (*b*) explain how mode coupling causes loss.

3.8 The mean optical power launched into an optical fiber is 2.25 μW and the fiber has an attenuation of 0.5 dB/km. Determine the maximum possible link length (with repeaters, when the minimum optical power level required at the detector is 3 μW.)

3.9 Discuss Rayleigh scattering losses in optical fiber.
An estimated value of the fictive temperature of silica is 1400 K with an isothermal compressibility of 7×10^{-11} met^2 N^{-1}. The refractive index and photo elastic coefficient of silica are 1.46 and 0.286 respectively estimate the theoretical attenuation in decibels per kilometer due to the fundamental Rayleigh scattering in silica at an optical wavelength of 0.63 μm.

3.10 A single mode step-index fiber gives a total pulse broadening of 95 ns over a5 km length. Estimate the bandwidth length product for the fiber when a nonreturn to Zero digital code (NRZ) is used.

3.11 A step index fiber operated at 1.3 mm is found to have dispersion factors of 10 ns/km intermodal, 0.5 ns/km material, and 0.8 ns/km waveguide. The fiber is 15 km long and is restricted so that only one mode propagate (*a*) what is the total effective dispersion (*b*) what maximum bit rate it will handle.

3.12 The refractive indexes of the core and cladding of a step index fiber in problem 3.11. are 1.5000 and 1.4980 respectively. Find (*a*) the intermodal dispersion factor for the fiber (*b*) the total dispersion in an 18 km length and (*c*) the maximum bitrate allowed assuming dispersion limiting. (6.68 ns/km, 120 ns, 1.66 M bits/second)

3.13 A step index fiber has a core index of 1.5000 and a fractional difference of 0.0005 with cladding refractive index. It is operated at a wavelength of 1.20 μm. What is the largest diameter of the core may have if it is to operate as a single mode fiber? (*b*) Which mode does a single mode fiber propagate (19.4 μm, HE_{11} mode).

CHAPTER 4

Light Emitting Diode: LED

CHAPTER FOCUS

- *Introduction: Requirements of an optical source*
- *Physical principle of operation of a p-n photo-diode*
- *The p-n junction on the basis of valence and conduction bands*
- *Fermi-dirac function*
- *Energy level diagram: Extrinsic semiconductor*
- *Energy band picture of unbiased p-n junction*
- *Energy band picture of a forward biased p-n junction*
- *Energy band and current in reverse biased p-n diode*
- *Direct bandgap semiconductors*
- *Indirect bandgap semiconductors*
- *Minority carrier lifetime*
- *Spontaneous emission: Electroluminescence*
- *Design of LED for optical communications*
- *Basic theory of a double heterojunction LED (DHLED)*
- *Power and internal efficiency of LED*
- *Different LED structures*
- *Double heterostructure surface emitter LED (SLED, Burrus type)*
- *Stripe geometry AlGaAs edge-emitter LED (ELED)*
- *Stripe geometry InGaAs edge-emitter LED (ELED)*
- *Super luminescent LED (SLD)*
- *Super luminescent AlGaAs contact stripe SLD*
- *Lens coupling to fiber*

- *Truncated spherical microlens coupling*
- *Output spectrum*
- *Modulation bandwidth*
- *Electrical bandwidth*
- *Optical bandwidth*
- *Factors determining modulation bandwidth*
- *Drive circuits for LED and laser*
- *Analog transmission drive circuit*
- *Solved examples and review questions*

4.1 INTRODUCTION: REQUIREMENTS OF AN OPTICAL SOURCE

In fiber communications, light sources produce an optical wave to act as a carrier wave in the transmitter and electrical signals either analog or digital serve to amplitude modulate the carrier light wave. The optical source for fiber communications must satisfy the following requirements.

1. Firstly, the light sources must be as nearly as possible monochromatic which means that the light consists of single frequency. In practice, most light sources do not emit radiation of single frequency, but do emit light of several frequencies. Of course, few light sources such as light emitting diodes (abbreviated as LED's) and laser diodes emit light over a narrow band of frequencies. In strict sense, they are not absolutely monochromatic, but emit over an extremely narrow band of frequencies.
2. Secondly, the intensity of light which is referred to as "radiance" should be high as far as possible over a narrow band of wavelengths in the range of 0.6 to 1.7 μm and as such, sufficient light can be transmitted to overcome the loss suffered by the light wave during transmission through the optical fiber. Radiance is a measure in watts into a unit solid angle from per unit area of the emitting surface.
3. Thirdly, the light sources must be capable of being amplitude modulated by the electrical signal to be transmitted through the fiber.
4. The device must be small in size, compatible with the size of the fiber, compact, inexpensive to manufacture and can easily to be coupled with the fiber without suffering excessive losses due to coupling with the hair sized fiber.

The light emitting diodes and semiconductor laser diodes approximately satisfy the above requirements. Both devices emit radiations almost at fixed frequency and find extensive applications in fiber communications.

In their operation the excited electrons which are free to recombine with holes (electron vacancy) generate photons of light in each recombination process. The photon of light is the result of converting potential energy into light energy.

Figure 4.1 shows the emission spectrum of few light waves used in fiber communications.

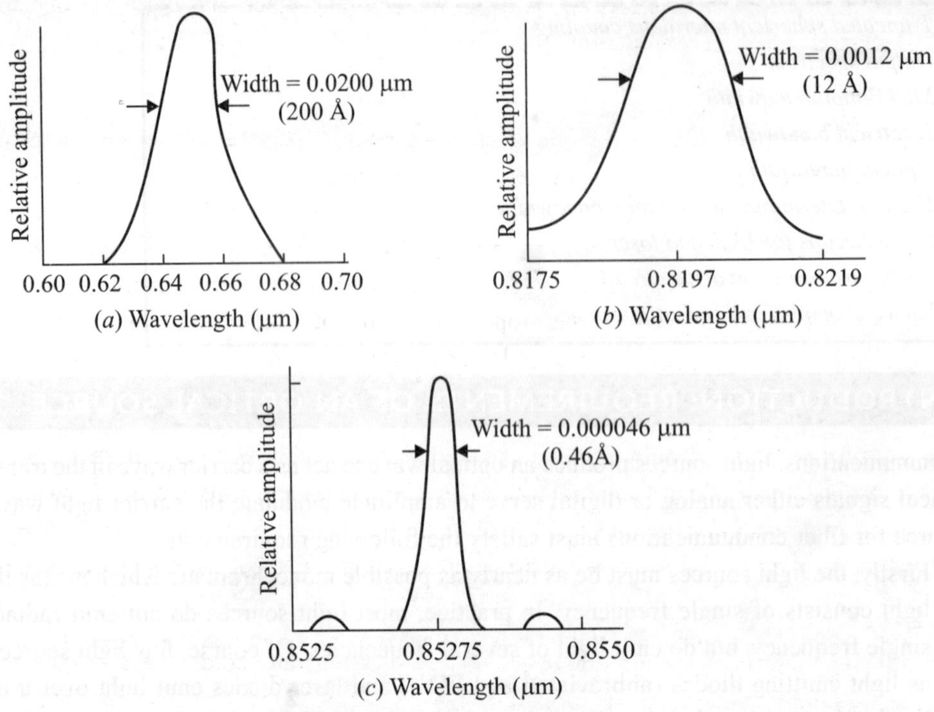

Fig. 4.1 The emission spectra of few light sources used in optical fiber communications (*a*) an LED operating at 0.65 μm (*b*) a broad spectrum laser diode operating at 0.8197 μm and (*c*) a narrow spectrum laser diode at 0.85275 μm.

4.2 PHYSICAL PRINCIPLE OF OPERATION OF A *p-n* PHOTO-DIODE

The light emitting diode which is basically a semiconductor *p-n* junction diode emits a coherent light, when a forward current is passed through it. The basic principle on which it work is the process of "sponanteous emission". In order to understand the operation of a light emitting diode, we shall review the working of a *p-n* junction diode.

The *p-n* Junction Diode

A perfect semiconductor crystal with no impurities or lattice defects is referred to as an "intrinsic semiconductor". Germanium, silicon and gallium arsenide are the examples of the semiconductor materials. One side of the single crystal semiconductor diode is develeped as the *p*-type semiconductor by adding a minute quantity (1 part in 10^8 parts) of the acceptor impurity atoms such as indium or aluminium which belong to Group-III of the periodic table. In the *p*-type region "holes" (or electron vacancies) are the majority carriers, while free electrons are the minority carriers. When a covalent bond structure is broken in a crystal, a hole is formed by the removal of an electron from the bond structure. Thus, a hole is nothing but an electron vacancy in a crystal bond. The other side of the crystal is developed as the *n*-type region by adding a minute quantity (1 part in 10^8 parts) of donor impurity atoms such as arsenic or antimony which belong to Group V of the periodic table. The *n*-region contains large number of free electrons as the majority carriers, while the holes are the minority carriers in the *n*-region of the crystal.

As soon as the *p-n* junction is formed, two phenomena take place and these are due to the properties of "Diffusion" and "Recombination".

Diffusion

At the junction free electrons and holes are at liberty to move inside the crystal. As soon as the junction is formed, some of the free electrons from the *n*-region are able to cross the junction into the *p*-region where the electron concentration is very low, whereas some of the holes from the *p*-region pass across the junction into the *n*-region where the concentration of the holes is extremely small. The phenomenon of movement of free electrons and holes from one region to other due to the difference in the concentration is known as "Diffusion". Thus, diffusion is the property by virtue of which the charge carriers move from a location of higher concentration to a that of lower concentration.

Recombination

After crossing the junction a free electron encounters a large concentration of holes and the mobile free electrons quickly disappear due to recombination with a hole. Similarly, the holes that cross the junction to move from the *p*-region to the *n*-region due to diffusion recombine with free electrons which have got a high concentration in the *n*-region and as a result, the holes disappear just after diffusion and recombination.

Depletion Layer

As a result of diffusion and recombination there are no free electrons and holes in the vicinity of the metallurgical junction, but there exist positively ionized donor atoms in the *n*-region near the junction and the negatively charged ions of the acceptor atoms in the *p*-region near the junction. These ions are immobile since they are tightly bound to the crystal structure. The whole region near the junction which does not contain any free electrons and holes but does contain immobile charges is known as the "Depletion Layer" (or transition region, or space charge region or uncovered charge region), because all mobile charges such as free electrons and holes are depleted from this region. The depletion layer has got usually a small thickness of the order of 0.5 μm or less. Figure 4.2 shows a *p-n* junction with charge carriers and potential barrier across the depletion region.

Potential Barrier

As a result of charge accumulation of opposite nature as shown in Fig. 4.2, a barrier potential V_B develops across the depletion layer making the potential of the *n*-region positive with respect to that of *p*-region. The barrier potential and hole electron energies of an unbiased *p-n* junction are shown in Fig. 4.2(*c*), (*d*) and (*e*) respectively. The barrier potential gradient is of such polarity as to oppose the further movement of free electrons and holes between the two regions.

The equilibrium condition will be established so that free carriers such as free electrons and holes with their highest possible energy are unable to cross the junction. The barrier potential V_B of an unbiased junction is given by

$$V_B = \frac{K_B T}{e} \ln \frac{n_n}{n_p} = \frac{K_B T}{e} \ln \frac{p_p}{p_n} = \frac{K_B T}{e} \ln \frac{n_n p_p}{n_i^2} \tag{4.2.1}$$

where n_n and p_n are the densities of free electrons and holes respectively in the *n*-region. p_p and n_p are the densities of holes and free electrons respectively in the *p*-region. n_i is the density of electron-hole pairs in the intrinsic semiconductor, *e* is numerically equal to the electronic charge = 1.6×10^{-19} C, K_B = Boltzmann constant = 1.38×10^{-23} J/K, *T* = absolute temperature for a germanium *p-n* junction diode the barrier potential is of the order of 0.2 to 0.3 volts, while that for a silicon device the barrier potential is of the order of 0.5 to 0.6 volts at room temperature.

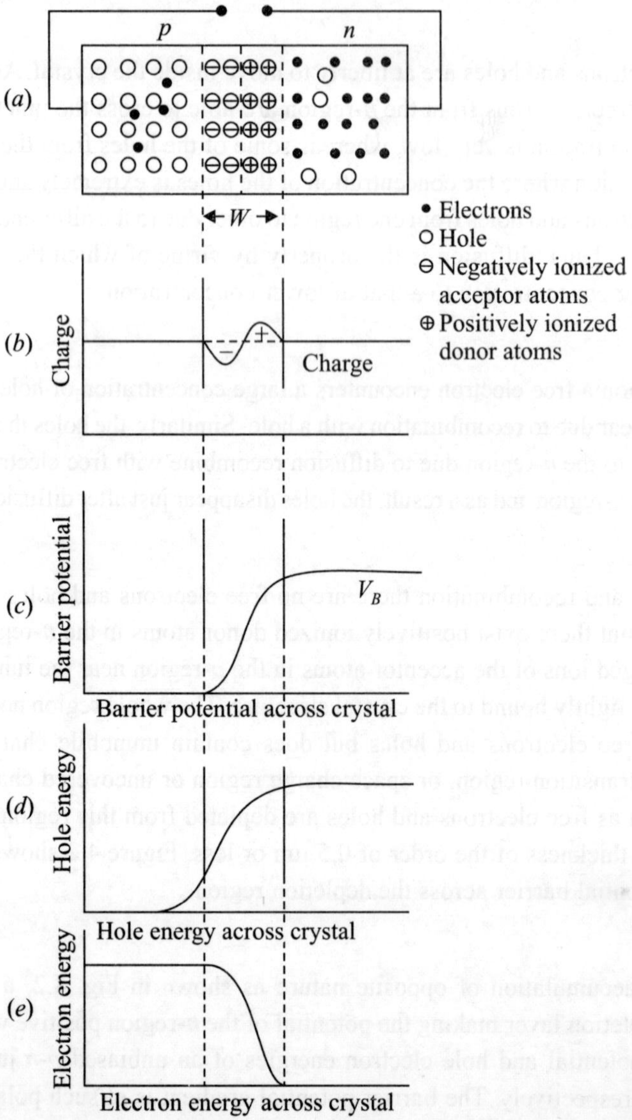

Fig. 4.2 (*a*) Distribution of charge carriers, such as electrons and holes across the length of an unbiased *p-n* junction diode (*b*) charge distribution across the depletion layer due to immobile charges due to negatively ionized acceptor atoms and positively ionized donor atoms (*c*) barrier potential V_B making *n*-region positive with respect to the *p*-region (*d*) hole energy and (*e*) electron energy across the length of crystal

Forward Bias of a *p-n* Junction

When a *p-n* junction is forward biased by voltage E_f by making the *p*-region positive with respect to the *n*-region [Fig. 4.3(*a*)], the effective potential barrier decreases considerably and will be equal atmost to $(V_B - E_f)$ volts, because all applied voltage develops across the junction and is in the opposite direction to V_B. Figure 4.3(*c*), (*d*) and (*e*) show the potential barrier hole and electron energies respectively of a forward biased *p-n* junction. The width of the depletion layer *W* decreases with the potential barrier and is given by

$$W = \left[\frac{2\varepsilon_r\,\varepsilon_o (V_B - E_f)}{e}\left(\frac{1}{N_d} + \frac{1}{N_a}\right)\right]^{1/2} \tag{4.2.2}$$

where $\varepsilon_r\,\varepsilon_o$ is the permittivity of the semiconductor material. ε_r = relative dielectric constant and $\varepsilon_o = (8.85 \times 10^{12})$ farad/m. N_d and N_a are the densities of donor and acceptor atoms per cubic metre in the *n*-type and *p*-type material regions respectively.

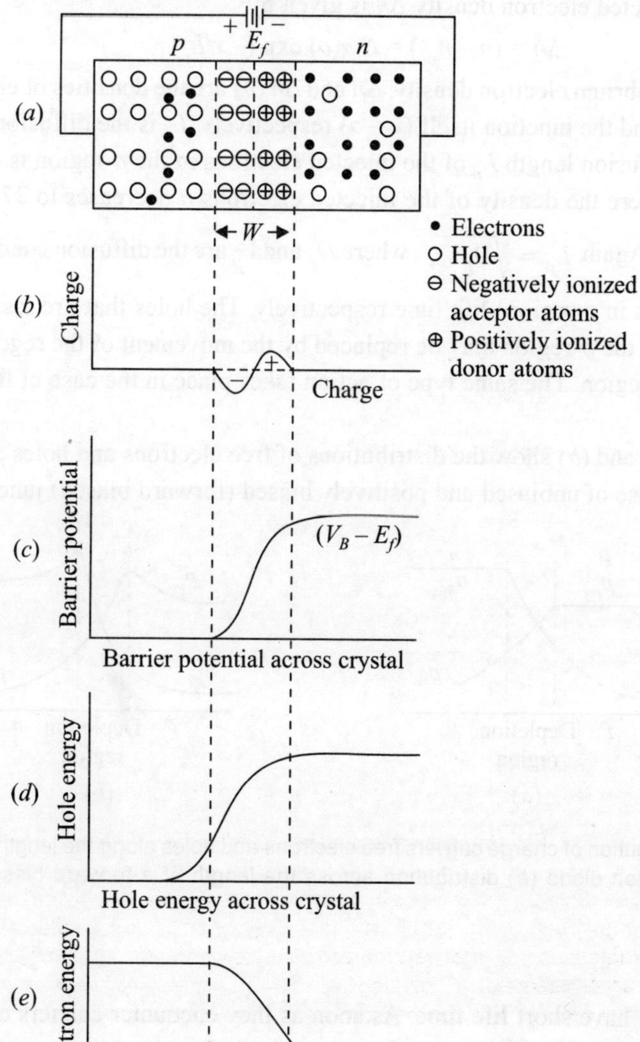

Fig. 4.3 (*a*) Distribution of charge carriers across the length of a forward biased *p-n* junction diode (*b*) charges in the depletion layer (*c*) reduced potential barrier aross the depletion layer (*d*) hole energy and (*e*) electron energy across the length of the crystal

As a result of reduction in the potential barrier, holes with comparatively higher energy will be able to move from the *p*-region to the *n*-region and the free electrons with comparatively higher energy will move from the *n*-region to the *p*-region. After entering into the *n*-region the holes become the minority carriers and in the *p*-region electrons become the minority carriers. Technically, we say that holes and free electrons are "injected" in the *n*-region and *p*-region respectively.

It may appear that the movement of the majority carriers from one region to the other may result in an increase in minority carriers in each region. However as soon as the free electrons are injected into the *p*-region, a small fraction of the same may recombine with holes of the *p*-region. Thus, as the stream of electrons moves through the *p*-region, their density decreases exponentially with the distance *x* from the junction. The injected electron density Δn is given by

$$\Delta n = (n - n_{po}) = \Delta\, n(o)\, \exp\,(-x/L_p) \quad (4.2.3)$$

where n_{po} is the equilibrium electron density. Δn and $\Delta n\,(o)$ are the densities of electrons at any distance *x* from the junction and the junction itself ($x = o$) respectively. L_p is the diffusion length of electrons in the *p*-region. The diffusion length L_p of the injected electrons in the *p*-region is defined as the distance from the junction where the density of the injected electrons *n* decreases to 37 per cent (times) of its value at the junction. Again $L_p = \sqrt{D_p \Gamma_p}$, where D_p and Γ_p are the diffusion coefficient and the minority carrier (free electrons in *p*-region) life time respectively. The holes that are lost due to recombination with free electrons in the *p*-region may be replaced by the movement of the regenerated holes from the further end of the *p*-region. The same type of action takes place in the case of the injected holes in the *n*-region.

Figure 4.4 (*a*) and (*b*) show the distributions of free electrons and holes across the length of the *p*-*n* junction in the case of unbiased and positively biased (forward biased) junctions respectively.

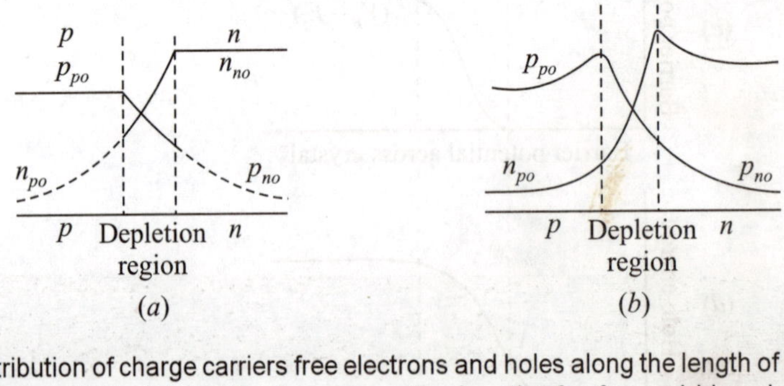

Fig. 4.4 (*a*) Distribution of charge carriers free electrons and holes along the length of an unbiased *p-n* junction diode (*b*) distribution across the length of a forward biased *p-n* junction diode

Photo Emission

The injected carriers have short life time. As soon as they encounter carriers of opposite polarity, the electrons and holes recombine. When an electron and a hole recombine in a suitable condition in the depletion region, they may emit light of certain energy which is equal to the forbidden energy gap of the semiconductor material used to fabricate the *p-n* junction. Figure 4.5 shows that the recombination may occur in any place in the diode structure.

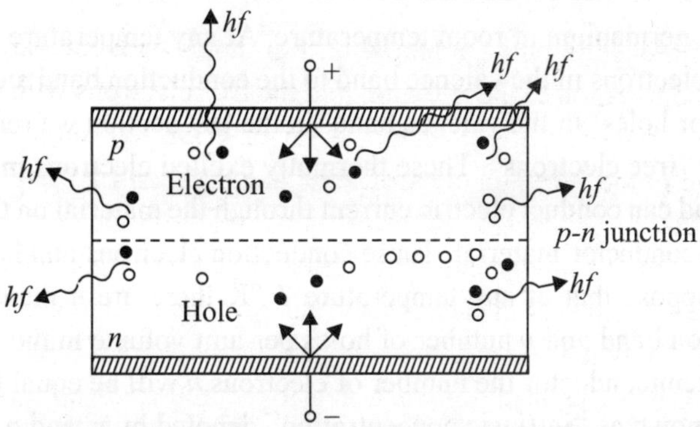

Fig. 4.5 Carrier recombination may occur at any place in the whole of a *p-n* junction structure although light is emitted from the carrier recombination sites close to the junction

Relationship Between Forbidden Energy Gap E_g and Emitted Wavelength

The energy of each photon E and the frequency f of the electromagnetic radiation are related by

$$E = hf \tag{4.2.4}$$

where E is the energy in Joule, f is the frequency in Hz and h is the Planck's constant ($h = 6.68 \times 10^{-34}$ J-s). The light spectrum is usually expressed in terms of wavelength λ instead of frequency f and the two have relationship

$$f = c/\lambda \tag{4.2.5}$$

where c is the velocity of the electromagnetic radiation (light) ($c = 3 \times 10^8$ m/s) and λ is the wavelength in metres.

When the energy E is expressed in electron-volts we have

$$E = eE_{ev} \tag{4.2.6}$$

Where e is numerically equal to electronic charge 1.6×10^{-19} C and E and E_{ev} are the energies in Joules and electron-volts respectively combining Eqs. (4.2.4, 4.2.5 and 4.2.6) the relationship between the forbidden energy gap E_g in electron volts and wavelength λ in μm (10^{-6} metres), we get

$$E_{ev} = \frac{hc}{e\lambda} = \frac{1.24}{\lambda} \tag{4.2.7}$$

4.3 THE *p-n* JUNCTION ON THE BASIS OF VALENCE AND CONDUCTION BANDS

In the last section, we have discussed the properties of *p-n* junction on physical basis and now we shall review the operation of a *p-n* junction in terms of valence and conduction bands. This knowledge is essential to understand the basic principle of operation of opto-electronic devices such as light emitting diode (LED), laser, and photo diode. We have already defined "Intrinsic Semiconductor" as one with no impurities or lattice defects.

Figure 4.6(*a*) shows the energy band structure of an intrinsic semiconductor and the figure shows that the top of the valence band and the bottom of the conduction band are separated by forbidden energy gap E_g. The value of E_g is different for different semiconductors. The value of E_g is 1.12 eV for silicon and 0.72 eV for germanium at room temperature. At any temperature higher than 0 °K thermal excitation raises some electrons in the valence band to the conduction band and leaves an equal number of electron vacancies (or holes) in the valence band. In the last section we referred the electrons in the conduction band as the "free electrons". These thermally excited electrons in the conduction band and the holes in valence band can conduct electric current through the material on the application of external voltage across the semiconductor material. These conduction electrons and holes are referred to as the "Current Carriers". Suppose that at any temperature T °K there are n number of electrons per unit volume in the conduction band and p number of holes per unit volume in the valence band. In the case of an intrinsic or pure semiconductor the number of electrons n will be equal to the number of holes, p. This concentration is known as "intrinsic concentration" denoted by n_i and n_i is given by

$$n = p = n_i = A \exp\left[\frac{-eE_g}{2K_BT}\right] \tag{4.3.1a}$$

where
$$A = 2\left[\frac{2\pi K_BT}{h^2}\right]^{3/2} [m_p\, m_n]^{3/4}, \tag{4.3.1b}$$

K_B is the Boltzmann constant = 1.38 × 10^{-23} J/K, h is the Planck's constant = 6.62 × 10^{-34} J-s and m_n and m_p are effective masses of an electron and hole respectively.

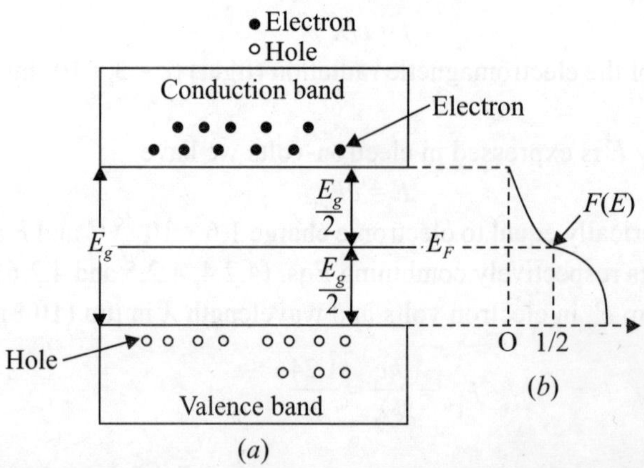

Fig. 4.6 (*a*) Energy band structure of an intrinsic semiconductor at any temperature above absolute zero illustrating that there are an equal number of holes and electrons in the valence and the conduction bands respectively. The Fermi level E_F is located at the middle energy gap region, (*b*) Fermi-Dirac probability corresponding to energy band structure of an intrinsic semiconductor and Fermi level is located at the middle of energy band structure.

The effective mass of an electron or a hole is a quantum mechanical concept and the effective mass of an electron or a hole is often smaller than the free electron rest mass by a factor of 10 or more. The rest mass of a free electron = 9.11 × 10^{-31} kg.

4.4 FERMI-DIRAC FUNCTION

The concept of free electrons and holes is essential to understand the operation of active semiconductor devices. Free electrons and holes are introduced profusely in a semiconductor crystal by the process of doping (adding minute quantity of impurity elements). Depending on the temperature of the crystal, the carriers are distributed in energy in the conduction band or valence band or in dopant energy levels. The number of carriers at any energy level depends on the energy distribution function and density of state function.

We know that in a semiconductor the occupancy of states by carriers either in the conduction band or in the valence band or in the impurity level is decided by Pauli's exclusion principle. The Pauli's principle states that no two electrons with the same spin are allowed to occupy the same energy state. The carriers are also indistinguishable. This stringent condition implies that Boltzmann distribution function cannot be used to determine the statistics of electrons and holes in a semiconductor.

Fermi-Dirac statistics describes the distribution of energy of the carriers in a semiconductor appropriately and expresses the occupation of the states in a semiconductor by Fermi-Dirac function $F(E)$. $F(E)$ is expressed as:

$$F(E) = \frac{1}{1 + \exp\left(\frac{E - E_F}{K_B T}\right)} \tag{4.4.1}$$

Where E is the energy of the state of level whose occupation we are interested in and could be an energy in the conduction or valence band or the energy of an impurity level within the forbidden energy gap. The energy E_F is referred to as Fermi level and used as the reference level. E_F is the characteristic of the material.

Figure 4.6 (*b*) shows the nature of the Fermi level (or function) E_F as a function of energy at 0°K and higher temperatures. These curves show several important characteristics of Fermi function $F(E)$.

1. The maximum value of $F(E)$ can be unity and therefore, the probability of occupation of an energy level can never exceed unity or not more than one electron can occupy the same quantum state.
2. At $E = E_F$ the value of function can go to unity at 0°K and at any temperature T, $F(E) = 1/2$ at $E = E_F$.

Thus the Fermi energy can be defined as that energy level up to which all levels are occupied and above which all levels are empty at 0°K and at any temperature $T >$ 0°K, the probability of occupation of Fermi level is 1/2.

The energy position of Fermi level in an extrinsic semiconductor is closely related to the type of doping (whether *n*-type of *p*-type) and level (the concentration of doping material). In an *n*-type semiconductor the Fermi level is situated near to the conduction band edge, while in a *p*-type one the Fermi level is close to the valence band edge. E_F does have a temperature dependence and its value decrease with the increase of temperature.

Figure 4.6(*a*) shows that the Fermi level is situated at the middle of the energy gap region for an intrinsic semiconductor where the number of electrons in the conduction band is equal to the number of holes in the valence band.

4.5 ENERGY LEVEL DIAGRAM: EXTRINSIC SEMICONDUCTOR

We already know that in order to prepare the extrinsic semiconductor, the pure semiconductor material is doped with (adding an extremely minute quantity) impurity atoms which either creates more free electrons (by adding donor impurity) or more holes (by adding acceptor impurity). These situations are illustrated in energy band picture of Fig. 4.7(*a*) which shows that the donor impurity forms the energy level just below the conduction band. Figure 4.7(*b*) shows that the acceptor impurity forms the acceptor impurity energy level just above the top of the valence band.

Fermi Level

In the case of an intrinsic semiconductor crystal the number of free electrons is equal to the number of holes so that the Fermi level is situated at the centre of the energy gap region as shown in Fig. 4.6(*a*). In an *n*-type crystal, the Fermi level shifts from the centre of energy gap region toward the conduction band. The shift of the Fermi level from the centre is the measure of relative abundance of the conduction electrons with respect to the valence holes in an *n*-type crystal. Figure 4.7(*a*) shows the Fermi level of an *n*-type crystal which is situated at an energy ΔE above the centre of the energy gap region.

Suppose the density of donor atoms in the *n*-type semiconductor is N_d and at room temperature the electron density n_n in the *n*-type material is equal to N_d, since all the donor atoms are ionized at room temperature. If the density of electron-hole pairs in the intrinsic semiconductor at the room temperature T°K is n_i, it can be shown that

$$n_n = N_d = n_i \exp\left(\frac{e\Delta E}{K_B T}\right) \tag{4.5.1}$$

Where K_B is the Boltzmann constant = 1.38×10^{-23} J/K; e is numerically equal to electronic charge = 1.6×10^{-19} C and ΔE is expressed in electron-volts.

Since in an *n*-type crystal the electron density n_n is greater than the density of electron-hole pair in the intrinsic semiconductor, the shift ΔE comes out to be positive so that the Fermi level moves toward the conduction band : At constant temperature the amount of shift toward the conduction band depends on the ratio (N_d/n_i).

Suppose N_a is the density of the acceptor atoms added to a *p*-type semiconductor so that the density of holes in the material p_p is equal to N_a, because all the acceptor atoms ionize at the room temperature. If $\Delta E'$ is the amount by which the Fermi level is displaced down from the centre of the energy gap to the valence band in the *p*-type crystal, (Fig. 4.7(*b*)) it can be shown that at the room temperature

$$p_P = N_a = n_i \exp\frac{e\Delta E'}{K_B T} \tag{4.5.2}$$

where $\Delta E'$ is expressed in electron, volts, e =1.6×10^{-19}, T is the absolute temperature and $K_B = 1.38 \times 10^{-23}$ J/K.

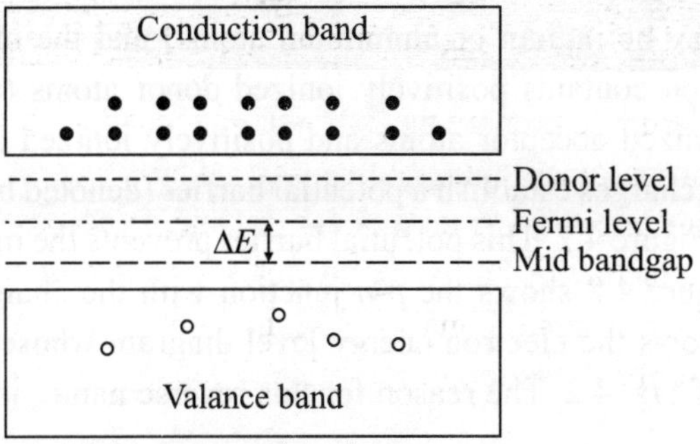

(*a*) *n*-type semiconductor

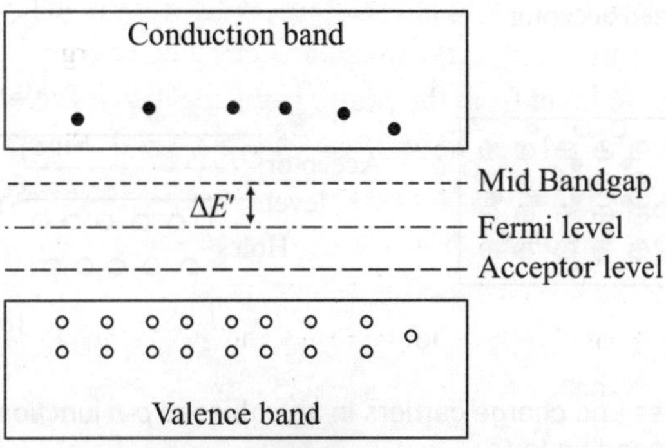

(*b*) *p*-type semiconductor

Fig. 4.7 Energy band structure of an *n*-type and *p*-type semiconductors (*a*) Energy band structure of an *n*-type semiconductor showing that the donor impurity energy level, and the Fermi level shifts from the middle of the energy gap region toward the conduction band by an amount ΔE. (*b*) Energy band structure of a *p*-type semiconductor showing the acceptor impurity level, and that the Fermi level shift from the middle of the energy gap toward the valence band by an amount $\Delta E'$

Degenerate Semiconductor

When the density of the donor atoms N_d in an *n*-type semiconductor is extremely high the value of ΔE (shift toward conduction band) may exceed half the value of the forbidden energy gap so that the Fermi level may enter into the conduction band. Again, when the density N_a of acceptor atoms in a *p*-type semiconductor is very high, the value of the shift $\Delta E'$ toward the valence band exceeds half the value of energy gap, the Fermi level in the *p*-type material may enter into the valence band. Both types of semiconductor crystals are referred to as the "Degenerate semiconductor crystals" due to excessive dopings.

4.6 ENERGY BAND PICTURE OF UNBIASED *p-n* JUNCTION

As discussed earlier a thin depletion layer is formed at the junction of a *p-n* diode due to diffusion and carrier recombination. All mobile charge carriers such as free electrons and holes are completely removed from the depletion layer. The portion of the depletion layer situated in the *p*-region contains negatively ionized acceptor atoms (may be indium or aluminium atoms) and the other portion of the depletion layer situated in the *n*-region contains positively ionized donor atoms (such as arsenic or antimony atoms). Both negatively ionized acceptor atoms and positively ionized donor atoms are fixed in the lattice and immobile. These charges establish a potential barrier (denoted by V_B) between the *n*-type and *p*-type regions as shown in Figure 4.2. This potential barrier prevents the inter-diffusion of carriers from their respective region. Figure 4.8 shows the *p-n* junction with the charge distribution in the *p*- and *n*-regions. Figure 4.8(*b*) shows the electron energy level diagram whose curvature is opposite to the potential barrier diagram of Fig. 4.2. The reason for this reverse nature is that the electronic charge is negative.

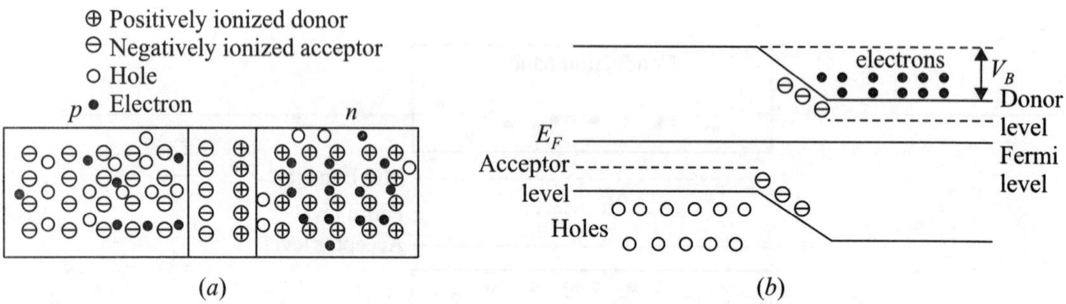

Fig. 4.8 (*a*) The impurities and charge carriers in an unbiased *p-n* junction (*b*) The energy band diagram corresponding to (*a*)

When a *p-n* junction is in equilibrium state with no external bias voltage at constant temperature, the Fermi energy levels in the *n*-type and *p*-type regions are in the same level as shown in Fig. 4.8(*b*). In the absence of any external potential, no current flows even when two ends are connected externally, because the barrier potential prevents the flow of charge carriers from one region to the other. The figure shows that there are more holes in the valence band of the *p*-region and more conduction electrons in the conduction band of the *n*-region. There are extremely small number of holes and electrons in the conduction band of the *n*-region and in the valence band of the *p*-region respectively. The energy difference between the top of the conduction band in the *p*-region and that in the *n*-region is equal to eV_B electron volts, where V_B is potential barrier and $e = 1.6 \times 10^{-19}$ C (Fig. 4.8*b*).

4.7 ENERGY BAND PICTURE OF FORWARD BIASED *p-n* JUNCTION

When a *p-n* junction is forward biased by applying positive potential E_f to the *p*-region with respect to the *n*-region, the effective potential barrier reduces to $(V_B - E_f)$ volts where V_B is the barrier potential of the unbiased *p-n* junction. This happens due to the fact that the barrier potential V_B is in the opposite direction to the applied forward biased voltage E_f and almost all applied voltage develops across the depletion layer in the junction. Due to the decrease of the effective barrier potential on the application of the forward biased voltage, the width of the depletion layer also decreases, since it depends on the square root of the effective barrier potential (Eq. 4.2.1).

Due to the reduction in the barrier potential, the electrons having comparatively higher energy will be able to cross the junction from the n-region to the p-region due to diffusion, while the holes with the comparatively higher energy will be able to cross the junction from the p-region to the n-region due to diffusion. As a result the injected electrons become the minority carriers in the p-region, while injected holes become the minority carriers in the n-region. Figures 4.9(a) shows a forward biased p-n junction with carrier distribution and Fig. 4.9(b) shows the electron energy level diagram of a forward biased p-n junction. The Fermi energy level in the n-region is higher than that in the p-region by $E_f eV$. The energy difference between the top of the conduction band of the p-region and that in the n-region is equal to $(V_b - E_f)\, eV$ (Fig. 4.9b).

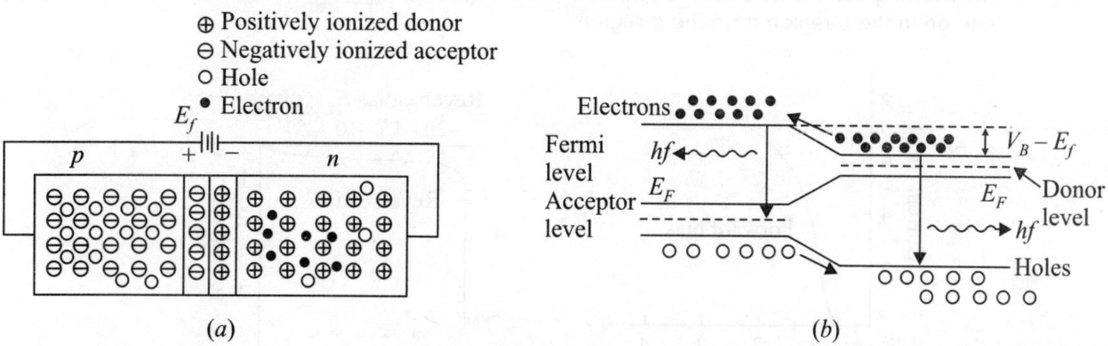

Fig. 4.9 (a) The p-n junction with forward bias giving (b) spontaneous emission of photon of energy hf

The carriers are effectively injected across the junction due to the forward biased voltage and form a current flow as the carriers continuously diffuse away from the interface. Figure 4.11(a) shows the nature of the forward current variation with applied forward voltage.

4.8 ENERGY BAND AND CURRENT IN REVERSE BIASED p-n DIODE

When a p-n junction diode is reverse biased by applying a positive potential E_R to the n-region with respect to the p-region, the effective barrier potential increases to $(V_B + E_R)$ volts. This happens due to the fact that the reverse voltage is in the same direction as the barrier potential and almost all the applied voltage develops across the depletion layer near the junction. As the reverse biased voltage increases, the effective barrier potential and the width of the depletion layer also increase (Eq. 4.2.2). Due to enhanced potential barrier electrons of the n-region having the highest energy will not be able to cross the junction from the n-region to the p-region. Similarly, holes having the highest possible energy cannot cross the junction from the p-region to the n-region due to diffusion. As a result no current flow takes place due to the majority carriers in each region. But the minority carriers in each region (such as holes in the n-region and electrons in the p-region) do not encounter the potential barrier and flow from each region as "drift" current. But the concentration of the minority carriers in each region is extremely small and as such, the reverse current in negligible. Figure 4.10 shows a reverse biased p-n junction with (b) energy level diagram. Figure 4.11(b) shows the nature of the variation of current with reverse voltage.

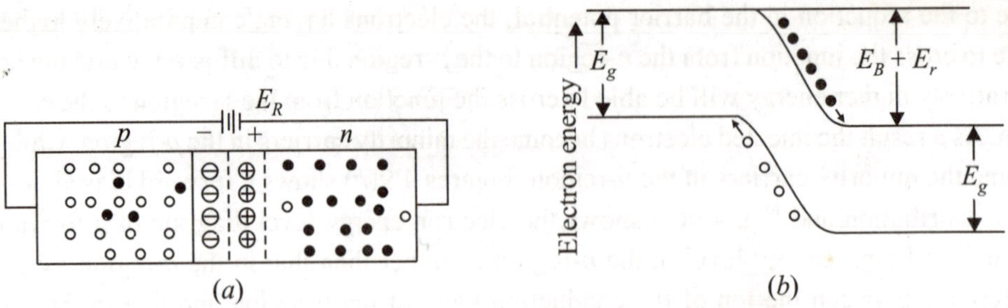

Fig. 4.10 (*a*) The *p-n* junction with reverse bias (*b*) The energy band diagram corresponding to (*a*) showing that the holes can come from the *n*-region to the *p*-region, whereas electrons can go to the *n*-region from the *p*-region

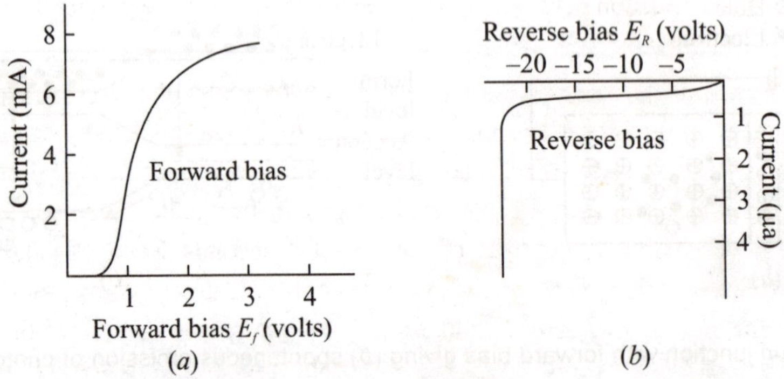

Fig. 4.11 (*a*) The volt-compere characteristics for a forward biased *p-n* junction, (*b*) The volt-ampere characteristics for a reverse biased *p-n* junction

4.9 DIRECT BAND GAP SEMICONDUCTORS

The most suitable materials for the emission of light are the direct band gap semiconductors. In the case of direct band gap semiconductors the electrons at the bottom of the conduction band and the holes at the top of the valence band on the either side of the forbidden energy gap have the same value of the crystal momentum. Therefore, the direct recombinations between the electrons in the conduction band and the holes in the valence band may take place. Figure 4.12 shows the diagram for the electron energy versus momentum for a direct band gap semiconductor and illustrates the process of recombinations between the conduction electrons and holes with the emission of a photon. We observe in the figure that the energy maximum in the valence band and the energy minimum in the conduction band occur at the almost same value of the electron crystal momentum. The crystal momentum p and the wave vector K are related by the equation

$$p = \frac{h}{2\pi} K \tag{4.9.1}$$

where h is the Planck's constant.

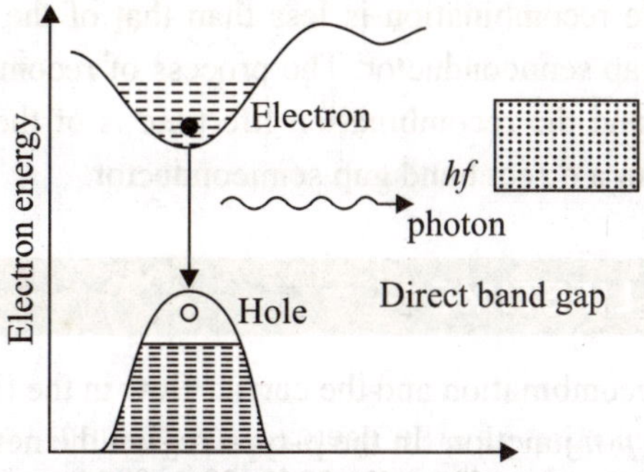

Fig. 4.12 Energy-momentum diagram for a direct bandgap semiconductor where the energy minimum in the conduction band occurs at the same value of momentum for the energy maximum in the valence band.

Thus, when the electron-hole recombination takes place, the momentum of both types of carriers, remains the same and the energy which corresponds to the band gap energy E_g may be emitted as light. The direct transition of an electron across the band gap gives the efficient process of photo emission. The average time during which the minority carrier remains in free state is known as the minority carrier life time which is very short of the order of 10^{-8} to 10^{-10} seconds. Some of the direct band gap semiconductors are gallium arsenide (GaAs), gallium antimonide (GaSb), indium arsenide (InAs) and indium antimonide (InSb) but silicon, germanium and gallium phosphide (GaP) are the indirect band gap semiconductors. The band gap energy E_g for GaAs, GaSb, InAs and InSb are 1.43, 0.73, 0.35 and 0.18 eV respectively.

4.10 INDIRECT BAND GAP SEMICONDUCTORS

In the case of an indirect band gap semiconductor, the energy maximum in the valence band and the energy minimum in the conduction band occur at different values of crystal momentum as shown in Fig. 4.13. In this case when an electron-hole recombination occurs, the electron must lose some momentum so that it has the same momentum corresponding to the energy maximum of the valence band. The conservation of momentum during the recombination process requires the emission of a third particle known as "phonon" with momentum Δp, where Δp is the difference in momentum between the energy minimum in the conduction band and the energy maximum in the valence band as illustrated in Fig. 4.13.

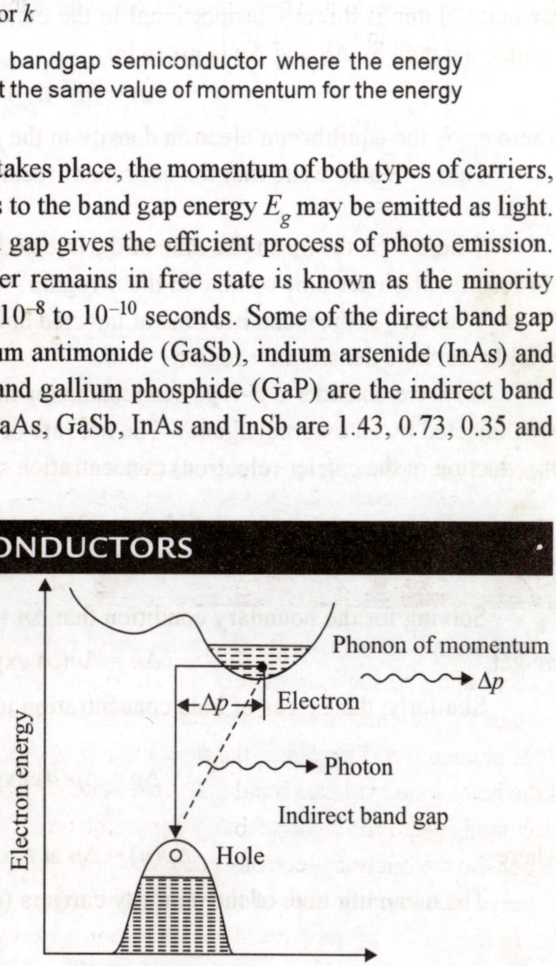

Fig. 4.13 Energy momentum diagram for an indirect bandgap semiconductor where the maximum energy at the valence band occurs at a different value of the energy minimum in the conduction band, the difference of the momentum being Δp

The probability of three-particle recombination is less than that of the two-particle recombination which occurs in a direct band gap semiconductor. The process of recombination in an indirect semiconductor is comparatively slow and the recombination life time is of the order of 10^{-2} to 10^{-3} second which is much larger than that in a direct band gap semiconductor.

4.11 MINORITY CARRIER LIFETIME

We shall now consider the process of recombination and the carrier flow in the field free region outside the depletion layer of a forward biased *p-n* junction. In the *p*-type region the net rate of recombination per unit volume is directly proportional to the concentration of the injected carrier electrons per unit volume denoted by Δn and Δn is given by

$$\Delta n = (n - n_{po}) \tag{4.11.1}$$

where n_{po} is the equilibrium electron density in the *p*-region.

Thus, we may write that the net rate of recombination per unit volume in the *p*-region = $\Delta n/\Gamma_p$ (4.11.2)

Where Γ_p is the mean life time of the injected carrier electron in *p*-region. Semilarly, the net rate of recombination per unit volume in the *n*-region = $\Delta p/\Gamma_n$ (4.11.3)

Where Γ_n is the mean life time of injected holes in the *n*-region and $\Delta p = p - p_{no}$, where p_{no} is the equilibrium hole density in *n* region (4.11.4)

Now we consider a *p*-type semiconductor in which uniform excess of carriers (electrons) has been injected by an external agency. The net rate of recombination of electron-hole is equal to the rate of reduction in the carrier (electron) concentration so that

$$-\frac{d(\Delta n)}{dt} = \frac{\Delta n}{\Gamma_p} \tag{4.11.5}$$

Solving for the boundary condition that $\Delta n = \Delta n(o)$ at $t = 0$

we get
$$\Delta n = \Delta n(o) \exp -t/\Gamma_p \tag{4.11.6}$$

Similarly, the excess of hole concentration in the *n*-region is

$$\Delta p = \Delta p(o) \exp \frac{-t}{\Gamma_n} \tag{4.11.7}$$

where $\Delta p(o) = \Delta p$ at $t = 0$

The mean life time of the minority carriers (electrons) in the *p*-region is

$$\langle t \rangle = \frac{\int_o^\infty t\Delta n(o) \exp\frac{-t}{\Gamma_p} dt}{\int_o^\infty \Delta n(o) \exp\frac{-t}{\Gamma_p} dt} = \Gamma_p \tag{4.11.8}$$

4.12 SPONTANEOUS EMISSION: ELECTROLUMINESCENCE

When a *p-n* junction is forward biased, holes and electrons are injected into the *n*-region and *p*-region respectively so that the concentration of the minority carriers in both regions increases. This gives rise to the recombinations of electrons and holes across the band gap as illustrated in Fig. 4.12 in the case of a direct band gap semiconductor. In a direct band gap semiconductor material the normally empty electron state in the conduction band of the *p*-type and the normally empty hole state in the valence band of the *p*-type material are populated by the "injection" of the carriers which recombine across the band gap.

The energy released by an electron hole recombination is approximately equal to the band gap energy E_g of the semiconductor. The excess of population of the minority carriers is reduced by the electron-hole recombinations. These recombinations may be either radiative or non-radiative. Some of the radiative recombinations may emit light. This type of emission of light within the diode structure is referred to as "Electro-luminescence". In the case of non-radiative recombinations the energy released is dissipated in the form of lattice vibrations producing heat. However in the band-to-band radiative recombination, the energy is released with the production of photons (light radiation).

The wavelength of the emitted light is given by

$$\lambda = \frac{1.24}{E_g} \tag{4.12.1}$$

where λ is the wavelength in μm and E_g is the forbidden energy gap of the semiconductor in electron-volts.

The electroluminescence may occur in the site of the recombination which is close to the junction in most cases. However, the recombinations may occur through the whole of the diode structure, because the minority carriers diffuse away from the junction and the concentration of the minority carriers decreases to 37 per cent of its value at the junction at a distance of "diffusion length" from its junction. The shorter the minority carrier life time, the lesser will be the diffusion length $[L_n = (D_n\Gamma_n)^{1/2}]$.

4.13 DESIGN OF LED FOR OPTICAL COMMUNICATIONS

Figure 4.14 shows the structure of a typical conventional light emitting diode (LED). For a source in visible region, gallium arsenic phosphide (GaAsP) or gallium phosphide (GaP) doped with either nitrogen (N) or Zinc oxide (ZnO) should be used. An alternative type known as the Burrus type provides small emissive area with high radiance. This is known as the Burrus type surface emitting LED and its essential features will be illustrated in Fig. 4.20. Comparison with the more conventional LED of Fig. 4.14, Fig. 4.20 shows that the emitted radiation is brought out through the back side of the diode and that means it is brought out through the substrate. This has been etched out in order to make the distance between the active layer and the emitting surface minimum. An insulating silicon oxide layer separates the rear contact from the semiconductor exc pt in the light emitting region. In this way the flow of the electric current is concentrated into a laterally well defined active layer.

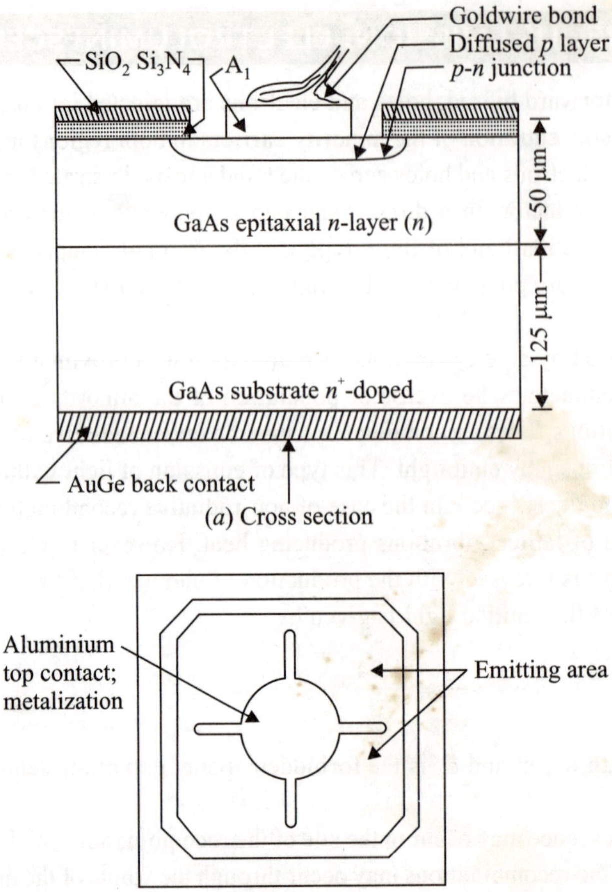

Fig. 4.14 (*a*) Structure of a typical LED (*b*) The front emitting surface of the LED

LEDs Emitting Different Colours

Figure 4.15 shows the cross-section of a red emitting LED. The device consists of two layers of GaAsP grown on GaAs substrate. The top layer $GaAs_{0\cdot6}P_{0\cdot4}$ and a bottom layer of graded GaAsP are grown on a gallium arsenide (GaAs) substrate. Some LEDs, such as an orange emitting, LEDs are grown on GaP substrate. The substrate material is grown in ingots by horizontal Bridgeman or *Czoharalski* method. For a *p*-phosphorous containing compound, such as GaP the resulting pressure at the melting point is extremely high of the order of 40 atmospheres. When Boron trioxide Br_2O_3 compound is used for encapsulation, then the boron impurities are introduced in the single crystal. For a high quality substrate, the defect density varies in the range of 10^2 to 10^5 per cm^2.

The epitaxial layers are grown by liquid phase or vapour phase epitaxial technique. For red light emitting LED, the *n*-layer is doped with tellurium and for orange emitting one, sulpher is used as a dopant. For *p*-type layer Zn is the best dopant.

NH_3 (ammonia) is used for creating *N* isoelectronic traps. If ZnO pairs are needed to create isoelectronic traps, then GaO_3 is passed through the material which has already been doped with Zn. GaAsP grown on GaAs substrate may produce mis-fit dislocation and therefore graded composition layers of graded n-$GaAs_{1-x}P_x : T_e$ are incorporated between the substrate and the active region of $GaAs_{0.6}P_{0.4}$ region.

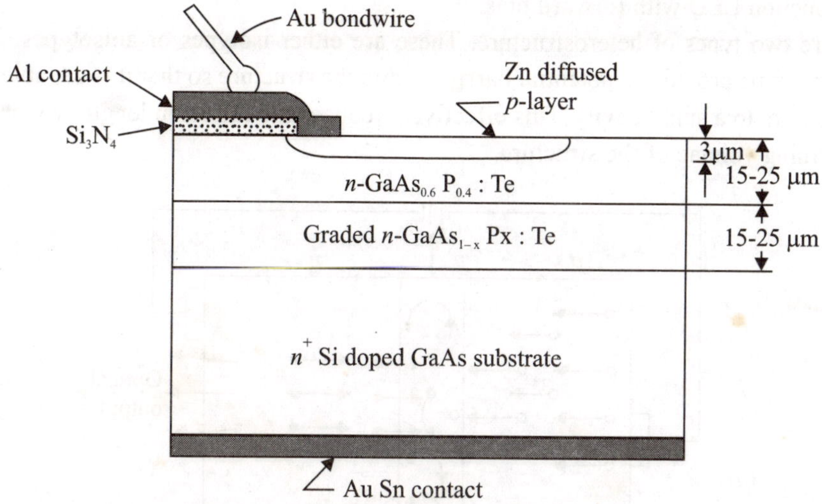

Fig. 4.15 LED emitting various colours of radiation by changing the substrate material

As shown in Fig. 4.15 an n-type substrate and top Zn diffused layer are the most common configuration because of the fact that Zn can easily be incorporated by diffusion. The diffusion is done selectively through a mask of Si_3N_4 to form p-n junction. Contact layers are usually formed of Al and AuSn on p- and n-sides respectively. Contact to the bias source is made via gold bond as shown in the figure. In order to make light scatter and radiate uniformly from the junction, the metalization on it is made so that the current distribution on the active region is uniform.

The completed LED is encapsulated in a dome as will be shown in Fig. 4.18(b) for display and counting applications.

Comparison of Conventional and BurrusType of LED

In the Burrus type the active layer is located to the heat sink so that the thermal impedance is small and that high current densities may be used without causing a high rise in the temperature. Excessive temperatures of the active layer have three detrimental effects:

1. The wavelength distribution of the emitted radiation changes with temperature.
2. The internal quantum efficiency decreases, because there is a rise in the rate of non-radiative recombinations.
3. The life of the light emitting diode falls rapidly with increasing junction temperature.

4.14 BASIC THEORY OF A DOUBLE HETEROJUNCTION LED (DHLED)

The LED discussed earlier has a very low value of quantum efficiency and low radiance (a low value of intensity of radiation) and as such they are not suitable for use in optical communications. In order to attain a high value of radiance and quantum efficiency, the LED structure must be capable of confining both charge carriers and optical power in the active region of the p-n junction. In the active region mostly the radiative recombination takes place. The carrier and optical confinements are achieved in the active region of a double heterojunction LED with forward bias Fig. 4.16 shows the layer structure of a double heterojunction LED with forward bias.

There are two types of heterostructure. These are either isotypes or anisotypes. The isotypes such as *p-p* or *n-n* type provides a potential barrier within the structure so that it is capable of confining the minority carriers to a small cavity. This effectively reduces the diffusion length, which is restricted within the confining volume of the structure.

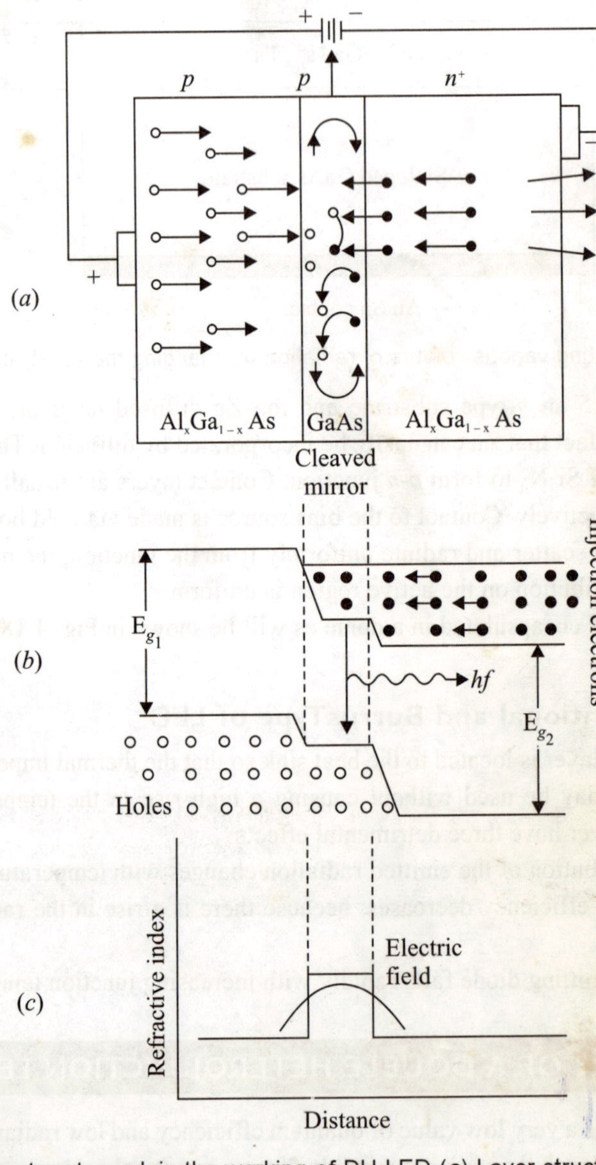

Fig. 4.16 LED structure to explain the working of DH LED (*a*) Layer structure (*b*) Energy band structutre (*c*) Refractive index profile

The DHLED structure consists of a p-type GaAs layer sandwiched between a p-type AlGaAs layer and an n^+-type AlGaAs layer (Fig. 4.16a). When a forward bias voltage is applied by making the p-region positive with respect to the n-region, the electrons from the n^+ layer are injected into the p-type layer. The injected electrons move away from the junction into the central GaAs layer due to diffusion and combine with holes which are the majority carriers in the central p-type GaAs layer.

As a result of recombination, photons (light quanta) are generated with energy corresponding to the band gap energy E_g of the central p-type GaAs layer. These injected electrons are prevented from diffusing into the p-type AlGaAs layer because of the potential barrier presented by the p-p heterojunction as shown in (Fig. 4.16b). Thus, the electro luminescence takes place only in the GaAs junction layer and provides good internal quantum efficiency and high radiance emission. Again, light is emitted from the device without any appreciable amount of reabsorption, because of the fact that the band gap energy of the p-type AlGaAs layer is larger than that of the central GaAs layer.

The difference between the bandgap energy of the adjacent layers of the p-p structure confines the charge carriers, while the difference between the refractive index of the adjacent layer (Fig. 4.16c) confines the optical power in the central GaAs layer due to total internal reflection. These double confinements give rise to both high quantum efficiency and high radiance. The parameters on which the operation of the device depend on are the heterostructure interfaces and the impurity concentration of the central GaAs active layer.

The central GaAs-layer into which electrons are injected from the top of the n^+ type AlGaAs-layer is made thin (1000 to 2000 Å) and the injected electrons are confined within this small thickness by the potential barrier of the p-type GaAs/p-type AlGaAs heterojunction.

The top AlGa As-layer serves several important functions. These are:

1. The band gap energy of this layer is larger and the layer acts as a window layer in which the emitted photons with energy close to the energy of the central GaAs are not reabsorbed.
2. The interface state density at the lattice matched junction is lower than the free surface state density and thus the non-radiative recombinations are drastically reduced.
3. The injection efficiency is higher.

Epitaxial growth method is used to fabricate such heterostructure LED's. Such heterostructure LED have also been fabricated with indium based materials (InP) such as InGaAsP/InP or InGaAs/InGaAsP/InP. These LED's emit radiation at longer wavelength.

4.15 POWER AND INTERNAL EFFICIENCY OF LED

We can calculate the power generated internally in an LED by considering the excess electrons Δn and excess holes Δp injected into the p-region and n-region respectively of the device. When a p-n junction is forward biased the excess electron density Δn is injected into the p-region and the excess holes density Δp into the n-region. The excess of electron density Δn and the excess hole density Δp are equal since the injected carriers are created and recombined in pairs such that the charge neutrality is maintained within the structure.

The excess of electron density Δn injected into the p-region at time t varies with the time according to the following relationship (Eq. 4.11.6)

$$\Delta n = \Delta n(o) \exp(-t/\Gamma_p) \tag{4.15.1}$$

where $\Delta n(o)$ and Δn are the initial injected electron density at $t = 0$ and the electron density at time t respectively. Γ_p is the minority carrier recombination life time in the p-region.

In equilibrium condition, the total rate at which the excess carrier increases is equal to the sum of the externally supplied carriers and the thermally generated carriers.

When a current density J ampere per square meter flows through the diode, the increase in electron density per unit time may be written as $\frac{J}{ed}$ electrons per cubic meter per second where d is the thickness of the recombination region and e is numerically equal to electronic charge in MKS system $= 1.6 \times 10^{-19}$.

Hence the rate equation for recombination in the diode is

$$\frac{d(\Delta n)}{dt} = \frac{J}{ed} - \frac{\Delta n}{\Gamma_p} \; (\text{m}^{-3}\text{s}^{-1}) \tag{4.15.2}$$

The equilibrium condition is obtained by putting $d(\Delta n)/dt = 0$ in Eq. (4.15.2). Thus

$$0 = \frac{J}{ed} - \frac{\Delta n}{\Gamma_p} \tag{4.15.3}$$

or

$$\Delta n = \frac{J\Gamma_p}{ed} \cdot (\text{m}^{-3}) \tag{4.15.4}$$

Equation (4.15.4) gives the increase in the steady state electron density Δn when a constant current density flows through the junction. It is also apparent from Eq. (4.15.4) that in a steady state the number of recombinations per second per unit volume of the active region (denoted by r_t) will be equal to

$$r_t = \frac{J}{ed} = r_r + r_{nr} \; (\text{m}^{-3}) \tag{4.15.5}$$

where r_r and r_{nr} are the radiative and non-radiative recombinations rates respectively per unit volume of the active region. When the forward currrent in an LED is I, then the total number of recombinations in the device (denoted by R_t) is given by

$$R_t = \frac{I}{e} \tag{4.15.6}$$

where I is the current in amperes and $e = 1.6 \times 10^{-19}$ coulomb.

The excess carriers can recombine in the device radiatively and non-radiatively. In the case of radiative recombination, a photon is generated, while in the case of a non-radiative recombination the energy is released in the form of heat (lattice vibration). The thickness of the active region of a double heterostructure junction LED is extremely small of the order of few microns (typical values 1000° to 2000° A) and the non-radiative recombination tends to be dominated by the surface recombinations at the heterojunction interfaces. Thus the radiative recombination is predominant in a double heterojunction LED.

Internal Quantum Efficiency (η_{int})

The internal quantum efficiency is defined as the ratio between the number of photon generated and the number of electron injected. The LED internal quantum efficiency can also be defined as the ratio of the radiative recombination rates to the total recombination rate. With the help of Eq. (4.15.5)

$$\eta_{int} = \frac{r_r}{r_t} = \frac{r_r}{r_r + r_{nr}} = \frac{R_r}{R_t} \tag{4.15.7}$$

where R_r and R_t are the number of radiative and total number of recombination rates respectively in the whole volume of the active region.

On combining Eq. (4.15.6) and Eq. (4.15.7) yields

$$R_r = \eta_{int} R_t = \eta_{int} I/e \tag{4.15.8}$$

or

$$\eta_{int} = \frac{eR_r}{I}$$

where radiative recombination R_r is equal to the number of photon generated per second in the active volume and each photon has an energy hf joules. Thus with the help of Eq. (4.15.8), the optical power generated internally P_{int} is given by

$$P_{int} = R_r hf = \eta_{int}(I/e)\, hf = (\eta_{int}\, Ih/e)\,(c/\lambda) \tag{4.15.9}$$

where c is the velocity of light, I is the drive current in amperes h is the Planck's constant (6.62×10^{-34} J.s). When the wavelength is expressed in metres, P_{int} comes out in watts. From Eq. (4.15.9) we find that the optical power produced in an LED is directly proportional to the drive current.

Quantum Efficiency and Minority Carrier Life Time

From the exponential decay of the excess minority carrier Δn with time as expressed by (4.15.1), the carrier life time Γ_p in the p-region is given by

$$\Gamma_p = \frac{\Delta n}{r_t} \quad \text{or} \quad r_t = \frac{\Delta n}{\Gamma_p} \tag{4.15.10}$$

and the radiative and non-radiative lifetime of the minority carriers are

$$\Gamma_r = \frac{\Delta n}{r_r} \quad \text{or} \quad r_r = \frac{\Delta n}{\Gamma_r} \tag{4.15.11}$$

and

$$\Gamma_{nr} = \frac{\Delta n}{r_{nr}} \quad \text{or} \quad r_{nr} = \frac{\Delta n}{\Gamma_{nr}} \tag{4.15.12}$$

Again we know that

$$r_t = r_r + r_{nr} \tag{4.15.13}$$

Substituting the values from Eqs. (4.15.10) (4.15.11) (4.15.12) (4.15.13), we get

$$\frac{\Delta n}{\Gamma_p} = \frac{\Delta n}{\Gamma_r} + \frac{\Delta n}{\Gamma_{nr}}$$

or

$$\frac{1}{\Gamma_p} = \frac{1}{\Gamma_r} + \frac{1}{\Gamma_{nr}} \tag{4.15.14}$$

where Γ_p, Γ_r and Γ_{nr}, are the total, radiative, and non-radiative recombination lifetime of the minority carriers respectively.

The internal quantum efficiency η_{int} is obtained from Eq. (4.15.7)

$$\eta_{int} = \frac{R_r}{R_t} = \frac{r_r}{r_t} = \frac{r_r}{r_r + r_{nr}} = \frac{1}{1+\frac{r_{nr}}{r_r}} = \frac{1}{1+\frac{\Gamma_r}{\Gamma_{nr}}} \tag{4.15.15}$$

Combining Eq. (4.15.11) and Eq. (4.15.12), we get

$$\eta_{\text{int}} = \frac{\Gamma_p}{\Gamma_r} \tag{4.15.16}$$

4.16 DIFFERENT LED STRUCTURES

There are five different types of LED structures. These are (1) Planer LED, (2) Dome LED (3) Double heterostructure surface Emitter (4) Double heterostructure Edge Emitter and (5) Super Luminescence LED.

1. Planer LED

Among the all light emitting diodes the planer LED has the simplest structure. It can be fabricated by either liquid or vapour phase epitaxial process or the junction is formed by diffusion technique. The device is so designed that the most of the radiative recombinations take place in the side of junction nearest to surface and therefore, the probability of reabsorption is greatly reduced. This is accomplished by arranging the most of the current flowing through the diode to be carried by those carriers which are injected to the surface layer. Figure 4.17(*a*) shows the planer LED structure made by epitaxial growth technique and Fig. 4.17(*b*) shows the planer surface emitting LED fabricated by diffusion technique. When a forward bias voltage is applied across the junction, the device emits light from all surfaces and only a small amount of light can escape due to total internal reflection. The radiance of the planer LED is low.

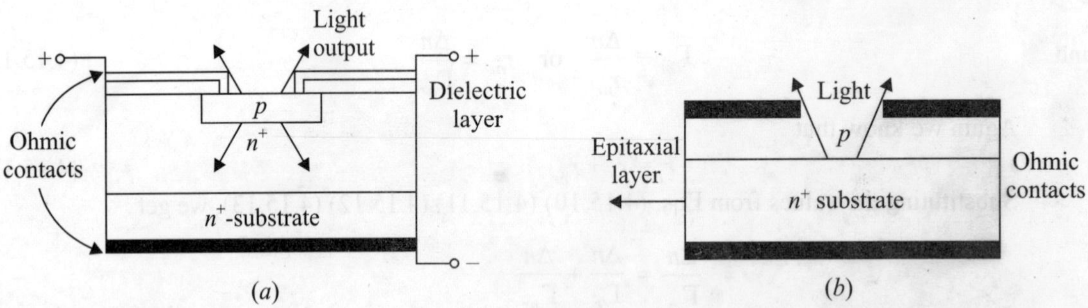

Fig. 4 .17 (*a*) A planer LED structure fabricated by epitaxial growth technique (*b*) A planer surface emitting LED prepared by diffusion technique

2. Dome LED

Figure 4.18(*a*) shows the structure of the dome LED in which the semiconductor surface is shaped in the form of a hemisphere. This is quite expensive and impractical for most applications, Fig. 4.18(*b*)

shows the second type which is comparatively cheaper. In fabricating this type a dielectric encapsulation shaped as a dome is used with a transparent material of high refractive index. Higher output can be obtained from this type of LED than that from a planer LED. Both types shown in Fig. 4.18(*a*) and (*b*) are called "Dome LEDs" and are useful for high volume low priced applications.

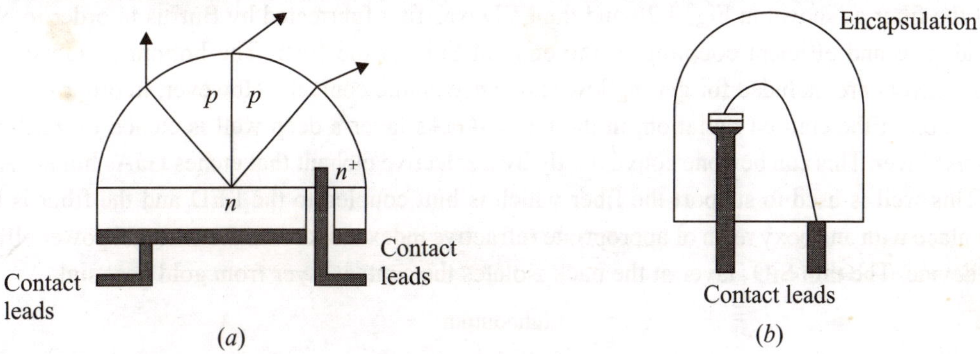

Fig. 4.18 (*a*) A dome LED with semiconductor surface is shaped in the form of hemisphere (*b*) A cheaper type of a dome LED

Planer and Dome LEDs are used as cheap alarm introducers, display and in calculators, TV channel changer and industrial counting.

In applications such as pocket calculators, parallel connections of several small area LEDs are used to display "alpha numerics". Two methods are generally used. Figure 4.19(*a*) shows the first method where a seven segment system is used. In this scheme each segment may consist of several LEDs. A digital logic control system is used to feed current to the desired segment. The negative terminals of all LEDs are connected to the common ground terminal.

Figure 4.19(*b*) shows the second scheme where a matrix of LEDs is used and logic control directs current to particular LEDs to produce a specific display. In both schemes the failure of one or small number of LED's will not ruin the display.

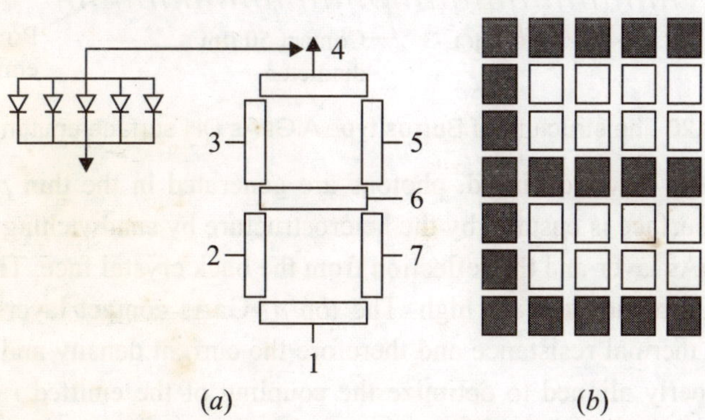

Fig. 4.19 Segment architecture (*a*) Matrix architecture of LEDs for alpha numeric display (*b*) Matrix structure where each segment structure consist of multiple LEDs

4.17 DOUBLE HETEROSTRUCTURE SURFACE EMITTER LED (SLED, BURRUS TYPE)

In a surface emitter LED the plane of the active light emitting region is oriented perpendicular to the axis of the fiber as shown in Fig. 4.20 and the LED was first fabricated by Burrus in order to obtain a high radiance and efficient coupling of the emitted light to the fiber. The bottom p^+-GaAs and top n^+-GaAs layers are included for getting low resistance ohmic contacts. However, in order to avoid the reabsorption of the emitted radiation, in the top n^+-GaAs layer a deep well is etched to reach the top n-AlGaAs layer. This can be done conveniently by a selective etchant that etches GaAs but not AlGaAs layer. This well is used to support the fiber which is butt coupled to the LED and the fiber is held in proper place with an epoxy resin of appropriate refractive index to increase the external power efficiency of the device. The thin SiO_2 layer at the back isolates the contact layer from gold heat sink.

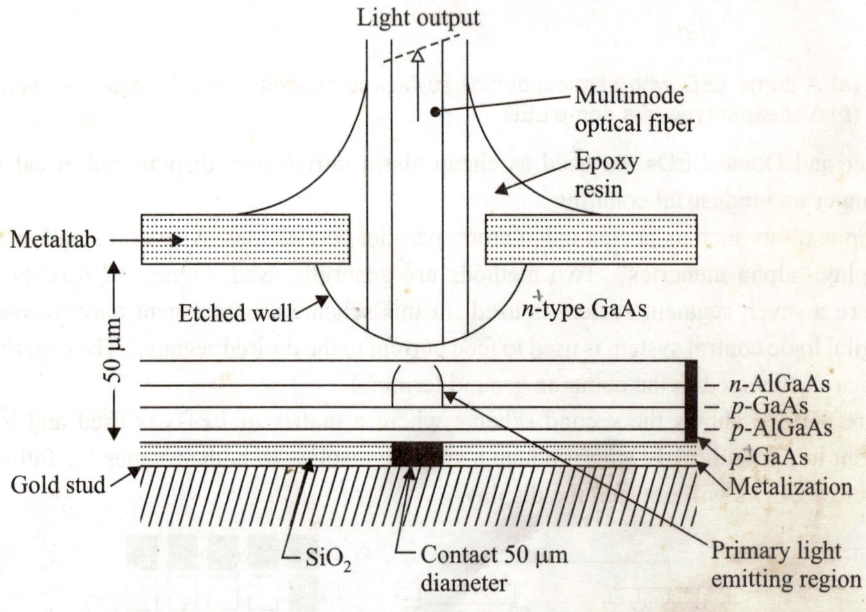

Fig. 4.20 The structure of Burrus type AlGaAs DH surface emitting LED

When the LED is forward biased, photons are generated in the thin p-GaAs region and the emission from the top surface is ensured by the heterostructure by sandwiching p-GaAs layer between n-AlGaAs and p-AlGaAs layer and the reflection from the back crystal face. Thus, the radiance of the device in the forward direction is very high. The top n^+-GaAs contact layer ensures low electrical contact resistance and thermal resistance and therefore the current density and radiation intensity are high. The fiber is properly aligned to optimize the coupling of the emitted radiation with the fiber. However, there is some loss due to the Lambertian distribution of radiation intensity.

The heterostructure of the LED provides a means for confining charge carriers and optical power in the active region of the p-n junction so that radiative recombinations occur in the active region. The internal absorption is low because of the large band gap of the adjoining layers.

The diameter of the circular active area in a practical surface emitter LED is approximately 40 to 50 μm and its thickness may be up to 25 μm. This LED usually operates in the wavelength range of 0.6 to 0.9 μm. The power coupled to the multimode step-index fiber may be estimated from the following relationship

$$P_c = \pi(1 - R) A R_d (NA)^2 \tag{4.17.1}$$

where R is the Fresnel reflection coefficient at the surface and R_d is the radiance of the source. Reflection coefficient R is given by

$R = \left(\frac{n_1 - n_2}{n_1 + n_2}\right)^2$ where n_1 and n_2 are the refractive indices of the device material and the medium in which the device is placed respectively, A is the fiber cross-section or the emission area of the device surface (whichever is less), (NA) is the numerical aperture of the fiber and R_d = radiance of the source. Radiance is defined as the measure of the optical power radiated per unit area of the surface into a unit solid angle.

However, the power coupled to the fiber depends on the distance and the alignment between the emission area and the fiber. For example, when epoxy resin is added to the etched well, it has the tendency to reduce the refractive index mismatch and increase the external efficiency of the LED.

Proton Bombardment

In Fig. 4.21 the light emitting region is prepared by the technique of proton bombardment on the semiconductor rather than the oxide isolation back contact. When a group III-V semiconductor such as GaAs or AlGaAs is bombarded with high energy protons, a severe damage is caused to the crystal lattice. This damage causes a large increase in the resistivity of the semiconductor. A dose of 3×10^{19} protons per square meter increases the resistivity of the semiconductor material by a factor of 1000. The penetration depth caused by the bombardment of proton is determined by the energy of the bombarding proton and is approximately 1 μm per 100 Kev of energy.

Besides the proton bombardment thc implementation of oxygen ions can also be used to produce semi-insulating material. But there are two advantages of proton bombardment. Silica has a low value of thermal conductivity so its removal increases the lateral heat losses from the active region and reduces the temperature rise for a given current density. The second advantage is that there remains a large area of inactive *p-n* junction which contribute to the depletion layer capacitance. This capacitance is not found in the case of proton bombardment.

4.18 STRIPE GEOMETRY AlGaAs EDGE-EMITTER LED (ELED)

An edge emitter LED is shown in Fig. 4.21. It consists of an active junction region which is the source of incoherent radiation and two guiding layer. The active region is lightly doped or undoped, and consists of thin *n*-AlGaAs layer which is sandwiched between a *p*-AlGaAs and an *n*-AlGaAs which act as carrier confinement layers. Above and below these carrier confinement layers there are two transparent p^+AlGaAs and *n*-AlGaAs layers working as optical guiding layers.

The both guiding layers have a refractive index which is lower than that of the active region but higher than the refractive index of the surrounding material. The structure forms a waveguide channel which directs the optical radiation toward the fiber core.

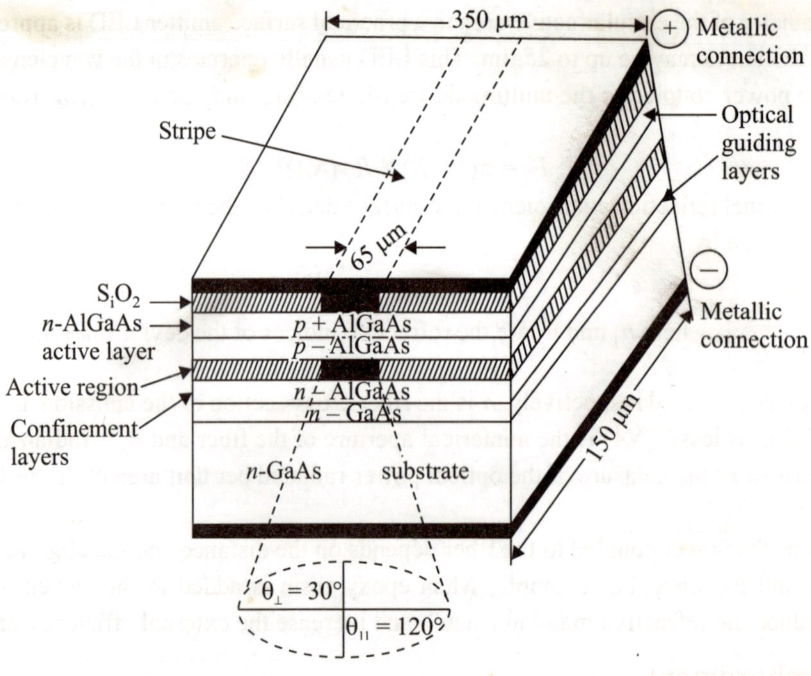

Fig. 4.21 The schematic diagram of the structure of stripe geometry DH AlGaAs edge emitting LED

The stripe geometry is introduced in the structure to provide optical confinement in the horizontal plane. The active region is within the stripe. Generally the stripe is formed by the creation of high resistance areas on either side by proton bombardment technique or by oxide isolation technique. The diameter of an optical fiber core usually ranges from 50 to 100 μm and the contact stripes for the edge emitters are usually 50 to 70 μm width. The device takes advantage of the transparent guiding layer with a very thin active layer (50 to 100 μm) in order the light produced in the active layer to spread into the transparent guiding layers reducing self-absorption in the active layer. The top face of the device should be connected to the positive terminal and the bottom face to the negative terminal of an external voltage source.

In order to match with the fiber core, the width of the active region of the LED is prepared ranging from 100 to 150 μm. The emission pattern of an edge emitter LED is more directional than that of the surface emitter LED (Burrus type) which was shown in Fig. 4.20. In the plane parallel to the junction where there is an waveguide effect, the beam is Lambertian (varying with cos θ) with half power beam width of $\theta_{11} = 120°$. In the plane perpendicular to the junction the half power beam width $\theta_\perp$ has been made as small as 25° to 35° by a proper choice of the waveguide thickness.

The enhanced waveguiding of the edge emitter enables to couple more power to a low (*NA*) fiber than a comparable surface emitter fiber. The stripe geometry of the edge emitter allow a very high carrier injection densities for a given drive current. Thus, it is possible to couple a milliwatt of optical power into a low *NA* (~ 0.14) multimode step index fiber with edge emitting LEDs operating with a high drive current of 500 mA.

Edge emitters have also been found to have a substantially better modulation bandwidth of the order of several megahertz.

Example 4.1: *The radiative and non-radiative recombination lifetimes of the minority carriers in the active region of double heterojunction LED are 50 ns and 100 ns respectively. Determine the total carrier life time and the optical power generated internally within the device when the peak emission wavelength is 0.85 μm at a drive current of 50 mA.*

Solution: The total carrier recombination life time is given by (Eq. 4.15.24). Data given $\Gamma_r = 50$ ns, $\Gamma_{nr} = 100$ ns, $I = 50$ mA, $\lambda = 0.85$ μm

$$T_p = \frac{\Gamma_r T_{nr}}{\Gamma_r + \Gamma_{nr}} = \frac{50 \times 100}{50 + 100} = \frac{100}{3} = 33.3 \text{ ns}$$

The internal equation efficiency is given by Eq. (4.15.26)

$$\eta_{int} = \frac{\Gamma_p}{\Gamma_r} = \frac{33.3}{50} = 0.666$$

The internal generated power P_{int} is given by Eq. (4.15.9)

$$P_{int} = \eta_{int} \frac{Ihc}{e\lambda} = \frac{0.66 \times 50 \times 10^{-3} \times 6.62 \times 10^{-34} \times 3 \times 10^{8}}{1.6 \times 10^{-19} \times 0.85 \times 10^{-6}} = 46 \text{ mW}.$$

Example 4.2: *A DH surface emitter LED is butt joined to a step index fiber with core diameter of 50 μm and numerical aperture of 0.18. The emission area of the device has a diameter of 80 μm. The radiance of the device is 30W Sr^{-1} cm^{-2} at constant operating drain current. Calculate the optical power coupled to the fiber. Assume that the Fresnel reflection coefficient at the index matched fiber is 0.01.*

Solution. Using Eq. (4.5.1) the optical power coupled to the fiber P_c is given by

$$P_c = \pi(1 - R)A\, R_D\, (NA)^2$$

In this case A = emission area of the source

$$A = \pi(25 \times 10^{-4})^2 = 1.96 \times 10^{-5} \text{ cm}^2$$

Thus $P_c = \pi(1 - 0.01) \times 1.96 \times 10^{-5} \times 30 \times (0.18)^2 = 69$ μW

4.19 STRIPE GEOMETRY InGaAs EDGE-EMITTING LED (ELED)

Figure 4.22 shows the layer structure of a InGaAs edge emitting LED which emits at longer wavelength region (1.3 μm). Generally, for a super-radiant LED a stripe geometry is incorporated to improve the injection efficiency as in a junction laser. The active layer is usually lightly doped or undoped and consists of n-$In_{0.53}Ga_{0.47}As$. A very large population of carriers for recombination in this region is injected by forward bias. The two p^+-InGaAsP and n^+-InGaAsP layers on both sides of the active layer serve as carrier confinement layers. At the outer sides of the confinement layers there are doped In P-layers. These layers serve as the cladding layers and the region in between two cladding layers forms an optical waveguide. Thus, the confining layers with wide band gap serves the dual purpose of optical and carrier confinements.

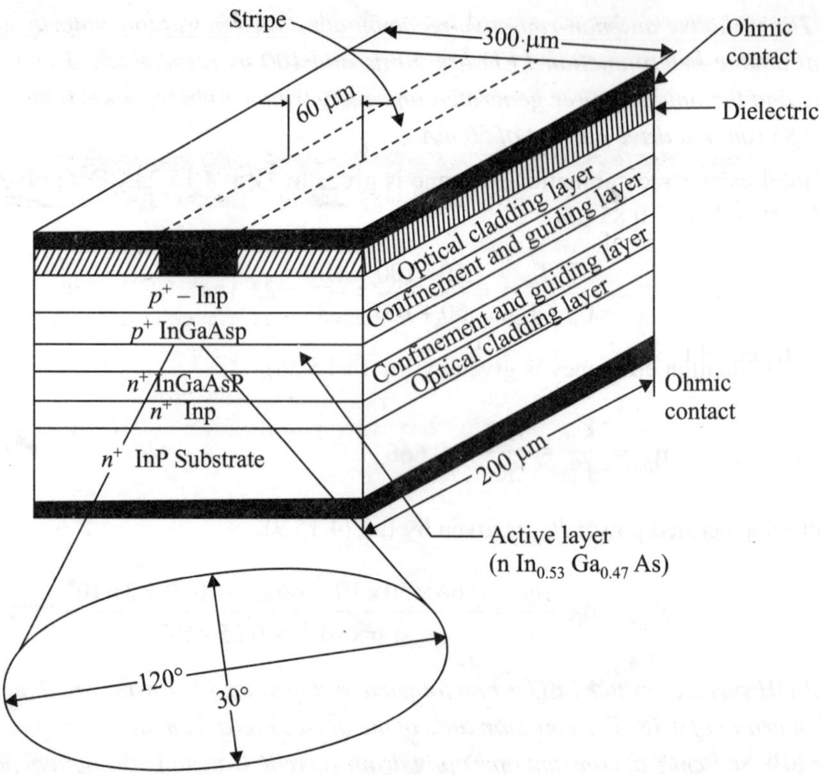

Fig. 4.22 Schematic diagram of the structure of stripe geometry DH InGaAs/InP edge emitting LED

The photons are generated in the very thin active layer when the LED is forward biased and spread into the guiding layers without reabsorption, because of their larger band gap. It is important to understand that the emitting structure allows the thin active region without sacrificing the radiating recombination efficiency in comparison with homojunction surface emitting LED.

The stripe geometry is produced by selective metallization on the top surface through a window opened in a SiO_2 layer and allows higher carrier injection density for the same drive current. It means that the optical power conversion efficiency is considerably improved.

4.20 SUPER LUMINESCENT LED (SLD)

Figure 4.23 shows the construction of super luminescent LED (abbreviated as SLD) and it exhibits a significant advantage over both SLED and ELED for communication applications. This type of device has the advantage of (1) high power output (2) more directional output beam and (3) narrow spectral beam width. However, the super radiant emission process within the SLD devices tends to increase device modulation bandwidth.

Figure 4.23 shows two forms of construction of super luminescent LEDs. It may be observed that the structure in both forms is similar to that of the Edge emitting LEDs (ELEDs). The super luminescence LED structure consists of a *p-n* junction either in the form of a long rectangular stripe or a narrow ridge waveguide or a buried structure. However, one end of the device is made optically lossy to avoid reflections and the laser action is suppressed. The output is provided by the other end of the device. When a forward bias voltage is applied across the device, the injected current is increased slowly until a stimulated emission takes place and thus amplification occurs. There is a high loss in one end of the device so that the "laser action" is suppressed due to the elimination of the optical feedback.

4.21 SUPER LUMINESCENT AlGaAs CONTACT STRIPE SLD

Figure 4.23 shows an earlier type of super luminescent LED which has contact stripes along with an absorbing region at one end so that the laser action is suppressed. Such devices are capable of providing a peak output of 60 mW at an wavelength of 0.67 μm in pulsed mode. Recently antireflection coatings are used in cleaved facets of the SLD so that Fabry Parrot resonance is suppressed. Such devices have been able to launch 350 μw of optical power in 50 μm diameter multimode graded index fiber at a drive current of 250 mA and 250 μw into a single mode fiber using a drive current of 100 mA. In both cases the line width of the device are in the range of 30 to 40 nm.

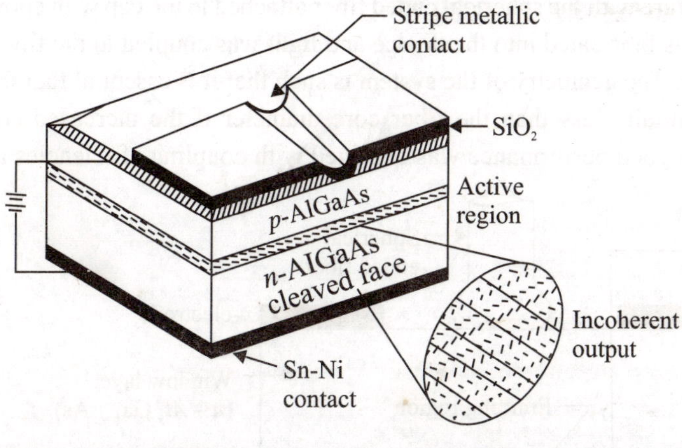

Fig. 4.23 Superluminescent (*a*) AlGaAs contact-stripe SLD.

Super Luminescent InGaAsP/InP type of SLD

Figure 4.24 shows the structure of a more recent type of InGaAs/InP type of super luminescent LED (SLD). The device emits light at the wavelength of 1.3 μm and consists of buried active layer within a *V*-shaped groove on a *p*-type InP substrate. The structure is suitable for high power operation because of its low value of leakage current. The device does not use antireflection (*AR*) coatings on both end facets to eliminate feedback, but a light diffusion surface is placed within this device. The surface which is applied diagonally on the active layer of length 350 μm serves to scatter the background light emitted from the active layer and thus decreases the feedback into the layer.

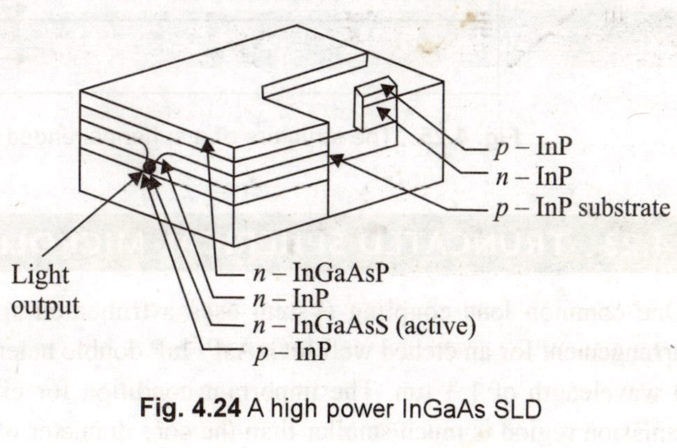

Fig. 4.24 A high power InGaAs SLD

The coupling of 1 mw of optical power into a spherically shaped lens end of a single mode fiber of 10 μm core diameter has been found to work at a drive current of 150 mA. The spectral distribution from SLD was observed to be smooth envelope with a (Full width half power) FWHP of 30 nm, while the device modulation bandwidth is 350 MHz at –1.5 dB point.

4.22 LENS COUPLING TO FIBER

It has been found in practical applications that the use of spherical ended fiber has increased coupling efficiency up to four times. The end of the fiber has been polished into a spherical form by controlled melting. This collects and collimates the divergent beam of radiation from the LED. A LED with 35 μm. emissive diameter has been coupled to a fiber with 85 μm core diameter and of numerical aperture of 0.14 and the end radius of the fiber has been 75 μm. Thereby the coupling efficiency has increased 4 times.

Figure 4.25 shows a GaAs/AlGaAs spherical ended fiber coupled LED. It consists of a planer surface emitting structure with the spherical ended fiber attached to the cap with epoxy resin. An emitting diameter of 35 μm was fabricated into the device and light was coupled to the fiber with core diameter of 75 μm and 110 μm. The geometry of the system is such that it is essential that the active diameter of the device be substantially less than the fiber core diameter if the increased coupling efficiency is desired. In this case a good performance was obtained with coupling efficiencies around 6 per cent.

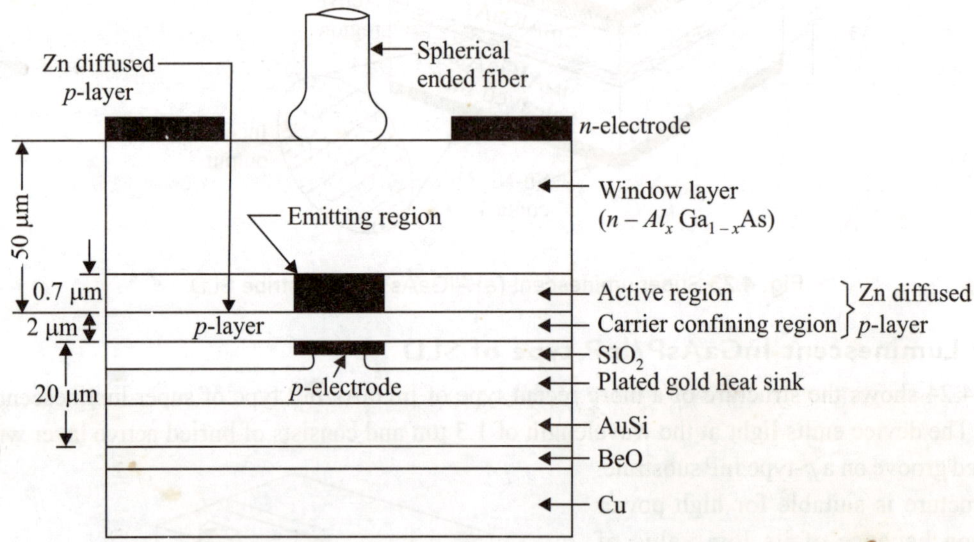

Fig. 4.25 The structure of a spherical ended fiber coupled with AlGaAs LED

4.23 TRUNCATED SPHERICAL MICROLENS COUPLING

One common lens coupling system uses a truncated spherical microlens. Figure 4.26 shows the arrangement for an etched well InGaAsP/ InP double heterostructure surface emitter LED operating at a wavelength of 1.3 μm. The important condition for efficient coupling is that the diameter of the emission region is much smaller than the core diameter of the fiber. In this case the most satisfactory results were obtained with a 14 μm active region diameter and an 85 μm core diameter of step index fiber having numerical aperture of 0.16. The coupling efficiency was found to increase by a factor of 13. The overall power conversion efficiency η_p which is defined as the ratio between the optical power coupled to the fiber P_c and the electric power applied at the terminals of the device P is given by

$$\eta_{pc} = \frac{P_c}{P} \tag{4.23.1}$$

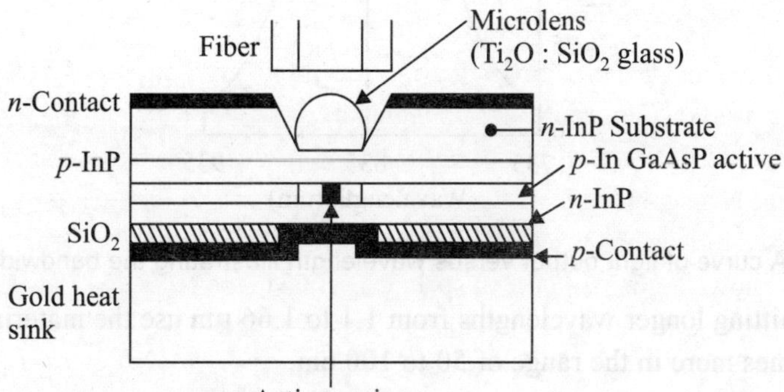

Fig. 4.26 The use of a truncated spherical microlens coupling for the emission from an InGaAsP surface emitter LED to the fiber

4.24 OUTPUT SPECTRUM

Figure 4.27(*a*) and 4.27(*b*) show the light output power versus drive current characteristics for good surface emitter and edge emitter LEDs respectively. It may be observed the surface emitter emits more optical power into air than the edge emitter at moderate drive currents. At low values of drive current the curves for both devices are almost linear.

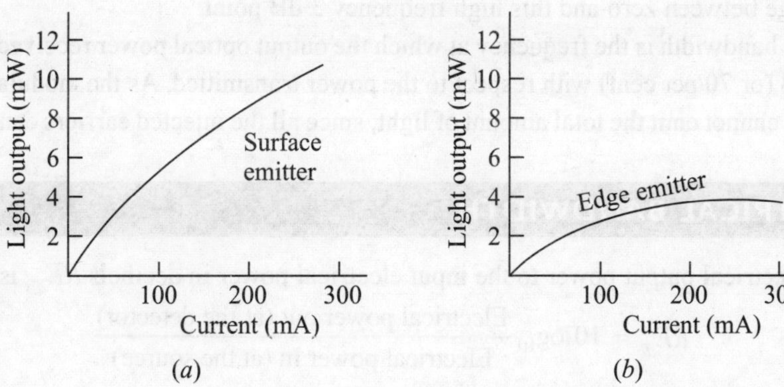

Fig. 4.27 A curve between light output versus dc drive current for typical good LEDs

The spectral response of the AlGaAs LED is shown in Fig. 4.28 which is the plot of relative light output intensity against the wavelength in nanometers. The peak intensity is approximate at 835 nm. The spectral linewidth measured in wavelength between the half intensity points is about 40 nm and it is termed as the full width half maximum (FWHM).

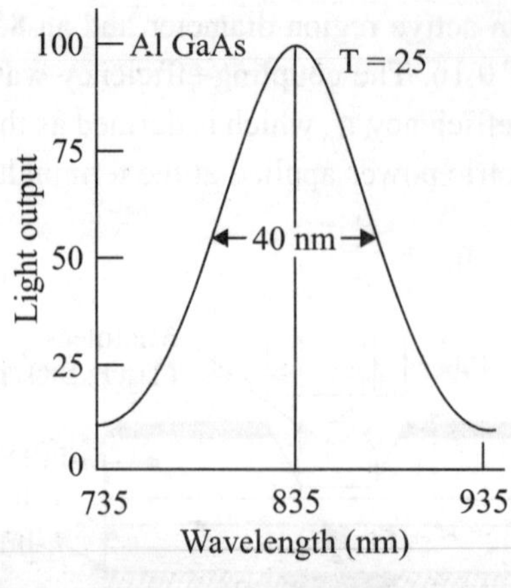

Fig. 4.28 A curve of light output versus wavelength illustrating the bandwidth output

The LEDs emitting longer wavelengths from 1.1 to 1.66 μm use the material InGaAsP/InP and their linewidth becomes more in the range of 50 to 100 nm.

4.25 MODULATION BANDWIDTH

The modulation bandwidth in optical communications may be defined in two ways: One on the basis of electrical power and the other on the basis of optical power.

In electrical definition, the bandwidth is defined as the frequency at which the electrical power has decreased to half of its constant value due to the modulated portion of the optical signal. This frequency corresponds to the electrical 3 dB point or the frequency at which the output electrical power is reduced by 3 dB with respect is the input electrical power. As optical sources operate down to dc extremely low frequency. We only consider the high frequency 3 dB point, the modulation bandwidth is the frequency range between zero and this high frequency 3 dB point.

The optical bandwidth is the frequency at which the output optical power received at the detector is reduced by 3 dB (or 70 per cent) with respect to the power transmitted. As the modulation frequency increases the LED cannot emit the total amount of light, since all the injected carriers cannot recombine.

4.26 ELECTRICAL BANDWIDTH

The ratio of the electrical output power to the input electrical power in decibels RE_{dB} is given by

$$RE_{db} = 10\log_{10} \frac{\text{Electrical power out (at the detector)}}{\text{Electrical power in (at the source)}} \tag{4.26.1}$$

$$= 10 \log_{10} \frac{I_{out}^2 \times R_{out}}{I_{in}^2 \times R_{in}} \tag{4.26.2}$$

$$= 10 \log_{10} \frac{I_{out}^2 R_{out}}{I_{in}^2 R_{in}} = 10 \log_{10} \left(\frac{I_{out}^2}{I_{in}^2} \right) \tag{4.26.3}$$

$$= 20 \log_{10} \left(\frac{I_{out}}{I_{in}} \right) \text{ (Assume that } R_{in} = R_{out}) \tag{4.26.4}$$

The electrical 3 dB point (70%) occurs when the electrical power ratio is 1/2.

$$\left(\frac{I_{out}}{I_{in}}\right)^2 = \frac{1}{2} \tag{4.26.5}$$

or

$$\frac{I_{out}}{I_{in}} = \frac{1}{\sqrt{2}} \tag{4.26.6}$$

Therefore, in electrical case the bandwidth is defined by the frequency at which the output current has dropped to $1/\sqrt{2}$ (or 0.701) of the input current in the system.

4.27 OPTICAL BANDWIDTH

The optical bandwidth is defined as the ratio of the output optical power to the optical input power, RO_{dB}. It is given by

Optical power out (at the detector)

$$RO_{dB} = 10 \log_{10} \frac{\text{Optical power out (at the detector)}}{\text{Optical power in (transmitted at the source)}} \tag{4.27.1}$$

$$= 10 \log_{10} \frac{I_{out}}{I_{in}} \tag{4.27.2}$$

This happen because the optical power is proportional to the current (due to the linear relationship between optical power and current at the source and at the detector). Hence the optical 3 dB points occur when the ratio

Thus

$$\frac{I_{out}}{I_{in}} = \frac{1}{2} \tag{4.27.3}$$

Hence in the optical case, the bandwidth is defined as the frequency at which the output current has dropped to 50 per cent of the input current in the system. As the optical source operates down to dc we consider the bandwidth as equal to the high frequency 3 dB point. Figure 4.29 illustrates two types of bandwidth concerning electrical and optical powers.

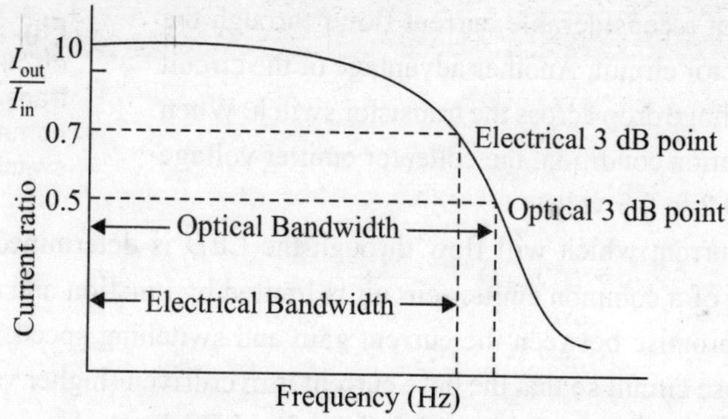

Fig. 4.29 The frequency response for an optical fiber system illustrating electrical and optical bandwidth

4.28 FACTORS DETERMINING MODULATION BANDWIDTH

The modulation bandwidth of LEDs is generally determined by the following mechanism:

1. The amount of impurity concentration in the active layer of the device.
2. The reduction in radiative life time Γ_r due to the injected carriers.
3. The parasitic capacitance present in the system.

Assuming negligible parasitic capacitance the speed at which an LED can be directly modulated is determined by the recombination life time Γ_i of the carriers. The optical power output $P_e(W)$ of the device with constant base current and the angular modulation frequency w is given by

$$\frac{P_e(W)}{P_{dc}} = \frac{1}{1+(\omega\Gamma_i)^2} \tag{4.28.1}$$

where Γ_i is the injected carrier life time in the recombination region and P_{dc} is the dc optical power for the same device current.

4.29 DRIVE CIRCUITS FOR LED AND LASER

The output power of an LED is much smaller than that of a laser, but the greatest advantage of an LED is that it is quite easy to operate an LED. We shall now discuss some of the simple circuits which can be used to convert the information voltage signals in the transmitter into a modulation current suitable for a light source.

Simple Common Emitter Drive Circuit for Digital Transmission

The operation of an LED binary system depends on switching on and switching off of current in the range of few to several milliamps. This switching operation must be done extremely rapidly in response to the logic voltage applied at the input of the drive circuit. Figure 4.30 shows a simple circuit for this current switching for an LED circuit. The circuit uses an *n-p-n* transistor in common emitter configuration. The transistor circuit is capable of providing an appreciable current gain of the circuit so that a considerable current flows through the LED placed in the collector circuit. Another advantage of the circuit is that there is a small voltage drop across the transistor switch. When the transistor is in saturation condition, the collector emitter voltage may be of the order of 0.2 to 0.3 volts.

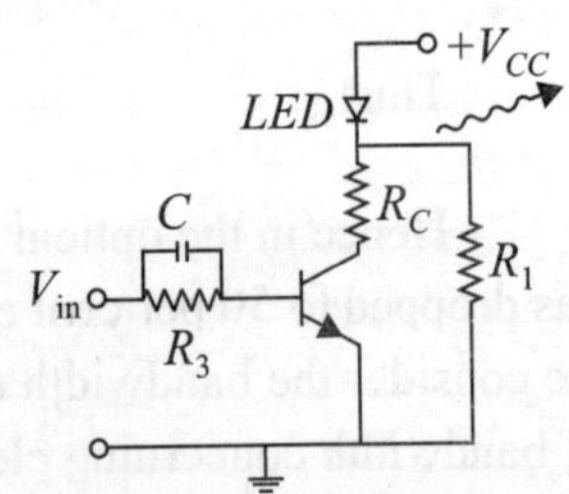

Fig. 4.30 A simple drive circuit for binary digital transmission consisting of a common emitter saturating switch

The maximum current which will flow through the LED is determined by the resistance R_c. However, the switching of a common emitter circuit is limited by junction and diffusion capacitances. In order to have a compromise between the current gain and switching speed, a pre-emphasis circuit (CR_3) is placed in the base circuit so that the base current is overdriven (higher value) during switching-on period. Figure 4.30 illustrates the pulse modulation of an LED by analog input voltage.

Emitter-follower Drive Circuit

When a low impedance drive circuit is used, it is possible to achieve the higher switching speed from an LED circuit without using a pulse shaping circuit or a speed up circuit. In the case of a low impedance drive circuit the charging of capacitances (space charge and diffusion) occurs very quickly (*CR* value is small). Figure 4.31 shows the emitter follower drive circuit which is capable of achieving comparatively higher speed.

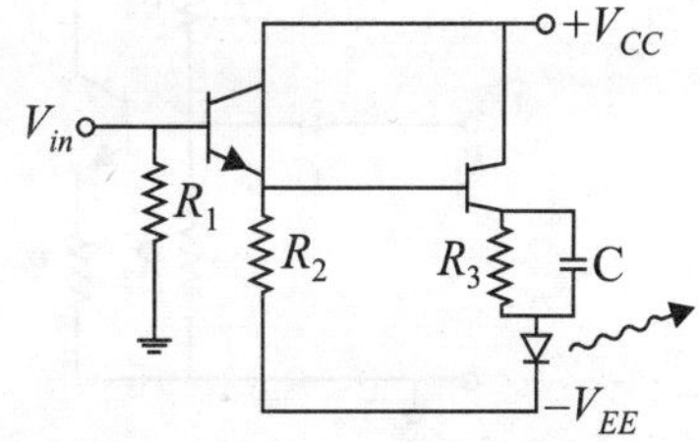

Fig. 4.31 A simple *n-p-n* common emitter LED circuit for current switch (making current on and off)

The advantage of this circuit of Fig. 4.31 with a compensating matching network (R_3C) allows fast modulation of LEDs with comparatively lower drive power. A circuit for optimum values for the matching network is capable of giving an optical rise time of 2.5ns with LEDs with capacitance of 180pf and therefore, the operation with 100 bits per second is possible.

Low Impedance Drive Circuit with Shunt Configuration

Figure 4.32 shows a low impedance drive circuit consisting of a simple shunt configuration. The switching transistor in this circuit is placed parallel to the LED providing a low impedance path for switching off the LED by shunting current around it. The switching-on time of the circuit is determined by the combination of resistance *R* and the LED capacitance. Stored space charge may be removed by applying a small reverse bias to the LED when the device is switched off. This is achieved by putting the emitter of the transistor V_{EE} below ground (V_{EE}-negative). In this case Schottky clamp (shown by dotted) may be placed to limit the amplitude of the reverse bias without introducing any minority carrier stored charge in the circuit.

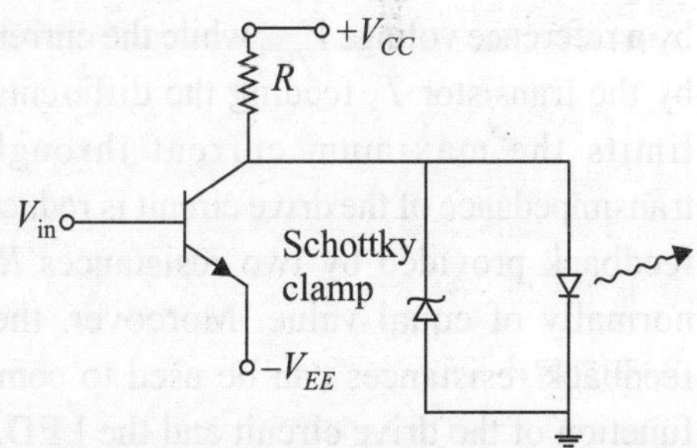

Fig. 4.32 A low impedance drive circuit consisting of an emitter follower with compensating matching network

4.30 ANALOG TRANSMISSION DRIVE CIRCUIT

Figure 4.33 shows two types of high speed drive circuit configurations for transmission of analog voltage. Figure 4.33 shows a driver circuit consisting of a common transconductance amplifier which converts the input base voltage into a collector current. This circuit is biased for a class *A* operation. The quiescent (operating) collector current is fixed at about half the peak value.

A similar transconductance circuit which uses a Darlington transistor pair in order to reduce the impedance of the source is shown in Fig. 4.34. This type of circuit is used to drive high radiance LED's at the frequencies around 70 MHz.

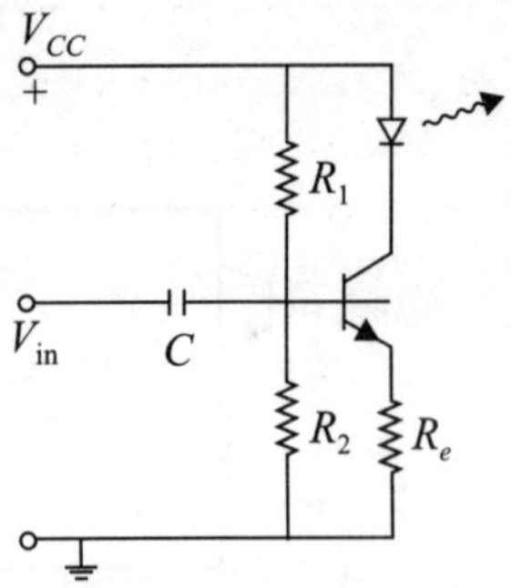

Fig. 4.33 Low impedance drive circuit consisting of a simple shunt configuration

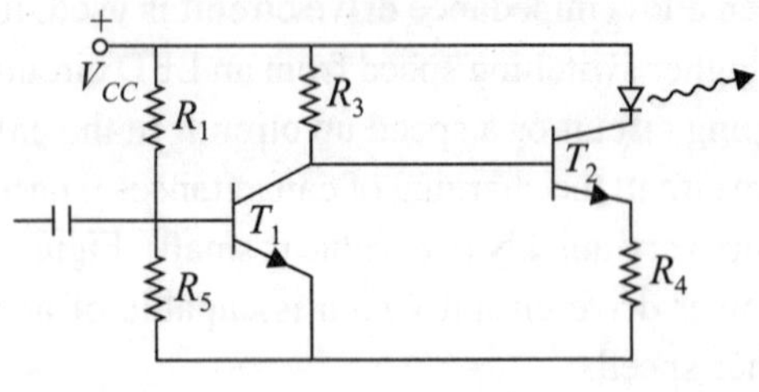

Fig. 4.34 (a) Transimpedance drive circuit for analog signal transmission: Common emitter configuration

Differential Amplifier Driver Circuit

Figure 4.35 shows a drive circuit using differential amplifiers operating over its linear region and it directly modulates the output of an LED. The operating point of the LED is controlled by a reference voltage V_{ref}, while the current generator provided by the transistor T_3 feeding the differential stage (T_1 and T_2) limits the maximum current through the device. The transimpedance of the drive circuit is reduced by a series current feedback provided by two resistances R_1 and R_2 which are normally of equal value. Moreover, the variation of these feedback resistances can be used to compensate the transfer function of the drive circuit and the LED.

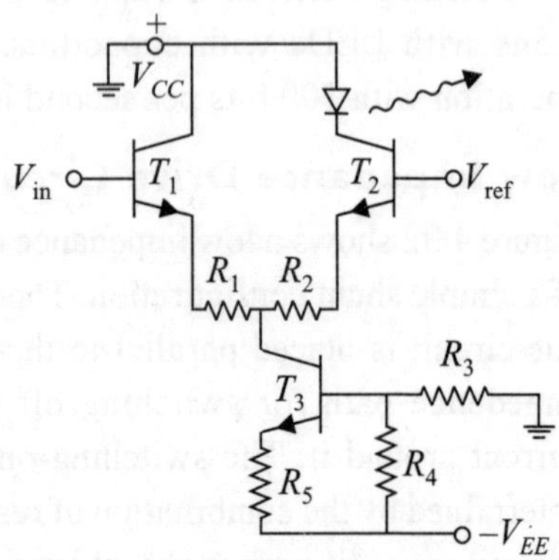

Fig. 4.35 A differential amplifier drive circuit

REVIEW QUESTIONS

4.1 Discuss the important requirements for an optical source for use in optical communication. Give the spectrum of a few of them.

4.2 Describe the operation of *p-n* junction diode on physical basis with reference to the formation of barrier potential and depletion layer. Explain the condition when it exhibits the electroluminescence.

4.3 Explain physically the Fermi Dirac statistics and Fermi function *F*(*E*). What will be the position of the Fermi level in energy band diagram for (*a*) intrinsic (*b*) *n*-type and (*c*) *p*-type semiconductors? What do you understand by "Degenerate semiconductor"?

4.4 Draw the energy band diagram for a (*a*) unbiased (*b*) forward biased and (*c*) reverse biased *p-n* junction illustrating the position of the Fermi level.
What do you mean by "impurity levels"? Draw the impurity level for both acceptor and donor impurities in the *p-n* junction energy level diagram.

4.5 What are the direct band gap and indirect band gap semiconductors? Illustrate with electron energy band diagrams. With suitable diagrams explain the mechanism of electronhole recombinations in each case. Mention the names of two materials in each type.

4.6 Explain with relevant diagrams the basic principle of confinement of carriers and optical power in the active region of a double heterojunction (DH) LED.

4.7 Estimate the power generated internally within a double heterojunction LED at a drive current of 60 ma and a peak emission wavelength of 1.31 μm. The radiative and nonradiative recombination life times of the minority carriers in the active region are equal to 100 μ second and 60 μ second respectively.

4.8 Estimate the optical power coupled to a step index fiber of 50 μm core diameter with a numerical aperture (*NA*) of 0.18 from a DH surface emitter LED with an emission area of diameter 75 μm and radiance of 60 W Sr^{-1} cm^2. The Fresnel reflection coefficient may be considered to be negligible.

4.9 Discuss the various techniques about the lens coupling to optical fibers and outline the various techniques employed.

4.10 Define electrical and optical modulation bandwidth for an optical fiber communication. Estimate the 3 dB optical bandwidth as well as 3 dB electrical bandwidth and relationship between them.
In a system of optical modulation the 3 dB electrical bandwidth is 5 MHz. Estimate the 3 dB optical bandwidth the assumption that the frequency response is Gaussian.

4.11 Discuss different modulation drive circuits for LEDs and explain their operations. Analog drive circuits are designed with Darlington pairs. Explain the operations of the circuits.

4.12 Estimate the electrical modulation bandwidth for an LED with carrier recombination lifetime of 8 ns. The frequency response of the device is assumed to be Gaussian.

4.13 Explain proton bombardment on semiconductor with the effect produced. Compare proton bombardment and implementation of oxygen ions techniques.

4.14 Compare the difference in performance characteristics between conventional LED used in fiber communications and super luminescent LED.

4.15 Describe with the help of a diagram the structure of a SLD used for longer wavelength region.

CHAPTER 5

Semiconductor Lasers

CHAPTER FOCUS

- *Introduction: Basic principle of laser action*
- *Einstein relationships for stimulated emission*
- *Population inversion*
- *Non-semiconductor laser system*
- *Condition of lasing (Laser action)*
- *Direct band gap semiconductors*
- *Indirect band gap semiconductors*
- *Recombination coefficient*
- *Stimulated emission: Intrinsic semiconductor*
- *Stimulated emission: p-n diode*
- *The salient points about the laser action*
- *Threshold current for lasing*
- *Relationship between J_{th} and g_{th}*
- *Homojunction laser with Fabry Perrot cavity structure*
- *Spectrum and fabrication*
- *Efficiency of a laser*
- *Double heterojunction laser*
- *Broad area double heterojunction laser (DH laser)*
- *Stripe geometry laser*
- *Laser modes*
- *Broadening of longitudinal mode peaks*
- *Single mode operation*
- *Gain guided laser*

- *Distributed feedback laser (DFB) and distributed Bragg reflector (DBR) laser*
- *Temperature dependence of threshold current of a laser*
- *Advantages of a semiconductor injection laser*
- *Solved examples and review questions*

5.1 INTRODUCTION: BASIC PRINCIPLE OF LASER ACTION

Spontaneous Emission, Stimulated Emission and Absorption

The term "laser" is the abbreviation of "light amplification by stimulated emission of radiation". Although the laser action has been achieved by using many different types of materials, only the semiconductor lasers are used in fiber communications. Despite their difference in sizes and materials used for their constructions, the basic principle of operation of all types of lasers is the same.

Absorption and Emission of Radiation in Two Level Systems

Laser action takes place due to three independent processes. These are: (1) absorption (2) spontaneous emission and (3) stimulated emission. Figure 5.1 illustrates the different processes in a simple two-energy level diagram. In the figure E_1 is the ground state energy level and E_2 is the excited state level. According to Planck's Law the transition between these two energy states gives emission or absorption of photon of energy given by

$$E = E_2 - E_1 = hf \tag{5.1.1}$$

where f is the frequency of radiation in Hertz and h is the Planck's constant = 6.626×10^{-34} J-s. The energy is in Joules. Normally the system is at the ground state. When a photon of energy hf impinges on an electron in the ground state of energy E_1, the electron absorbs the energy of the impinging photon and it is excited to the electron higher energy state E_2 as shown in Fig. 5.1(*a*). But this is an unstable state and the electron will shortly return back to the ground state E_1 emitting a light (radiation) of energy hf. This phenomenon happens automatically without any external stimulation. This emission is isotropic and of random phase. It appears as a narrow Gaussian output.

The electron can be induced to make the downward transition from the excited state E_2 by an external stimulation. As shown in Fig. 5.1(*c*), if a photon of energy hf (where $hf = E_2 - E_1$) impinges on the system while the electron is still in the excited state, the electron is immediately stimulated to fall on the ground state E_1 and radiates a photon of energy hf along with the impinging photon.

The emitted photon is in the same phase with the incident photon and the resultant emission is known as the "Stimulated emission".

Spontaneous emission process is of random nature and light is emitted by electronic transitions of large number of atoms and is of incoherent in nature. Spontaneous emission is the basic mechanism of light generation within the light emitting diode (LED). But the stimulated emission is the mechanism of light emission from the "laser". Firstly, a photon produced by the stimulated emission is generally of identical energy to one which caused it and hence, the light associated with them is of the same frequency.

Secondly, the light associated with the stimulating and stimulated photons are in the same phase and have the same polarization. Therefore, the coherent radiation is obtained in the case of the stimulated emission.

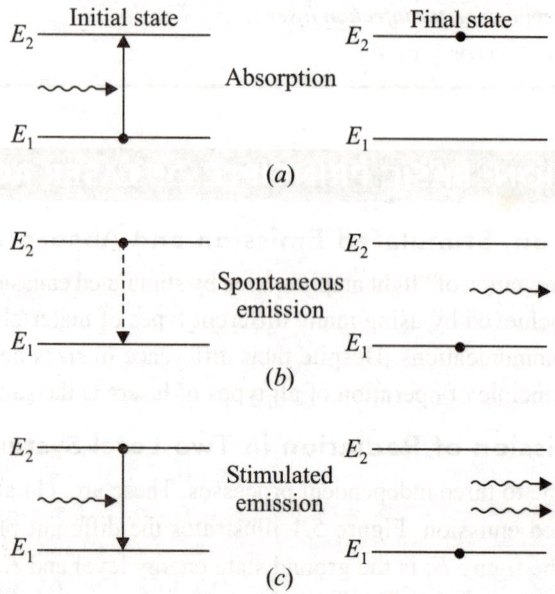

Fig. 5.1 Energy-state diagram illustrating three processes in two energy level system (*a*) absorption (*b*) spontaneous emission and (*c*) stimulated emission. The black dots indicate the state of the atom before and after a transition occurs.

5.2 EINSTEIN RELATIONSHIPS FOR STIMULATED EMISSION

In 1917 Einstein showed that the rates of three transition processes of absorption, spontaneous emission and stimulated emission are related mathematically. He considered the atomic system was in thermal equilibrium so that the rate of upward transition must be equal to the rate of downward transition. Let us consider a two energy-level system as shown in Fig. 5.1 inside an enclosure in thermodynamical equilibrium. The population in two energy-level system can be expressed by Boltzmann statistics and may be given by

$$\frac{n_1}{n_2} = \frac{g_1 \exp\left(\dfrac{-E_1}{K_B T}\right)}{g_2 \exp\left(\dfrac{-E_2}{K_B T}\right)} = \frac{g_1}{g_2} \exp \frac{(E_2 - E_1)}{K_B T} \tag{5.2.1}$$

Substituting the value of $(E_2 - E_1)$ in Eq. (5.2.1), we get

$$\frac{n_1}{n_2} = \frac{g_1}{g_2} \exp\left(\frac{hf}{K_B T}\right) \tag{5.2.2}$$

where n_1 and n_2 denote the densities of atoms in the energy level E_1 and E_2 respectively. T = absolute temperature, g_1 and g_2 are the degeneracies in the levels E_1 and E_2 respectively. K_B is the Boltzmann constant.

Since the density of atoms in the ground state E_1 is n_1 and the spectral density of radiation energy at the frequency f is P_f, the rate of the upward transition from E_1 to E_2 level is obviously proportional to the product of n_1 and P_f. The rate of upward transition, denoted by R_{12} from E_1 to E_2 level is given by

$$R_{12} = B_{12}\, n_1\, P_f \tag{5.2.3}$$

where B_{12} is the constant of proportionality and is referred to as "Einstein coefficient of absorption".

Again, atoms in the higher excited energy state can undergo transition from E_2 to E_1 level either spontaneously or stimulated by an external radiation field. In the case of spontaneous emission, the average time during which an electron exists in the excited state before the occurrence of the downward transition is known as the Spontaneous lifetime (denoted by Γ_{21}). If the density of atoms in the E_2 state is n_2, then the spontaneous emission rate is equal to the product of n_2 and $1/\Gamma_{21}$. Denoting $1/\Gamma_{21}$ by A_{21}, the rate of spontaneous emission rate may be written as $A_{21}\, n_2$ where A_{21} is referred to as *"Einstein coefficient of spontaneous emission"* and is given by the reciprocal of the spontaneous lifetime Γ_{21} so that

$$A_{21} = 1/\Gamma_{21} \tag{5.2.4}$$

The rate of stimulated downward transition of an electron from the energy level E_2 to E_1 is proportional to the product of n_2 and P_f. The rate of stimulated downward transition is given by $B_{21}\, n_2\, P_f$ where B_{21} is known as *"Einstein coefficient of stimulated emission"*. The total transition rate from energy level E_2 to E_1 (denoted by R_{21}) is the sum of spontaneous and stimulated emission rates. Hence,

$$R_{21} = B_{21}n_2 + B_{21}n_2P_f \tag{5.2.5}$$

In the case of a system in thermal equilibrium, upward and downward transition rates are equal so that

$$R_{21} = R_{12} \tag{5.2.6}$$

Substituting the values from Eq. (5.2.3) and Eq. (5.2.5) into Eq. (5.2.6), we get

$$B_{12}n_1P_f = A_{21}n_2 + B_{21}n_2P_f \tag{5.2.7}$$

Rearranging properly, we get

$$P_f = \frac{A_{21}n_2}{B_{12}n_1 - B_{21}n_2} \tag{5.2.8}$$

Dividing both numerator and denominator of right hand side of Eq. (5.2.8) by $B_{21}\, n_2$, we get

$$P_f = \frac{\dfrac{A_{21}}{B_{21}}}{\dfrac{n_1}{n_2}\exp\dfrac{B_{12}}{B_{21}} - 1} \tag{5.2.9}$$

on substitution of the value of n_1/n_2 in Eq. (5.2.9) from Eq. (5.2.2) yields

$$P_f = \frac{\dfrac{A_{21}}{B_{21}}}{\dfrac{g_1}{g_2}\dfrac{B_{12}}{B_{21}}\exp\left(\dfrac{hf}{K_BT}\right) - 1} \tag{5.2.10}$$

However, the atomic system under consideration is in thermal equilibrium and therefore, it produces radiation density which is identical with the black-body radiation. Planck has shown that the radiation spectral density P_f for a black-body radiation within a frequency range of f and $f + df$ is given by

$$P_f = \frac{8\pi h f^3}{c^3} \frac{1}{\exp\frac{hf}{K_B T} - 1} \tag{5.2.11}$$

where c is the velocity of light; $c = 3 \times 10^8$ m/s.

When we compare Eq. (5.2.10) and Eq. (5.2.11), we get two Einstein relationships:

$$B_{12} = \frac{g_2}{g_1} B_{21} \tag{5.2.12}$$

and

$$\frac{A_{21}}{B_{21}} = \frac{8\pi h f^3}{c^3} \tag{5.2.13}$$

Let us consider a case where the degeneracies of two levels are equal ($g_1 = g_2$), the probability of absorption and emission are equal ($B_{12} = B_{21}$).

Relationship between Stimulated and Spontaneous Emission

$$\frac{\text{Stimulated emission rate}}{\text{Spontaneous emission rate}} = \frac{B_{21}}{A_{21}} P_f$$

Comparing Eq. (5.2.9) and Eq. (5.2.11), we get

$$\frac{B_{21}}{A_{21}} P_f = \frac{c^3}{8\pi h f^3} P_f = \frac{1}{\exp\frac{hf}{K_B T}} \tag{5.2.14}$$

Example 5.1: *A lamp is operating at a temperature of 727°C and emitting radiation of an wavelength of 0.5 μm. Calculate the ratio between the stimulated and spontaneous emission rate.*

Solution: With the help of Eq. (5.2.9) and (5.2.11), we get

$$\frac{\text{Stimulated emission rate}}{\text{Spontaneous emission rate}} = \frac{B_{21}}{A_{21}} P_f = \frac{c^3}{4\pi h f^3} P_f = \frac{1}{\exp\frac{hf}{K_B T} - 1}$$

$$\text{frequency } f = \frac{c}{\lambda} = \frac{3\times10^8}{0.5\times10^{-6}} = 6\times10^{14} \text{ Hertz.}$$

$$T = 727°\text{C} + 273 = 1000°\text{K}, h = 6.626 \times 10^{-34} \text{ J-S}$$

$$K_B = 1.38 \times 10^{-23} \text{ J/K}$$

Substituting the values, we get

$$\frac{B_{21}}{A_{21}} P_f = \frac{1}{\exp\frac{6.626\times10^{-34} \times 6\times10^{14}}{1.38\times10^{-23}\times1000} - 1}$$

$$= \frac{1}{\exp 28.8 - 1} = \exp(-28.8) = 3.1\times10^{-13}$$

The ratio between the stimulated and spontaneous emission rate is 3×10^{-13}.

5.3 POPULATION INVERSION

According to Boltzmann distribution law the number of atoms n_1 in the lower energy level E_1 is much greater than the number n_2 in the upper energy level E_2 under the condition of thermal equilibrium. Figure 5.2(*a*) shows the condition at room temperature when the system is in thermal equilibrium, ($n_1 > n_2$).

In order to get optical amplification it is necessary to produce the non-equilibrium distribution of atoms so that the population in the upper energy level E_2 is greater than that of the lower energy level E_1 (*i.e.*, $n_2 > n_1$). This condition is referred to as "population inversion" and is illustrated in Fig. 5.2(*b*). Since this is not the equilibrium condition, the population inversion is achieved by the "pumping technique". There are different methods for pumping techniques.

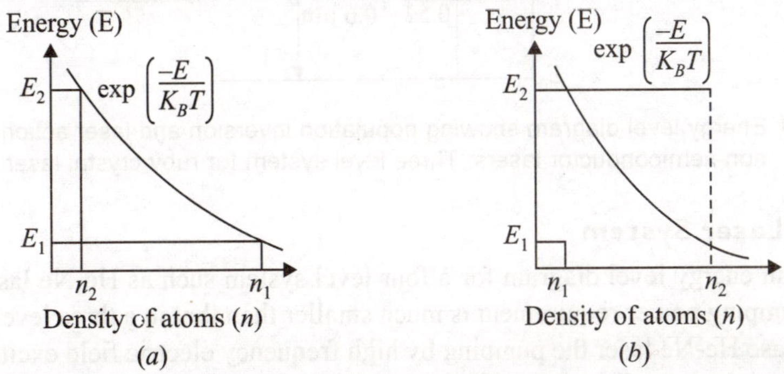

Fig. 5.2 Populations in a two energy level system (*a*) Boltzmann distribution for system in thermal equilibrium and (*b*) a non-equilibrium distribution showing population inversion

In a semiconductor laser, the population inversion is achieved by injecting electrons into the material at the device contacts to fill the lower energy states in conduction band. In the case of a *p-n* junction diode the population inversion is produced by a forward bias voltage applied to the diode so that electrons are injected into the conduction band of the *p*-region or holes injected into the valence band of the *n*-region.

5.4 NON-SEMICONDUCTOR LASER SYSTEM

The common methods for pumping in a non-semiconductor lasers (such as a ruby laser or helium-neon laser) involves the application of an intense light from a flash light tube or a high frequency electric field.

Ruby Crystal Laser

In the case of a ruby laser an intense light from a flash light tube is used for the population inversion. Figure 5.3 illustrates a three level laser system which consists of a ground energy level E_o, a metastable energy level E_1 and a third level E_2 above the metastable level. In equilibrium condition, atomic distribution obeys Boltzmann distribution law. With the help of a suitable pumping on the system such as an intense light from a flash light tube, the electrons in some of the atoms may be excited from the ground state E_o to the higher energy state E_2. Since E_2 is a normal level, the number of electrons will rapidly decrease by non-radiative process either to E_1 or directly to E_o. The metastable state has a much longer lifetime than that in E_2, so large number of atoms are accumulated in the energy level E_1. During a certain period of time, the density, n_1 in the metastable state increases and becomes more than that in the ground state E_o (denoted by n_o) and a population inversion is achieved by the levels E_o and E_1.

Stimulated emission and Lasing can take place creating radiative transitions between energy levels E_1 and E_0. The drawback of the three level system is that it requires a high value of pump power.

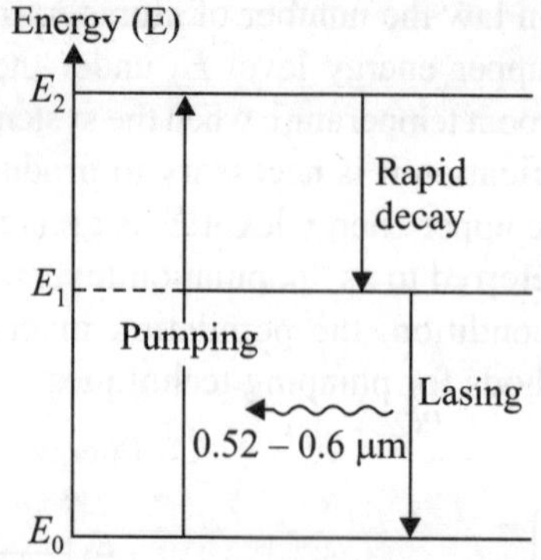

Fig. 5.3 Energy level diagram showing population inversion and laser action for two non-semiconductor lasers. Three level system for ruby crystal laser.

Helium-Neon Laser System

Figure 5.4 shows an energy level diagram for a four level system such as He-Ne laser. In this case of He-Ne laser the pumping power requirement is much smaller than that in a three level system such as a ruby laser. In the case He-Ne laser the pumping by high frequency electric field excites the atoms from the ground state E_0 to the state E_3 and they decay rapidly to a metastable state E_2. Thus, the population levels E_3 and E_1 remain unchanged and a small increase in the number of atoms in the level E_2 creates a population inversion so that lasing occurs between levels E_2 and E_1.

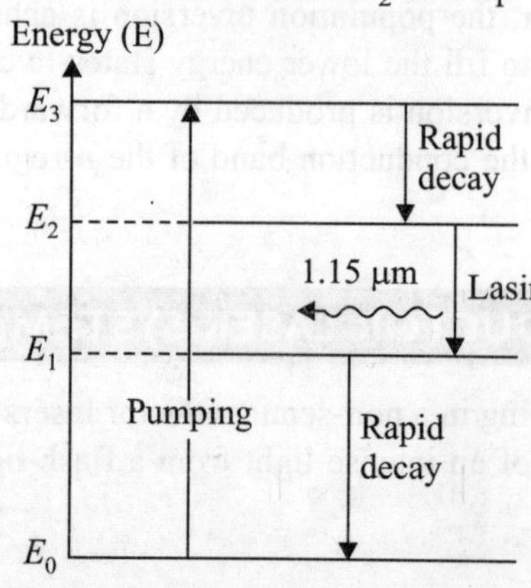

Fig. 5.4 Non-semiconductor laser: four level system of He-Ne (Helium-Neon) gas laser.

5.5 CONDITION OF LASING (LASER ACTION)

We consider a beam of radiation passing in the z-direction through an assembly of atoms. We also assume that it carries a power density P expressed in Watts/m^2 and extends over a narrow frequency range around f. The beam of radiation suffers an absorption and a net loss of power as it travels. We assume that all the losses except those due to transmission through the mirror is denoted by loss coefficient per unit length $\bar{\alpha}$ per cm and L is the length of the amplifying medium between two mirrors as shown in Fig. 5.5. The reflectivities of the two mirrors are R_1 and R_2 respectively. Thus, we can write

$$\bar{\alpha} = -\frac{1}{P}\frac{dP}{dz} \tag{5.5.1}$$

where $\bar{\alpha}$ is independent of z. The loss coefficient $\bar{\alpha}$ is defined as the loss in power per unit power of the wave as it travels through unit distance.

On solving Eq. (5.5.1) yields

$$P(z) = P(0)\exp(-\bar{\alpha} Z) \tag{5.5.2}$$

where $P(0)$ and $P(z)$ are the power densities at $z = 0$ and $z = z$ respectively.

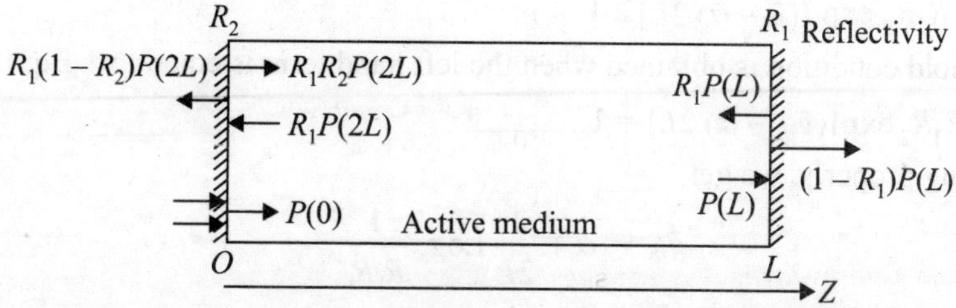

Fig. 5.5 Schematic illustration of the condition in a laser: $P(z) = P(0)\exp(g - \alpha)z$

When there are external means available, thereby the upper energy levels may be populated profusely, it becomes possible to create a situation in which the population of the upper energy level (n_2) is made greater than that in the lower energy level (n_1). In the case of a *p-n* junction laser, it is accomplished by forward biasing the junction and as a result, the conduction band in the *p*-region is populated by the "Carrier injection". Thus a "population inversion" is said to exist. The direct consequence is that the state of stimulated emission exceeds the absorption. The beam is supplemented (amplified) as it passes through the medium producing an optical amplifier. Let $\bar{g}$ per cm represent the gain coefficient per unit length produced by the stimulated emission. Thus,

$$\bar{g} = \frac{1}{P}\frac{dP}{dz} \tag{5.5.3}$$

Solving the equation, we get

$$P(z) = P(0)\exp(\bar{g}z) \tag{5.5.4}$$

where $P(0)$ and $P(z)$ are the power density at $z = 0$ and $z = z$ respectively.

Producing an optical gain by creating a population inversion is the first of the two important steps required for the laser action. The second step is to provide some positive feedback to transform the optical amplifier in a "laser oscillator". This is accomplished by two mirrors at two ends which reflect the amplified light back and forth through the active region as illustrated in Fig. 5.5. Then an optical cavity is formed. The optical power density at the resonant frequencies of the cavity comes into an equilibrium when the power loss during the transit of the radiation through the medium is equal to the optical power gain. The power loss occurs due to the reflection through the mirror.

Self-oscillation cannot start before the gain exceeds the loss. This requirement sets a threshold population inversion (denoted by n_{th}) which must be exceeded before the laser action can take place.

When both loss and gain are considered the variation of the power of the optical beam with distance z becomes

$$P(z) = P(0)\exp[(\bar{g} - \bar{\alpha})z] \tag{5.5.5}$$

In each round trip transit the beam moves through a distance $2L$ (Fig. 5.5) so that the gain in the round trip transit is found by putting $z = 2L$ in Eq. (5.5.5) as

$$P(2L) = P(0)\exp[(\bar{g} - \bar{\alpha})2L] \tag{5.5.6}$$

where L is the effective length of the active region as shown in Fig. 5.5. At the mirrors the fractions R_1 and R_2 of the incident light are reflected. Thus, the condition of self-oscillation in the laser cavity becomes

$$R_1 R_2 P(2L) \geq P(0)$$

or

$$R_1 R_2 \frac{P(2L)}{P(0)} \geq 1 \tag{5.5.7}$$

Combining Eq. (5.5.6) and Eq. (5.5.7), we get

$$R_1 R_2 \exp\,[(\bar{g} - \bar{\alpha})\, 2L] \geq 1 \tag{5.5.8}$$

The threshold condition is obtained when the left hand term is equal to 1 Eq. (5.5.8) becomes

$$R_1 R_2 \exp[(\bar{g}_{th} - \bar{\alpha})\, 2L] = 1 \tag{5.5.9a}$$

Rearranging properly, we get

$$\bar{g}_{th} = \bar{\alpha} + \frac{1}{2L} \text{Log}_e \frac{1}{R_1 R_1} \tag{5.5.9b}$$

when $R_1 = R_2 = R$, we get

$$\bar{g}_{th} = \bar{\alpha} + \left[\frac{1}{L} \text{Log}_e \frac{1}{R}\right] \tag{5.5.10}$$

5.6 DIRECT AND INDIRECT BAND GAP SEMICONDUCTORS

Direct Band gap Semiconductor

We have already mentioned that the most useful materials for electroluminescence are the direct band gap semiconductors. In a direct band gap semiconductor the energy maximum of the valence band and the energy minimum in the conduction band occur at the almost same value of electron crystal momentum as shown in Fig. 5.6(a) so that the direct recombination between an electron and a hole is possible. When the recombination occurs the momentum of the electron remains virtually constant and the energy released corresponding to the band gap energy E_g may be emitted as light. The direct transition of an electron across the energy gap provides an efficient mechanism of photon emission. The average time during which the minority carriers remain the free state before recombination is known as "minority carrier life time". Direct band gap semiconductors have a short value of the lifetime of the order of 10^{-8} to 10^{-10} seconds.

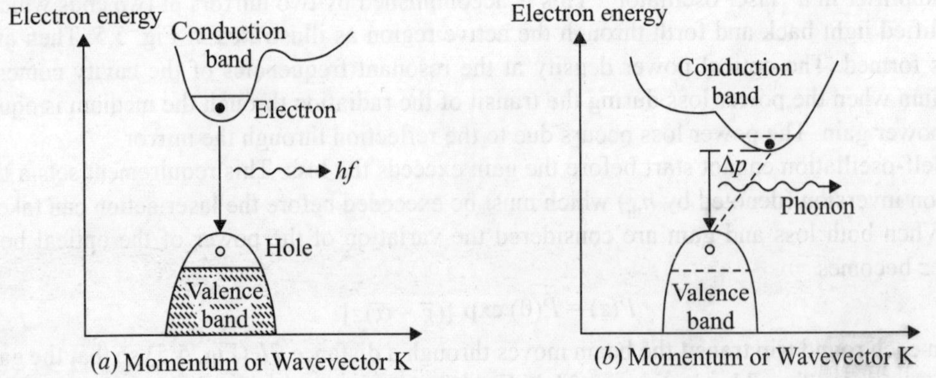

Fig. 5.6 Energy band structure of (a) a direct band gap semiconductor (b) an indirect band gap semiconductor

5.7 INDIRECT BAND GAP

In the case of an indirect band-gap semiconductor the energy maximum in the valence band and the energy minimum in the conduction band occur at different values of electron crystal momentum as shown in Fig. 5.6(*b*). For the electron-hole recombination to occur, it is essential that as electron loses momentum so that the electron has a momentum corresponding to the maximum energy in the valence band. The principle of conservation of momentum requires the emission of a third particle which is known as "phonon" with a momentum of Δp as shown in Fig. 5.6(*b*). The three particle recombination is less probable than the two particle recombination process exhibited by a direct band gap semiconductor. Hence the recombination in an indirect band gap semiconductor is relatively slow and the minority carrier life-time is of the order of 10^{-2} to 10^{-4} seconds in an indirect band gap semiconductor.

5.8 RECOMBINATION COEFFICIENT

The recombination coefficient B_r can be obtained by measuring the absorption coefficient of the semiconductor and for low injected minority carrier density relative to the majority carriers, the relationship between the recombination coefficient B_r and radiative minority carrier lifetime Γ_r is given by

$$\Gamma_r = [B_r(n_n + p_n)]^{-1} \tag{5.8.1}$$

where n_n and p_p are the respective majority carrier concentrations in the n and p type region respectively. Γ_r is the radiative minority carrier lifetime and can be defined as the average time a minority carrier can exist in a free state before radiative recombination takes place.

Table 5.1 Gives the value of Bandgap energy E_g and the recombination coefficient of some commonly used semiconductor materials

Semiconductor material	*Energy band gap in eV*	*Recombination coefficient B_r in cm³ per sec*
GaAs	Direct 1.43	7.21×10^{-10}
GaSb	Direct 0.73	2.39×10^{-10}
InAs	Direct 0.35	8.5×10^{-11}
InSb	Direct 0.18	4.58×10^{-11}
Si	Indirect 1.12	1.79×10^{-15}
Ge	Indirect 0.67	5.25×10^{-14}
GaP	Indirect 2.26	5.37×10^{-14}

5.9 STIMULATED EMISSION: INTRINSIC SEMICONDUCTOR

Population inversion is an essential condition for the occurrence of the stimulated emission. The inversion in the carrier population is achieved in an intrinsic semiconductor by the "injection" of electrons in the conduction band of the semiconductor material as shown in Fig. 5.7 which illustrates the electron energy and the corresponding filled state. At absolute zero the conduction band contains no electrons as shown in Fig. 5.7(*a*). Now electrons are injected into the material to fill the lower energy states in the conduction band up to the injection energy (or up to the quasi Fermi level of electrons). Since the charge neutrality should be maintained within the material, an equal density of holes is created at the top of valence band (by the absence of electrons) as shown in Fig. 5.7(*b*).

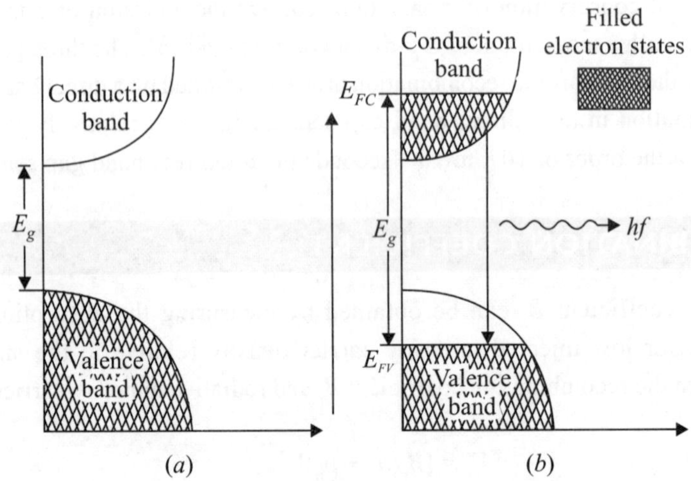

Fig. 5.7 The filled energy state for an intrinsic direct band gap semiconductor at absolute zero temperature (*a*) in equilibrium and (*b*) high carrier injection

The incident photons have energy E_g which is less than the separation energy E_q of the quasi Fermi level where $E_q = E_{FC} - E_{FV}$. These photons can not be absorbed because the necessary conduction states are occupied. But these photons are capable of inducing a downward transition of electrons from the filled conduction band to the empty valence band state and the emission of another photon is stimulated. Thus the basic condition of stimulated emission depends on the quasi-Fermi level separation E_q and also on the band gap energy E_g. The condition of the stimulated emission is given by

$$(E_{FC} - E_{FV}) > hf > E_g \tag{5.9.1}$$

where hf is the energy of the emitted photon, and $E_q = E_{FC} - E_{FV}$ = quasi-Fermi level separation in a *p-n* junction.

However, we have described the ideal situation, while at normal operating temperatures, the distribution of electrons and holes is less well defined, but the condition for stimulated emission is largely maintained.

5.10 STIMULATED EMISSION: *p-n* DIODE

The population inversion in a *p-n* junction diode is achieved by adding excessive quantity of impurity atoms (heavy doping) to both *n*-type and *p*-type regions of the diode. The heavy doping is, therefore, referred to as degenerative doping. Heavy *p*-type doping with acceptor impurities causes a considerable amount of lowering of the Fermi level into the boundary between the filled and the empty states in the valence band. Similarly, the degenerative *n*-type doping with donor impurity atoms causes the raising of the Fermi level into the conduction band of the *n*-type region as shown in Fig. 5.8(*a*). The figure illustrates the energy band diagram of a degenerate *p-n* junction, with the portion of the Fermi level and electron occupation when no external voltage applied across the diode. Since the *p-n* junction is in thermal equilibrium, the Fermi level will have the same value throughout the crystal.

Now the *p-n* junction is forward biased by an external voltage which is almost equal to the band gap energy E_g volts of the semiconductor. Figure 5.8(*b*) shows the band diagram of the forward biased *p-n* junction which has both *p*- and *n*-regions degenerated. When it is forward biased, electrons move from the *n*-region to the *p*-region and holes move from the *p*-region to the *n*-region due to the property of diffusion. This is known as the "injection" of minority carriers in respective region. At high carrier injection in such a junction, there exists an active region near the depletion region. The active region contains simultaneously the degenerate population of electrons and holes which is sometime referred to as "doubly degenerate" population.

For this region the condition of stimulated emission is satisfied for the electromagnetic wave of frequency *f* when

$$\frac{E_g}{h} < f < \frac{E_{FC} - E_{FV}}{h} \tag{5.10.1}$$

Therefore, any radiation of frequency *f* as given by Eq. (5.9.1) will be confined in the active regions near the depletion layer and gets amplified.

5.11 THE SALIENT POINTS ABOUT THE LASER ACTION

1. In general, the population inversion in a *p-n* junction is achieved by forward biasing the junction when the injection of the minority carriers occurs in both the regions.
2. Comparatively heavy doping which degenerates the semiconductor distinguishes a *p-n* junction which provides a stimulated emission from the one which provides the spontaneous emission as in the case of an LED. Thus a *p-n* junction made of degenerate semiconductors is responsible for the stimulated emission as in the case of a laser.
3. A further requirement of the *p-n* junction diode for establishing lasing is that there must be an arrangement for optical feedback which transform an optical amplifier into an optical oscillator. This may be achieved by constructing an optical cavity within the structure and by polishing the end faces of the junction diode to act as mirrors. Each end face of the junctions diode is polished and cleaved and the sides are roughened to prevent the undesirable light emission.
4. It is noticed that a high impurity concentration within the semiconductor gives rise to a slight difference in the value of band gap energy E_g in comparison with an intrinsic semiconductor. As a result, the lasing wavelength in the case of a GaAs *p-n* junction may be varied between 0.85 to 0.95 μm although the best performance can be achieved in the wavelength ranging from 0.88 to 0.91 μm.

5.12 THRESHOLD CURRENT FOR LASING

We are interested to determine the value of the threshold current in a *p-n* junction laser diode for the occurrence of "Lasing action". Two equations concerning the electron density *n* and the photon density ϕ are given below:

$$\frac{dn}{dt} = \frac{J}{ed} - \frac{n}{\Gamma_{sp}} - Cn\phi(\text{m}^{-3}\ \text{s}^{-1}) \tag{5.12.1}$$

and

$$\frac{d\phi}{dt} = Cn\phi + \frac{\delta \cdot n}{\Gamma_{sp}} - \frac{\phi}{\Gamma_{ph}}(\text{m}^{-3}\ \text{s}^{-1}) \tag{5.12.2}$$

J is the current density in the diode expressed in ampere per square meter, *e* is numerically equal to the electron charge in MKS system = 1.6×10^{-19} C, *d* is the thickness of the active region where the recombination takes place. Γ_{sp} is the spontaneous emission lifetime, Γ_{ph} is the photon lifetime, *C* is a coefficient and δ is a small fractional value. In Eq. (5.12.1), the first term in the right hand side represents the increase in the electron density in the conduction band of the *p*-region when the current density *J* flows in the forward biased diode, the second and third terms denote the electron lost from the conduction band by the spontaneous and stimulated emissions respectively.

In Eq. (5.12.2) the first term indicates that the stimulated emission from the diode is the source of photon. A small fraction of photons produced by spontaneous emission which combine with the energy in the lasing mode is represented by the second term. Since δ is small, this term is often neglected. The third term indicates the decay in the number of photons resulting from the losses in the optical cavity. Both the rate equations may be used to study both transient and steady state conditions of a semiconductor laser and here we are interested to make steady state solutions.

Steady State Condition

The steady state is obtained by putting left sides of Eq. (5.12.1) and Eq. (5.12.2) equal to zero when *n* and ϕ have non-zero values. In addition, the optical field in the cavity must build up from a small initial value and $(d\phi/dt)$ must be equal to a positive value when ϕ is small. By neglecting spontaneous emission we put $\phi = 0$ in Eq. (5.12.2) and it will be clear that for any value of ϕ, $d\phi/dt$ will have to be positive when

$$Cn\phi - \frac{\phi}{\Gamma_{ph}} \geq 0 \tag{5.12.3}$$

or

$$n - \frac{1}{C\Gamma_{ph}} \geq 0 \tag{5.12.4}$$

Therefore, there exists a threshold value of *n* satisfied by Eq. (5.12.4). If *n* is larger than this threshold value, then ϕ can increase. From Eq. (5.12.4) the threshold value of the electron density n_{th} is given by

$$n_{th} = \frac{1}{C\Gamma_{ph}}(\text{m}^{-3}) \tag{5.12.5}$$

The threshold current density J_{th} required to maintain $n = n_{th}$ in the steady state when $\phi = 0$ may be obtained from Eq. (5.12.1) by putting $\frac{dn}{dt} = 0$ as

$$0 = \frac{J_{th}}{ed} - \frac{n_{th}}{\Gamma_{sp}} - Cn_{th}\phi \tag{5.12.6}$$

When $\phi = 0$ we get threshold electron density in term of threshold current density

or
$$\frac{J_{th}}{ed} = \frac{n_{th}}{\Gamma_{sp}}(\text{m}^{-3}\text{s}^{-1}) \tag{5.12.7}$$

Hence, Eq. (5.12.7) gives the value of the current required to sustain an excess electron density in the laser cavity. Spontaneous emission provides the only decay mechanism. The steady state photon density ϕ_s is obtained by substituting Eq. (5.12.7) in Eq. (5.12.1).

$$0 = \frac{J}{ed} - \frac{n_{th}}{\Gamma_{sp}} - Cn_{th}\phi_s$$

$$0 = \frac{J}{ed} - \frac{J_{th}}{ed} - Cn_{th}\phi_s \tag{5.12.8}$$

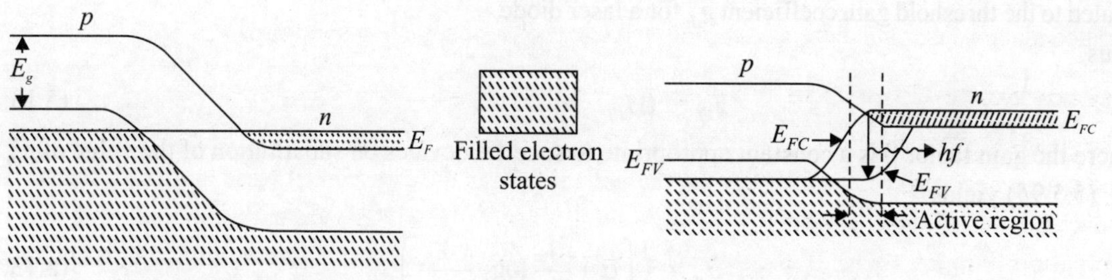

Fig. 5.8 The energy band diagram for a degenerate *p-n* junction (*a*) with no applied voltage and (*b*) with strong bias voltage which is approximately equal to the band gap energy E_g

Rearranging properly, we get

$$\phi_s = \frac{1}{Cn_{th}}\left(\frac{J - J_{th}}{ed}\right)(\text{m}^{-3}) \tag{5.12.9}$$

Substituting for Cn_{th} from (5.12.5), we get

$$\phi_s = \frac{\Gamma_{ph}}{ed}[J - J_{th}](\text{m}^{-3}) \tag{5.12.10}$$

The photon density ϕ_s cannot be a negative quantity as it is meaningless and for ϕ_s to be greater than zero, the current must exceed its "threshold value". Moreover, ϕ_s is proportional to the amount by which J is greater than J_{th}. As each photon has an energy hf, it is possible to estimate the optical power density in W m^{-2} (watts per sq. metre) with the assumption that half the photons are travelling in each of two directions.

5.13 RELATIONSHIP BETWEEN J_{th} AND g_{th}

An idealized characteristic between optical power and current in a *p-n* junction is shown in Fig. 5.9. The solid line represents the laser characteristics where the dashed line is the Eq. (5.12.9) showing the threshold value of current. It may be noticed that the device gives small output in the region below the threshold current which corresponds to the region of spontaneous emission within the structure. However as soon as the threshold current density is reached, the output light power increases considerably for a small increase in current in the device. This corresponds to the region of stimulated emission when the laser is working as an amplifier of light.

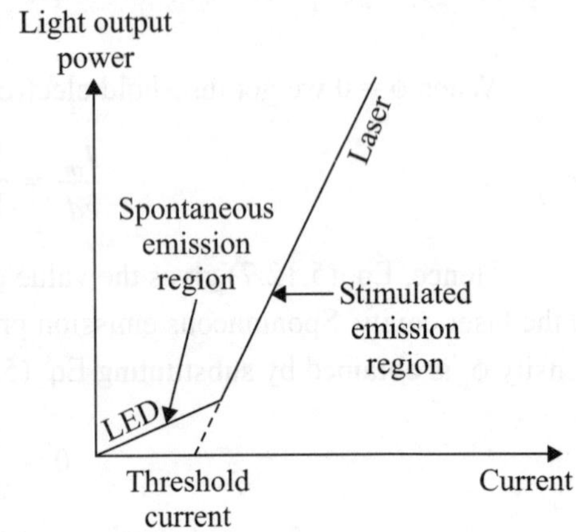

Fig. 5.9 An ideal output power versus current characteristic for injection laser

For a strongly confined structure, the threshold current density J_{th} for a stimulated emission is approximately related to the threshold gain coefficient g_{th} for a laser diode. Thus,

$$\bar{g}_{th} = \bar{\beta} J_{th} \tag{5.13.1}$$

where the gain factor β is a constant appropriate to specific devices on substitution of the value of g_{th} in Eq. (5.5.9*b*) yields

$$J_{th} = \frac{1}{\beta}\left(\bar{\alpha} + \frac{1}{2L}\log_e \frac{1}{R_1 R_2}\right) \tag{5.13.2}$$

When $R_1 = R_2 = R$, Eq. (5.13.2) takes form

$$J_{th} = \frac{1}{\beta}\left(\bar{\alpha} + \frac{1}{L}\log_e \frac{1}{R}\right) \tag{5.13.3}$$

The reflectivity for normal incident of a plane wave on the semiconductor-air layer interface may be obtained as (Fresnel law)

$$R_1 = R_2 = \frac{(n-1)^2}{(n+1)^2}$$

where n is the refractive index of the semiconductor material of the laser (say GaAs)

Example 5.2: *A gallium arsenide (GaAs) injection laser has an optical cavity of length 200 μm and width of 100 μm. The refractive index of GaAs is 3.8. At the normal operating temperature the gain factor β is 20 × 10⁻³ A cm⁻³ and the loss coefficient α per cm is 10. Estimate the threshold current density and the threshold current required for the device. Assume that the cleaved mirrors are uncoated and the current is wholly restricted within the cavity.*

Solution: The reflectivity for the normal incidence of a plane wave in GaAs-air interface is

$$R_1 = R_2 = \left(\frac{n-1}{n+1)}\right)^2 = \left(\frac{3.8-1}{3.8+1}\right)^2 = 0.34$$

$$\beta = 20 \times 10^{-3}\, A\, \text{cm}^{-3}\ \bar{\alpha} = 10\ \text{cm}^{-1}\ L = 200\ \mu\text{m} = 200 \times 10^{-4}\ \text{cm}$$

$$W = 1000 \times 10^{-4}\ \text{cm.}$$

Using Eq. (5.13.2)

$$J_{th} = \frac{1}{20\times10^{-3}}\left(10 + \frac{1}{200\times10^{-4}}\log_e\frac{1}{0.340}\right)$$

$$= \frac{1}{20\times10^{-3}}\left(10 + \frac{1.4\times10^{4}}{200}\right) = \frac{1}{20\times10^{-3}}[80]$$

$$= 4 \times 10^3\, A\, \text{cm}^{-2}$$

$$I_{th} = J_{th} \times L \times W = 4 \times 10^3 \times 200 \times 100 \times 10^{-8}$$

$$I_{th} = 800\ \text{mA.}$$

Example 5.3: *Show that the approximate value of radiative minority carrier lifetimes in GaAs is 1.4 ns when the minority carriers are electrons injected into the p-type region which has a hole concentration of 10^{18} cm⁻³. The injected electron density is small compared to the majority carrier hole density ($B_r = 7.21 \times 10^{-10}$ and $p_n = 10^{18}$ cm⁻³).*

Solution. Equation (5.8.1) gives the radiative minority carrier lifetime

$$\Gamma_r = [B_r(n_n + p_n)]^{-1} = (B_r p_n)^{-1}$$

In the p-type region the hole concentration determines the radiative carrier lifetime as $p >> n$.

Hence

$$\Gamma_r = (B_r p_n)^{-1} = [7.21 \times 10^{-10} \times 10^{+18}]^{-1}$$

or

$$\Gamma_r = \frac{1}{7.21\times10^{-8}} = 1.4\ \text{ns}$$

The minority carrier lifetime 1.4 ns

5.14 HOMOJUNCTION LASER WITH FABRY PERROT CAVITY STRUCTURE

Figure 5.10 shows the basic structure of a homojunction injection laser in the form of a Fabry Perrot Cavity. This was fabricated with GaAs which is lattice matched alloy of group III-V compound semiconductor having electro-luminescence property. In this laser structure the cleaved ends of the crystal act as partial mirrors to encourage the stimulated emission in the cavity when electrons injected into the *p*-region.

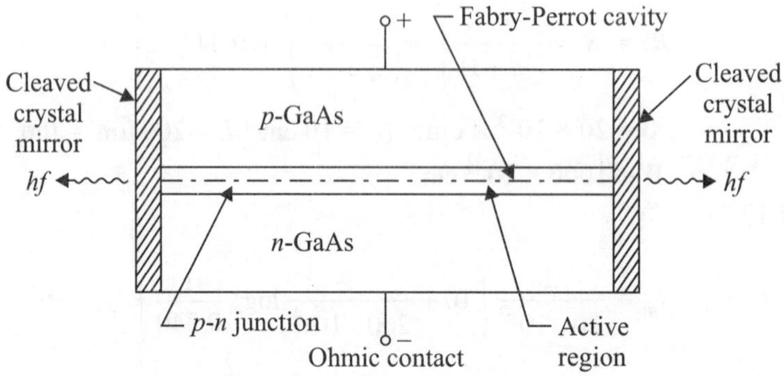

Fig. 5.10 A schematic diagram of GaAs homojunction laser with Fabry Perrot cavity

Disadvantage

The greatest shortcoming of the device is that it requires a very high value of threshold current density J_{th} (greater than 10^4 A cm^{-2}) to achieve stimulated emission. This happens because there is not any arrangement for carrier and optical confinements and the device proved to be an inefficient light source. However, in spite of this shortcoming the device is sometime used in pulsed mode to minimize high junction temperature so that the damage of the device is eliminated.

5.15 SPECTRUM AND FABRICATION

Semiconductor materials with direct band gap are used for fabricating the active layer of the optical source. In the case of a direct band gap semiconductor, an electron and a hole can combine directly across the band gap without requiring a third particle for maintaining the conservation of momentum. In the case of a direct bandgap semiconductor, the radiative recombination is considerably high to produce an adequate level of optical emission.

None of the single element semiconductors such as germanium or silicon is a direct band gap semiconductor. Many binary compounds which are the alloys of Group III and group V materials are found to be direct band gap semiconductors. They are prepared from compounds of Group III elements (such as aluminium or gallium or indium) and a compound of Group V elements (such as phosphorus or arsenic or antimony). Several ternary and quaternary combinations of binary compounds of these elements are also direct band gap materials and are used for the fabrication of optical sources.

1. Gallium arsenide (GaAs) diodes can be fabricated with the forbidden energy gap E_g ranging from 1.5 to 2.0 eV and the corresponding wavelengths of the emitted radiation range from 0.62 to 0.83 μm.

2. Gallium Indium arsenide (GaIn As) diodes are capable of covering the band gap energy range from 0.5 to 1.5 eV with the corresponding wavelengths ranging from 0.83 to 2.4 μm.
3. In addition to these materials the most important materials used for the fabrication of the LED and laser for the operation in 0.8 to 0.9 μm wavelength region is the ternary alloy of $Al_xGa_{1-x}As$. The ratio between the aluminium arsenide and gallium arsenide determines the band gap energy of the alloy and corresponding to the wavelength of the peak emitted radiation. This is illustrated in Fig. 5.11. The figure shows a curve between the band gap energy in electron-volts and the ratio x. The value of x for the active area material is chosen to give the emission wavelength ranging from 0.8 to 0.85 μm.

   For example, the value of x is 0.03 for $Al_xG_{1-x}As$ for active region of an LED and determines the peak value of the emitted wavelength of the radiation. From the curve of Fig. 5.11 we find that when $x = 0.08$, the value of the band gap energy E_g comes out to be 1.49 eV. The corresponding value of wavelength

$$\lambda = \frac{1.24}{1.49} = 0.81\ \mu\text{m}.$$

4. At still longer wavelengths the quaternary alloy $In_{1-x}Ga_xAs_yP_{1-y}$ is one of the primary materials used. By varying the ratio x and y in the material of the active layer of LED with the peak wavelength lying between 1.0 and 1.7 μm can be fabricated.

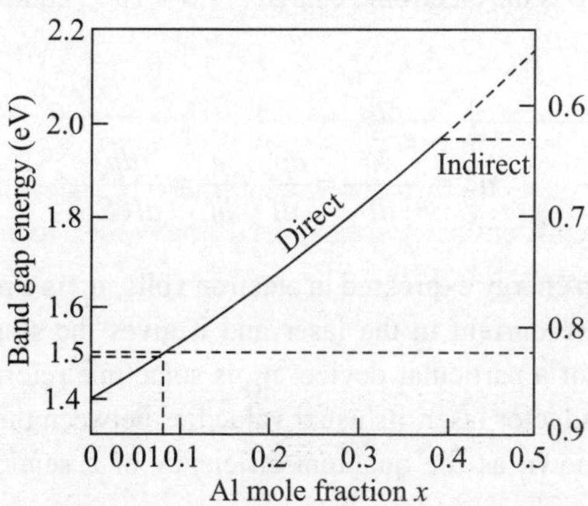

Fig. 5.11 Curve between band gap energy E_g and output wavelength as a function of Aluminium mole–fraction x for an $Al_xGa_{1-x}As$ at room temperature. When $x = 0.07$, $E_g = 1.51$ eV and $\lambda = 0.82$ μm.

For example, in the case aluminium gallium arsenide we take the value of $x = 0.07$ therefore the formula for the compound is $Al_{0.07}Ga_{0.93}As$ and $E_g = 1.51$ eV from the graph. The corresponding wavelength with peak emission comes out as $\lambda_m = 1.24/1.51 = 0.82$ μm.

In the case of aluminium gallium arsenide the band gap energy E_g in eV for the value of x lying between 0 and 0.37 (the direct band gap region) can be found from the empirical relationship.

$$E_g = 1.424 + 1.266x + 0.266x^2$$

where the value of E_g is expressed in eV. For example, $Al_{0.07}Ga_{0.93}As$ has the value of $E_g = 1.51$ eV from the equation and corresponding wavelength is 0.82 μm.

Table 5.2 Some common materials for the fabrication of electroluminescence sources

Material systems	*Useful wavelength*	*Substrate*
$GaAs/Al_xGa_{1-x}As$	0.8 – 0.9	GaAs
$GaAs/In_x Ga_{1-x}P$	0.9	GaAs
$Aly\ Ga_{1-y}As/Al_xGa_{1-x}As$	0.65 – 0.9	GaAs
$In_y\ Ga_{1-y}As/In_x\ Ga_{1-x}P$	0.85 – 1.1	GaAs
$GaAs_{1-x}Sb_x/Ga_{1-y}Al_y\ As_{1-x}Sb_x$	0.9 – 1.1	GaAs
$Ga_{1-y}Al_y\ As_{1-x}Sb_x/GaSb$	1.0 – 1.7	GaSb
$In_{1-x}Ga_xAs_y\ P_{1-y}/InP$	0.92 – 1.7	InP

5.16 EFFICIENCY OF A LASER

There are several ways in which the operational efficiency of a semiconductor laser may be defined. A useful definition is that of "Differential external quantum efficiency" (denoted by η_D).

The differential external quantum efficiency is defined as the ratio of the increase of photon output rate for a given increase in the number of injected electrons. If p_e is the optical power emitted from the laser, I is the current, e is the electronic charge = 1.6×10^{-19} coulomb and hf is the energy of a photon, then

$$\eta_D = \frac{e\dfrac{dp_e}{hf}}{dI} = \frac{dp_e}{dI} \cdot \frac{e}{hf} = \frac{dp_e}{dI(E_g)} \tag{5.16.1}$$

where $hf/e = E_g$ is the band gap energy expressed in electron volts, η_D is a measure of the rate of change of optical output with the drive current in the laser and it gives the slope of output characteristics (Fig. 5.9) in the lasing region of a particular device. η_D is sometime referred to as the slope quantum efficiency. For a C.W. semiconductor laser, its usual value lies between the range 40 and 60%.

Second definition is known as the quantum efficiency of a semiconductor laser η_i which is defined as

$$\eta_i = \frac{\text{Number of photons produced in the laser cavity}}{\text{Number of injected electrons}} \tag{5.16.2}$$

η_i may have a high value ranging from 50 to 100%. This parameter is related to the differential quantum efficiency η_D by the expression:

$$\eta_D = \eta_i \left[\frac{1}{1 + \dfrac{2\bar{\alpha} L}{\ln \dfrac{1}{R_1 R_2}}} \right] \tag{5.16.3}$$

where $\bar{\alpha}$ is the average loss coefficient of the laser cavity, L is the length of the laser cavity and R_1, R_2 are the reflectivities of the cleaved mirror. Another parameter is the total efficiency η_T which is defined as

$$\eta_T = \frac{\text{Total number of output-photons}}{\text{Total number of injected electrons}} = \frac{p_e(hf)}{I/e} = \frac{p_e}{IE_g} \quad (5.16.4)$$

when the emitted power p_e is directly proportional to the drive current where I is greater than the threshold current I_{th}, then

$$\eta_T = \eta_D\left(1 - \frac{I_{th}}{I}\right) \quad (5.16.5)$$

when I the injection current is much higher than the threshold current I_{th} η_T is almost equal to η_D.

The external power efficiency (or device efficiency) η_{ep} in converting electrical input into optical output is given by

$$\eta_{ep} = \frac{p_e}{P} \times 100\% = \frac{p_e}{IV} \times 100\% \quad (5.16.6)$$

where $P = IV$ is the electrical input power. Combining Eq. (5.16.4) and Eq. (5.16.5), we get

$$\eta_{ep} = \eta_T(E_g/V) \times 100\% \quad (5.16.7)$$

5.17 DOUBLE HETEROJUNCTION LASER

If a single *p-n* junction diode is fabricated from suitable single crystal semiconductor material and it exhibits photo emissive properties, it is known as "homojunction" *p-n* diode. However, the emission properties of a junction diode can be improved considerably by the use of "heterojunction". A heterojunction is an interface between two adjoining single crystal semiconductors having different values of band gap energies. Devices which are fabricated with heterojunctions are said to have heterostructures. There are two types of heterostructures. These are (1) isotypes such as *n-n* or *p-p* type. (2) anisotype such as *p-n* type. The isotype *p-p* junction has a potential barrier within the structure. The structure is capable of confining the minority carriers to small active region, which is called "Cavity". It effectively reduces the diffusion of the carrier within the confining volume of the structure where radiative recombination may occur. Figure 5.12 shows the schematic layer structure, energy band diagram and refractive index profile for a double heterojunction injection laser diode with biasing. The laser oscillations takes place in the central *p*-type GaAs region which is known as *active layer.* There are heterojunctions at the both sides of the active layer. A forward bias voltage is applied by connecting the positive electrode of the supply voltage to the *p*-side of the structure and the negative electrode to the *n*-side. When a voltage which is almost equal to the band gap energy, E_g of the active layer is applied, a large number of electrons are injected into the active *p*-layer from the *n*-layer and the laser oscillations start. These carriers are confined to the active layer by potential barrier provided by the heterojunction (Fig. 5.12*b*).

The heterojunctions are used to provide potential barrier in the injection laser. In a double heterojunction structure it is possible to obtain both carrier and optical confinements and thereby the threshold current is reduced considerably and sometime it may be reduced by a factor of 100. It becomes possible to achieve stimulated emission with a considerably small threshold current (50 to 200 mA). It may be noticed from Fig. 5.12(*c*) that there exists a refractive index step (difference of 5 to 10 per cent) at the heterojunctions which provides an optical confinement to the active layer.

5.18 BROAD AREA DOUBLE HETEROJUNCTION LASER (DH LASER)

Figure 5.13 shows the layer structure of a broad area DH GaAs/AlGaAs injection laser diode. A *p*-GaAs layer acts as the active layer which is sandwiched between a *p*-type AlGaAs and an *n*-type AlGaAs layers and these two layers act as the confinement layers. Then *n*-AlGaAs layer is developed on a substrate of n^+-GaAs layer, while there exist a *p*-type AlGaAs layer light is emitted from the central GaAs active layer through the front and back side of the device.

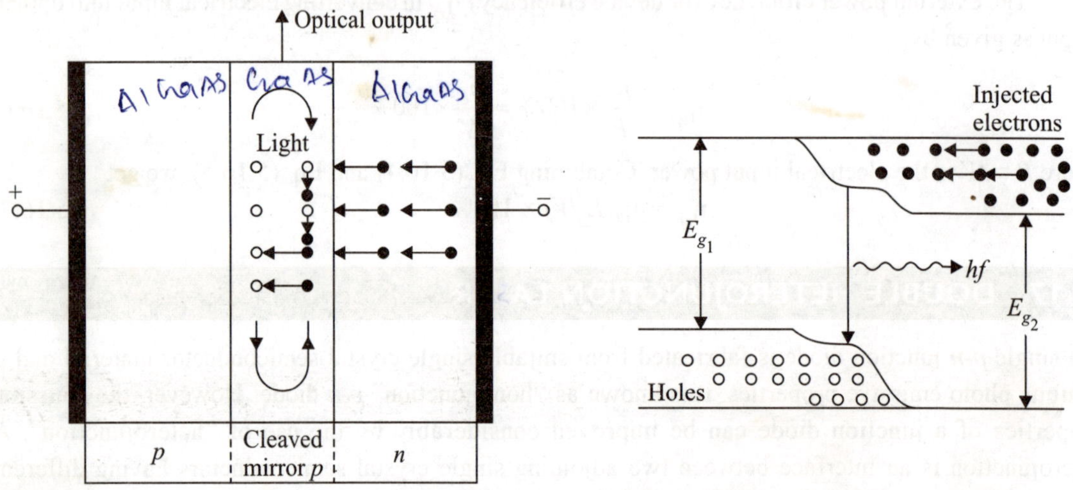

Fig. 5.12 (*a*) layer structure (*b*) energy band diagram showing a *p-p* heterojunction on the left and *p-n* heterojunction on the right (*c*) the corresponding refractive index profile and electric field distribution of a double heterojunction laser.

In the case of the DH broad area laser structure, the optical confinement in the vertical direction is achieved by the refractive index steps (by 5 to 10%) at the heterojunction interfaces between the active layer and the confinement layers, but the laser action takes place across the whole width of the device. In a broad area laser the sides of the cavity are formed by simple roughening the ends of the device. This is done in order to reduce the unwanted emission in these directions and to reduce the number of transverse mode.

However, a broad area laser possesses several drawbacks. These are (1) difficult heat sinking (heat radiation of the device) (2) unsuitable geometry of light output for efficient coupling to a cylindrical fiber (3) considerably high value of the threshold current.

Layer structure with width of each layer is shown in Fig. 5.13.

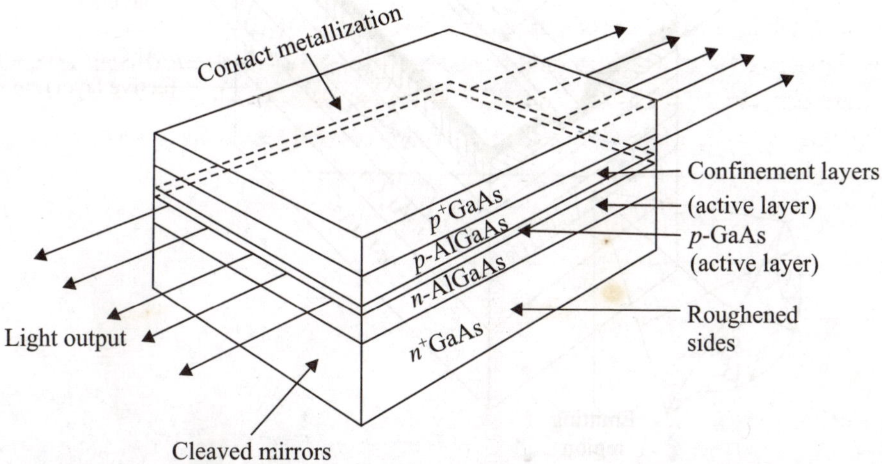

Fig. 5.13 The layer structure of a broad area GaAs/AlGaAs DH injection laser

5.19 STRIPE GEOMETRY LASER

In order to overcome the above difficulties in a broad area laser structure, the stripe geometry laser structure has been developed and in this structure the active area does not extend up to the edges of the device. A common method used is to introduce the stripe geometry to the structure which provides the optical confinement in the horizontal plane. Figure 5.14 shows the structure of a DH stripe contact laser geometry in which the active region lies within the stripe. Major portion of the smaller drive current flows through the stripes.

The stripe geometry is usually formed by creating a high resistance area on the either side of the stripe by *"proton bombardment"* technique or *"by oxide isolation"*. The stripe usually acts as a guiding mechanism which avoids all major difficulties encountered in the case of a broad area laser. However, although the width of the active area is reduced considerably, the light output is not collimated due to isotropic emission from a small active region and diffraction within the structure. Figure 5.14 illustrates the optical output for field emission pattern. The output beam divergence is typically 45° perpendicular to the plane of junction and 9° parallel to it. However, this is a considerable improvement over the broad area laser.

The contact stripe devices provide the balance of guiding single transverse mode operation in a direction parallel to the junction whereas the broad area devices allow multiple mode operation in this horizontal plane. The width of the stripe generally ranges from 2.0 to 65 μm and the stripe lasers find wide applications in fiber communications.

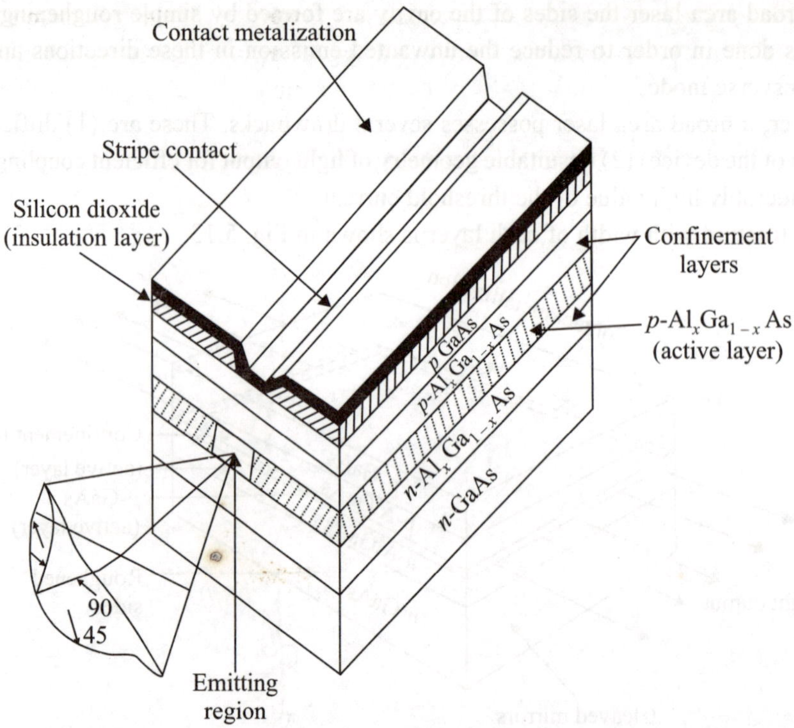

Fig. 5.14 The structure of an oxide DH stripe contact AlGaAs DH injection laser in which the active region lies within the stripe. (Silicon dioxide is used as isolation layers)

5.20 LASER MODES

Figure 5.15 shows a typical output spectrum from a broad area injection laser. The output spectrum does not consist of a single wavelength but a series of wavelength peaks corresponding to different longitudinal modes in the plane of the junction along the optical cavity within the structure of the laser. The wave-length of these modes depends on the length of the optical cavity. Each mode corresponds to the integral number of half wavelength.

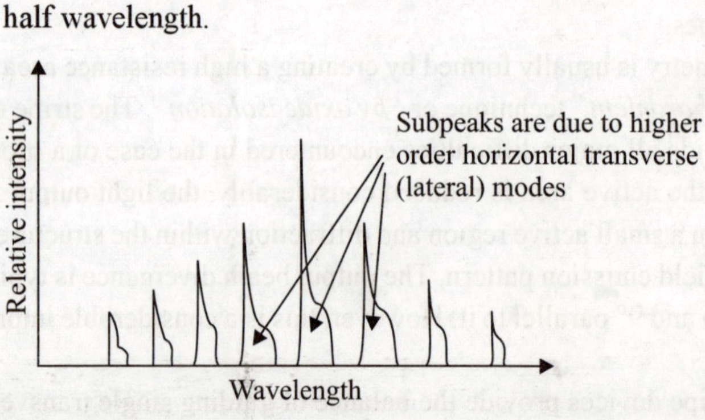

Fig. 5.15 A typical output spectrum of multimode injection laser

When the optical length of the active region is n_1L where n_1 is the refractive index and L is the actual length of the cavity, the resonance condition along the axis of the cavity with the formation of the standing wave in the active region is that n_1L must be the multiple of the half wavelength. Thus

$$n_1L = K\lambda/2 \quad (5.20.1)$$

where λ is the emission wavelength and K is the integral number giving the number of modes. Thus, the frequency is given

$$f = \frac{c}{\lambda} = \frac{Kc}{2n_1L} \quad (5.20.2)$$

where c is the velocity of light.

The separation between two adjacent modes δf is obtained by setting $K = 1$ in Eq. (5.20.2)

$$\delta f = \frac{c}{2n_1L} \quad (5.20.3)$$

when we assume that $\delta f << f$ the mode separation in terms of the free wavelength is given by (using Eqs. (5.20.2) and (5.20.3))

$$d_\lambda = \frac{\lambda df}{f} = \frac{\lambda^2}{c}\delta f = \frac{\lambda^2}{2n_1L} \quad (5.20.4)$$

The separation of wavelength is usually equal to few tenths of a nano-metre and the laser is said to be a multimode device.

5.21 BROADENING OF LONGITUDINAL MODE PEAKS

We observe in Fig. 5.14 that there exists some broadening of the longitudinal mode peaks. These sub peaks are caused by higher order horizontal "transverse modes". The higher order transverse or lateral modes may exist in the broad area device due to unrestricted width of the active region. The choice of the correct value of width of the active region in the stripe geometry laser can eliminate the occurrence of higher order lateral modes. The width of the optical cavity is so chosen that it gives only one "single" lateral mode. Figure 5.16 shows the output spectrum with a single lateral mode. The figure represents the typical output spectrum of an excellent multimode injection laser.

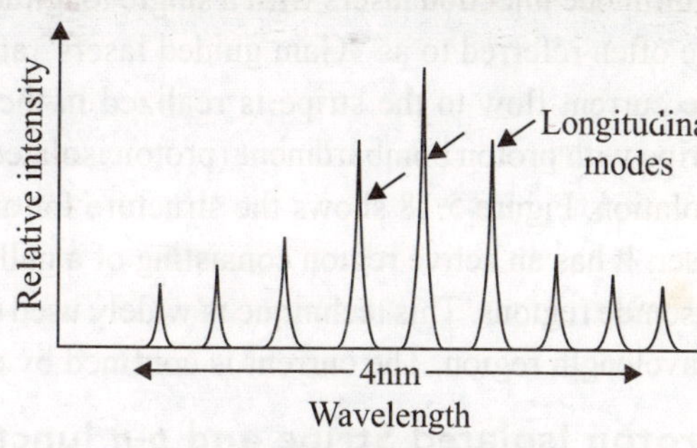

Fig. 5.16 Stripe geometry device with single transverse mode

5.22 SINGLE MODE OPERATION

In the case of single mode operation, the typical optical output spectrum of the laser must contain one longitudinal and one lateral (transverse) mode. One method of achieving single longitudinal mode operation is to reduce the length of the cavity L so that the frequency separation δf between two adjacent mode is given by

$$\delta f = \frac{c}{2nL} \tag{5.22.1}$$

where δf is larger than the laser line width c = Velocity of light and n = retractive index of the semiconductor. Then only the single mode which falls within the transition line width can oscillate within the laser cavity. However, the rigid control of parameters is essential to provide stabilization so that the single mode is attained and maintained.

The correct DH structure will restrict the vertical width of the waveguide region to less than 0.4 μm allowing only the fundamental transverse mode to be supported and removing any interference from the higher order transverse modes on the emitted longitudinal modes. In order to achieve required longitudinal modes, tight current confinement and good waveguiding are essential and the modes are formed between the mirror facets in the plane of the junction. Figure 5.17 shows a typical longitudinal mode output spectrum of a single mode injection laser. However, it is extremely difficult to handle an injection laser having a short cavity lengths of the order of 50 μm.

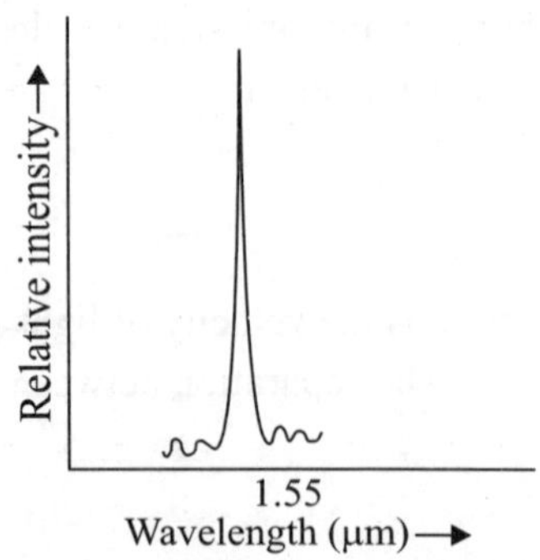

Fig. 5.17 Typical single longitudinal mode output spectrum from a single mode injection laser

5.23 GAIN GUIDED LASER

Multimode injection lasers with a single longitudinal mode or a small number of lateral modes geometry are often referred to as "Gain guided lasers" and achieved by the stripe geometry. The constriction on the current flow to the stripe is realized in the structure either by implanting the regions outside the stripe with proton bombardment (proton isolated type) to make highly resistive or by oxide *p-n* junction isolation. Figure 5.18 shows the structure for an aluminium gallium arsenide oxide isolated stripe DH laser. It has an active region consisting of a gallium arsenide bounded on both sides aluminium gallium arsenide regions. This technique is widely used especially for multimode laser structures used in shorter wavelength region. The current is confined by etching a narrow stripe in a silicon dioxide film.

Proton Isolated Stripe and *p-n* Junction Isolated Stripe Lasers

Two other basic techniques for the fabrication of gain guided laser structures are shown in Fig. 5.18(*a*) and 5.18(*b*). Figure 5.18(*a*) shows the proton isolated stripe structure type of laser. In this structure the highly resistive region is made by proton bombardment and the device provides better current confinement than the simple oxide stripe. It possesses a superior thermal properties due to the absence of silicon dioxide layer.

Figure 5.18(*b*) illustrates a *p-n* junction isolated stripe structure type of laser. *p-n* junction isolation structure involves a selective diffusion through the *n*-type surface region so as to reach the *p*-type layer as shown in Fig. 5.18(*b*). None of the structures is capable of confining the current and radiation to the stripe region fully and a spreading takes place on the both sides of the stripe. When the stripe width is

10 μm or less, such planer stripe lasers provide highly efficient coupling into the multimode fiber, but considerably lower coupling efficiency is obtained into the singlemode fibers with small core diameters.

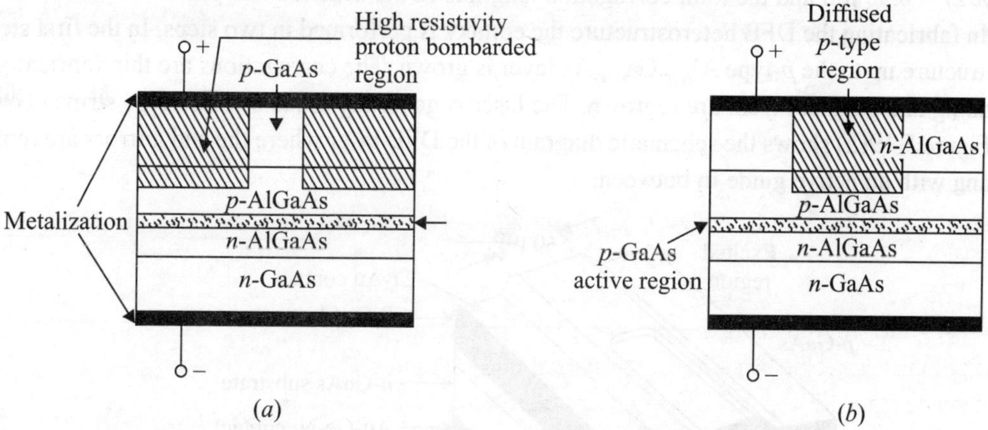

Fig. 5.18 Schematic representation of structure for stripe geometry injection laser (*a*) proton isolated stripe GaAs/AlGaAs laser (*b*) *p-n* junction isolated (diffused planer stripe) GaAs/AlGaAs laser.

5.24 DISTRIBUTED FEEDBACK LASERS [DFB]

In the case of a junction laser, feedback is an essential condition for the occurrence of lasing and obtained from the cavity facets formed by cleaving. Spectral emission of higher single frequency can be obtained by using a "Corrugated structure" or "Grating". In such structure the periodic variation of the refractive index within the cavity along the direction of the wave propagation is produced. For lasing feedback is essential. The feedback in the cavity occurs due to the energy propagating in the forward direction is being continuously feedback into the opposite direction by Bragg diffraction at the corrugation or grating. Lasers using such corrugations as feedback elements are known as "Distributed feedback" (DFB) lasers or "Distributed Bragg Reflector" (DBR) Lasers.

The first DFB lasers made use of thin film of dyed gelatin on a glass substrate and were investigated by Kogelnik and Shank.[1]

Figure 5.19 shows the schematic of a practical GaAs/AlGaAs DFB laser. The active layer consists of a *p*-type gallium arsenide (GaAs) and the thickness of the layer is as small as 0.2 μm. The active layer is confined between an *n*-$Al_{0.3}Ga_{0.7}As$ layer and a *p*-$Al_{0.17}Ga_{0.83}As$ layer. The optical confinement and gain are provided within the *p*-GaAs layer. The feedback is provided by the corrugation at the interface between the *p*-$Al_{0.07}Ga_{0.93}As$ and *p*-$Al_{0.3}Ga_{0.7}As$ layers where the main refractive index discontinuity for the optical guiding occurs. Such corrugations are made by ion-milling or reactive ion beam etching technique (RIBL). Thus, the thickness of the *p*-$Al_{0.17}Ga_{0.3}As$ and that the *n*-$Al_{0.3}Ga_{0.7}As$ are of the order of 0.1 μm. The roles of thin *n*- and *p*-layers are two folds. These are: (1) it helps to confine the guided modes the *p*-GaAs layer by corrugation and (2) it ensures the leakage of the optical field into the corrugated layer.

[1] H. Kogelnik and C.V. Shank, App. Phys Letter 18, 152, 1971, *Journal of Applied Physics*, 43, 232, 1972.

Since a small fraction of the combined mode power reaches the corrugated layer, the distributed feedback coefficient is obviously very small. In a typical DFB laser the corrugation period is $\Lambda = m(\lambda/2) = 0.35$ μm and the total corrugation length is of the order of 700 μm.

In fabricating the DFB heterostructure the epitaxy is performed in two steps. In the first step the heterostructure up to the *p*-type $Al_{0.07}Ga_{0.93}As$ layer is grown. The corrugations are thin fabricated and the cladding and contact layers are regrown. The laser is generally made as mesa type stripe geometry diode. Figure 5.19(*a*) shows the schematic diagram of the DFB laser where the end mirrors are replaced by grating with an active guide in between.

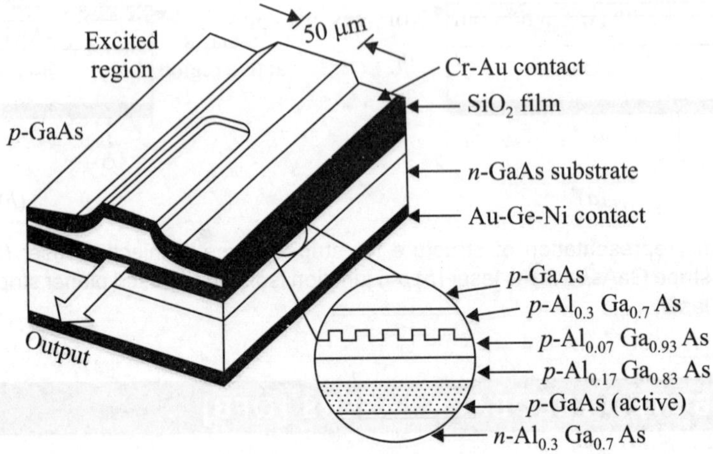

(*a*)

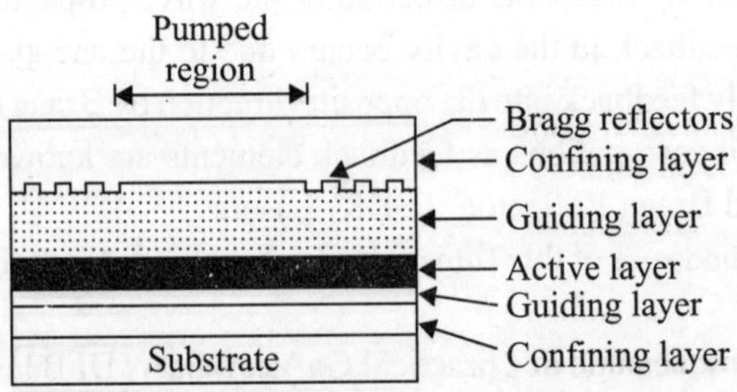

(*b*)

Fig. 5.19 (*a*) Schematic of distributed feedback (DFB) laser (*b*) Distributed Bragg reflector (DBR) laser.

5.25 DISTRIBUTED BRAGG REFLECTOR (DBR) LASER

Figure 5.19(*b*) shows the distributed Bragg Reflector (DBR) laser which also utilizes the corrugated interface. In DBR laser the periodic reflecting corrugated interface is placed outside the active lasing (pump) region as shown in the figure. The refractive index of the corrugating region is also alterable by current injection. The corrugated region is responsible for providing the reflection of light emitted from the active region. The wavelength λ_B produced by the DBR laser depends on the grating period Δ and is given by the equation

$$\lambda_B = \frac{2n_e\Delta}{L} \tag{5.25.1}$$

where n_e is the mode propagation constant, L represents an integer that indicates the order of the grating (usually $L = 1$). The corrugated region is situated outside the current pump active region. Both DBR and DFB lasers will oscillate at a single frequency λ_B. The shortcoming of both lasers are that they have usually a higher value of threshold current and susceptible to frequency chirp (very high frequency sharp notes made by small birds) when pulse voltages are applied to the lasers.

Example 5.4: *A GaAs injection DH laser has the length of cavity equal to 200 μm. The peak emission wavelength of the device is 0.85 μm. (a) Calculate the number of longitudinal mode emitted and their wavelength separation. The refractive index of GaAs is 3.6. Using equation (5.20.1), the number of modes can be obtained.*

Solution: From Eq. (5.20.4) seperation wavelength between two modes

$$\Delta\lambda = \frac{\lambda^2}{2n_1 L}$$

$$\Delta\lambda = \frac{(0.85)^2 \times 10^{-12}}{2 \times 3.6 \times 200 \times 10^{-6}}$$

$$= 0.5 \times 10^{-9} = 0.5 \text{ nm}.$$

SOLVED EXAMPLES

Example 5.5: *The total efficiency of an injection laser with GaAs active region is 18%. The band gap energy for GaAs is 1.43 eV. The voltage applied to the device is 2.5 Volt. Calculate external power efficiency of the device.*

Solution: The external efficiency n is given by Eq. (5.16.6)

$$\eta_e = \eta_T \frac{E_g}{V} \times 100\% = 0.18 \frac{1.43}{2.5} \times 100\% = 10\%$$

where η_e = external power efficiency and η_t = total efficiency.

5.26 TEMPERATURE DEPENDENCE OF THRESHOLD CURRENT OF A LASER

Although there is a very efficient heat sinking in a laser diode, the temperature of the laser diode may increase with time because of large injection current density. It is desirable that the lasing characteristics particularly the threshold current should remain constant independent of ambient temperature as shown in Fig. 5.20. The figure shows the variation of the threshold current with temperature for two gain guided lasers. Both devices have stripe width of approximately 20 μm, but were fabricated from different materials for emission wavelength of 0.85 μm and 1.55 μm for AlGaAs and In GaAsP devices respectively. The variation of the threshold current J_{th} can be expressed by the empirical equation

$$J_{th} = J_{th} \exp (T/T_o) \quad (5.26.1)$$

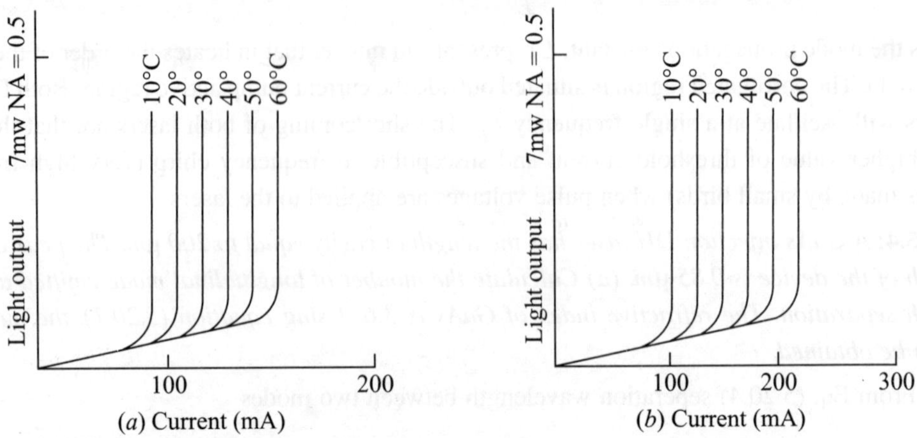

Fig. 5.20 The variation of threshold current of a laser with temperature *a*, *b*

Where T is the temperature of the device, T_o is the threshold temperature coefficient—as observed in Figure 5.20 the threshold current increases with the increase in temperature J_{th} is a constant. The amount of temperature dependence depends on the value of T_o. T_o depends on quality of material and structure of the device. For AlGaAs device, T_o is usually in the range of 120 to 190 K whereas for In GaAsP devices, it is between 40 and 75 K. This expresses the stronger temperature dependence for In GaAs devices. It will be clear from the Example 5.2.

Example 5.5 (A): *The total efficiency of an injection laser with GaAs active region is 20%. The band gap energy for GaAs is 1.43 eV. The voltage applied across the device is 2.5V. Calculate the external power efficiency of the device.*

Solution: The external efficiency η_e is given by Eq. (5.16.5) as

$$\eta_e = \eta_T \frac{E_g}{V} \times 100\% = 0.20 \frac{1.43}{2.5} \times 100\% = 11.44\% .$$

Example 5.6: *Compare the ratio of threshold current densities at 20°C and 80°C for an AlGaAs injection laser diode with threshold temperature coefficient T_o = 160 K and the similar ratio for an InGaAsP device with T_o = 55 K.*

Solution: From Eq. (5.26.1) we find the expression for the variation of threshold current density with temperature.

$$\text{For AlGaAs} \quad J_{th}(20°C) = J_{th} \exp \frac{273+20}{160} = J_{th} \exp \frac{293}{160} = J_{th} 6.24$$

$$J_{th}(80°C) = J_{th} \exp \frac{273+80}{160} = J_{th} \frac{353}{160} = J_{tho} \frac{353}{160} = J_{th} 9.08$$

Hence the ratio is

$$\frac{J_{th}(80°)}{J_{th}(20°)} = \frac{9.08}{6.24} = 1.46$$

For AlGaAsP device

$$J_{th}(20°C) = J_{tho} \exp\frac{273+20}{55} = J_{tho} 205.88$$

$$J_{th}(80°C) = J_{tho} \exp\frac{273+80}{55} = J_{tho} 612.89$$

Hence ratio $\dfrac{J_{th}(80°)}{J_{th}(20°)} = \dfrac{612.89}{205.88} = 2.98$.

Example 5.7: *A GaAs injection DH laser has 1700 longitudinal modes emitted. The peak emission wavelength of the device is 0.85 μm (a) calculate the length of the cavity in the laser and (b) the wavelength separation between two modes. The refractive index of GaAs may be assumed to be 3.6.*

Solution: (*a*) Using Eq. (5.20.1) the number of modes *K* is given by

$$K = \frac{2n_1 L}{\lambda}$$

or

$$L = \frac{K\lambda}{2n_1} = \frac{1700 \times 0.85 \times 10^{-6}}{2 \times 3.6} = 200\ \mu\text{m}$$

(*b*) from Eqn. (5.20.4) separation wavelength between two modes

$$d\lambda = \frac{\lambda^2}{2n_1 L} = \frac{(0.85 \times 10^{-6})^2}{2 \times 3.6 \times 200 \times 10^{-6}} = 0.5\ \mu\text{m}.$$

5.27 ADVANTAGES OF A SEMICONDUCTOR INJECTION LASER

The semiconductor injection lasers possess several advantageous features over the light emitting diodes (LEDs)

1. High values of radiance (intensity of radiation) due to the amplifying action in the case of stimulated emission. Injection lasers are able to provide several millivolts of optical power.
2. The line width of the radiation is considerably narrow of the order of 1 nm (10 Å) or can be less which is helpful in reducing the understandable material dispersion.
3. The modulation capabilities which depend on the highest frequency of the electrical signal extends up to sever all mega Hertz.
4. It possesses a good spatial coherence which allows the output to be focussed by an optical lens into a spot which has a greater light intensity than a dispersed unfocussed emission. This is helpful in obtaining efficient coupling of the optical output power into a fiber of low value of numerical aperture.

REVIEW QUESTIONS

5.1 Explain with necessary diagrams three process of absorption, spontaneous emission and stimulated emission in a two-level energy system. In what type optical sources are these emission properties utilized?

5.2 Derive Einstein relationships connecting absorption, stimulated emission and spontaneous emission coefficients.

Calculate the ratio of the stimulated emission and spontaneous emission rates of a tungsten lamp which radiates at an average wavelength of 0.6 μm at an average temperature of 1300°K.

5.3 What do you mean by population inversion? How is the population inversion accomplished in semiconductor and non-semiconductor lasers?

5.4 Derive the condition of lasing in terms of gain and loss co-efficients per unit length and the reflectivity of mirrors placed at the end facets of the laser crystal.

A GaAs injection laser has an optical cavity of 250 mm and a width of 100 mm. At operating temperature gain factor β is 21 × 10^{-3} A cm^{-3} and the loss co-efficient per cm. is 10 the refractive index of GaAs is 3.6.

(*a*) Determine the threshold current density required for lasing.

(*b*) Estimate the total current required for lasing.

5.5 Illustrate with suitable diagrams to explain the properties of a direct band gap and an indirect band gap semiconductor. What are the order of their recombination lifetimes?

Give the names of two semiconductors of each type with their band gap energies.

5.6 An electroluminescent *p-n* junction diode is fabricated using gallium antimonide (GaSb) for the use as a light source. Determine the required concentration of the holes in the *p*-region in order that the radiative minority carrier lifetime is 1ns. The recombination coefficient in GaSb is 2.39 × 10^{-10} cm^3 s.

5.7 Explain the operating principle of a double heterostructure laser with diagrams for layer structure, energy band diagram and the refractive index profile. Explain how does the active layer confine charge carriers and optical power.

CHAPTER 6

Photo Detectors

CHAPTER FOCUS

- *Introduction: Photo detector materials*
- *Basic principle for optical detection*
- *The p-n junction photodiode*
- *The p-i-n photodiode*
- *Diffusion length and lifetime*
- *Expressions for current, quantum efficiency and responsivity of a photodiode*
- *Quantum efficiency*
- *Responsivity R*
- *Practical p-i-n photodiode structures*
- *The p-i-n avalanche diode*
- *Reach through avalanche photodiode (RAPD)*
- *Operation of reach through photodiode (RAPD)*
- *Electron and hole ionization rates*
- *Silicon reach through avalanche photodiodes*
- *Practical avalanche photodiode*
- *Noise in photodetectors*
- *Photodiode materials with band gap energy*
- *Phototransistors*
- *Photoconducting detector*
- *Solar cell*
- *Review Questions*

6.1 INTRODUCTION: PHOTO DETECTOR MATERIALS

A number of photosensitive devices have been examined for the use as a detector in optical communication system. In the initial stage, these were silicon photodiodes, photoresistors and phototransistors. Unfortunately, most of them do not possess the characteristics of fast response and high sensitivity as required in fiber communication system. However, the *p-i-n* photodiode and avalanche photodiode possess the above characteristics.

Silicon photodiodes possess a good sensitivity over the 0.8 μm to 0.9 μm wavelength band, possess a considerable speed, a negligible shunt conductance, low value of dark current and long term stability. In the initial state, they were used for the fabrication of photodetectors. But the main drawback of silicon is that it is an indirect band gap semiconductor with a considerable response time. Germanium photodiodes have a considerable dark current due its narrow band gap (E_g = 0.7 eV). The direct band gap semiconductors of III-V alloys have been investigated in the larger wavelength region. Ternary alloys, such as indium gallium arsenide (InGaAs) and gallium aluminium antimonide (GaAlSb) deposited in indium phosphide (InP) and gallium antimonide (GaSb) substrates respectively have been used to fabricate photodiodes for longer wavelength band.

6.2 BASIC PRINCIPLE FOR OPTICAL DETECTION

Figure 6.1(*b*) shows a reverse biased *p-n* photodiode and due to reverse biasing, a thick depletion layer develops on the either side of the junction. The large potential barrier across the depletion layer has the effect of preventing the majority carriers of both the regions crossing the junction in the opposite direction due to the field of the barrier potential.

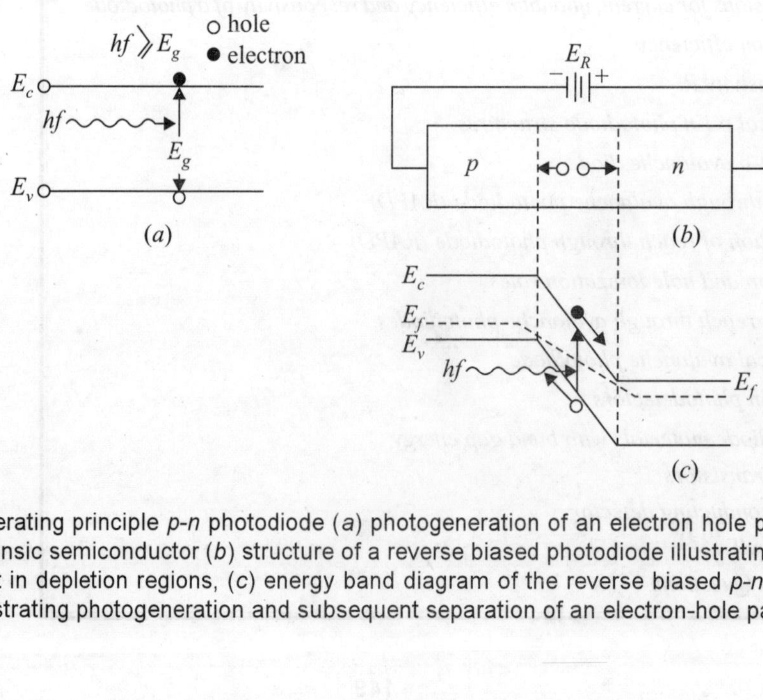

Fig. 6.1 Operating principle *p-n* photodiode (*a*) photogeneration of an electron hole pair in an intrinsic semiconductor (*b*) structure of a reverse biased photodiode illustrating carrier drift in depletion regions, (*c*) energy band diagram of the reverse biased *p-n* junction illustrating photogeneration and subsequent separation of an electron-hole pair.

Suppose a photon of light is incident in or near the depletion region and if the energy of the incident photon (hf) is equal to or greater than the band gap energy (E_g) of the semiconductor material of the *p-n* junction, the photon will excite an electron in valence band to the conduction band. This process will generate an "electron-hole" pair as shown in Fig. 6.1(*a*) when

$$hf \geq E_g \tag{6.2.1}$$

The photo-generated carriers "electron-hole" pairs are separated in the depletion layer and are swept away by the electric field due to the applied reverse biased voltage and a leakage current flows in the external circuit (Fig. 6.1*b*). The energy level diagram is shown in Fig. 6.1(*c*). The width of the depletion layer must be sufficiently thick so as to allow large portion of the incident light to be absorbed and thus the maximum carrier pair generation is obtained. However, the long carrier drift times in the depletion region decrease the speed of operation of the photodiode and the thickness of the depletion layer should be restricted. Thus, a compromise between the number of photons absorbed and the speed of response of the device should be made.

6.3 THE *p-n* JUNCTION PHOTODIODE

An ordinary *p-n* photodiode may be used as a "photodetector" with satisfactory speed of response, but its sensitivity is quite low. Figure 6.2 shows the basic structure of a photodiode showing depletion (carrier drift) and diffusion regions. The photodiode consists of a p^+ type thin layer deposited on an *n*-type substrate and light enters into its *p*-type layer. When the diode is reverse biased, larger portion of the depletion layer occupies the *n*-layer around the junction due to smaller impurity concentration in the *n*-layer. The photons of light enter into the depletion zone to produce electron-hole pairs. The electrons are attracted and drawn by the positive terminals, while the holes are drawn by the negative terminal of the applied reverse biased voltage. Thus, the electron-hole pairs constitute an electric current by the reverse voltage as in Fig. 6.2(*a*) and the current is known as "photocurrent" (or leakage current) having origin from the minority carriers.

Figure 6.2(*b*) shows the operating characteristics of a *p-n* photodiode which are curves between the photocurrent versus reverse biased voltage with light intensity as parameter. Figure 6.1(*c*) shows the energy-band diagram of a reverse biased photodiode illustrating photogeneration and the separation of electron-hole pairs. Since the photons are absorbed in the depletion layer which is made as wide as possible by decreasing the impurity concentration in the *n*-region so that the major portion of the depletion layer is situated in the *n*-region near the junction. The typical value of the depletion layer in a *p-n* photodiode is in the range of 1 to 3 μm. In the silicon devices the detected wavelengths fall in the visible spectrum (0.4 to 0.7 μm) and for the germanium devices in the near infrared region (0.7 to 0.9 μm).

Disadvantage

The disadvantage of a photodiode is that many of the photons entering into the depletion layer in the junction may enter into the *n*-region. These electron-hole pairs are not affected by the junction field and do not contribute to the photocurrent, because all reverse voltage develops across the depletion layer (Fig. 6.2*a*). The conversion efficiency is quite low. But the electron and hole will recombine in moving through the thin depletion layer and their lifetime is very short. Therefore, the *p-n* photodiode will respond very quickly to any change of light intensity.

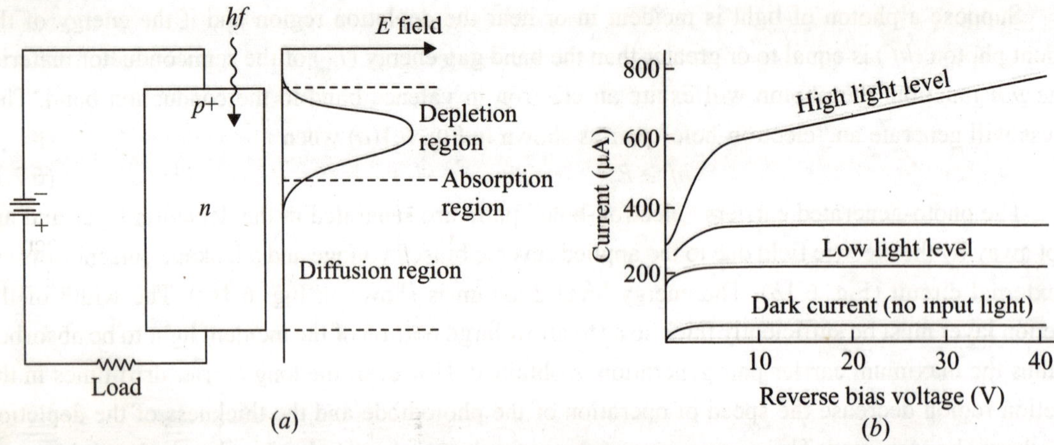

Fig. 6.2 (*a*) A typical *p-n* photodiode is shown with the depletion and diffusion regions (*b*) characteristic curves between the diode current in μa and reverse voltage in volts with light level as the parameter

6.4 THE *p-i-n* PHOTODIODE

The conversion efficiency of a *p-n* photodiode can be improved considerably by including a lightly doped (nearly intrinsic) *n*-layer between two heavily doped n^+ and p^+ layers. Figure 6.3(*a*) shows the basic structure of a *p-i-n* photodetector which shows that p^+- and n^+-regions are separated by a light *n*-doped (nearly intrinsic) *i*-region. In normal operation of the device, a sufficient reverse voltage is applied across the device so that the central *i*-region is fully devoid of mobile charge carriers. In this region the electron-hole concentrations are negligibly small in comparison with impurity concentrations. The depletion region is comparatively thicker.

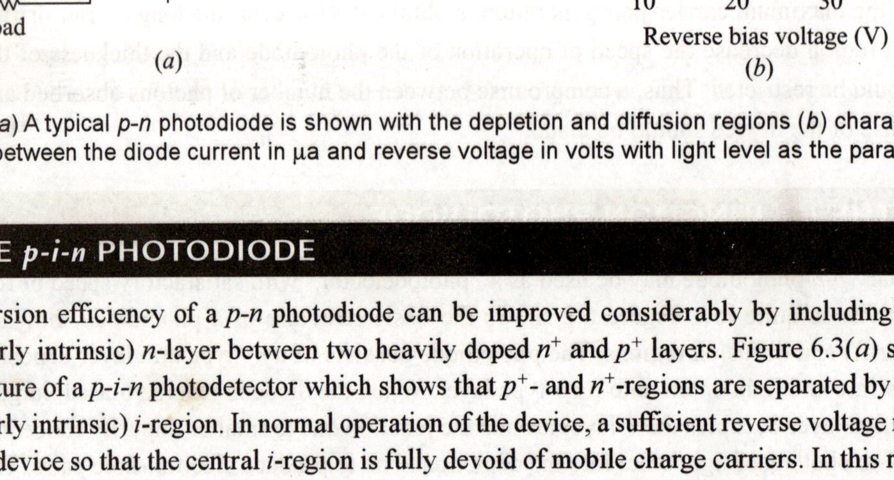

Fig. 6.3 (*a*) Schematic of a typical *p-i-n* photodiode showing absorption and depletion regions (*b*) energy level diagram

When an incident photon has an energy greater than or equal to the band gap energy E_g of the semiconductor material, the energy of the photon excites an electron from the valence band to the conduction band. This process produces a free electron and hole. The carriers generated by photons are known as "photocarriers". Figure 6.3(*b*) illustrates the energy band diagram of a *p-i-n* photodiode with the formation of an electron in the conduction band and a hole in the valence band.

The photodiodes are so designed that these carriers are generated in the depletion region (depleted intrinsic region) where most of the impinging light is absorbed. The high electric field developed across the depletion region causes the carrier to separate and to be collected by the reverse biased voltage. This causes a current flow in the external circuit with an electron flowing for every electron-hole pair generated and the current is referred to as "photocurrent". As the charge carriers move through the material some electrons and holes recombine to disappear and produces the photocurrent in the external circuit with an electron flowing for every electron-hole pair generated.

6.5 DIFFUSION LENGTH AND LIFETIME

On the average the free electrons and holes move through the distance L_n and L_p respectively before recombination. These distances are known as "diffusion length". The average times it takes for an electron or a hole to recombine are known "carrier lifetimes" and is represented by Γ_n and Γ_p for the electron and hole respectively.

The relationship between the carrier lifetime and diffusion length are given by

For electron $$L_n = (D_n \Gamma_n)^{1/2} \tag{6.5.1}$$

and for hole $$L_p = (D_p \Gamma_p)^{1/2} \tag{6.5.2}$$

where D_n and D_p are the electron and hole diffusion coefficients respectively and Γ_n and Γ_p are the lifetimes of electron and hole respectively. D_p and D_n are expressed in units of cm^2/sec.

6.6 EXPRESSIONS FOR CURRENT, QUANTUM EFFICIENCY AND RESPONSIVITY OF A PHOTODIODE

When an optical radiation is incident on the semiconductor surface of a photodiode, it gets absorbed as it penetrates into the material. Let P_o and P_x represent the incident optical power level and the optical power absorbed in a distance x respectively. The relationship between P_x and P_o is given by

$$P_x = P_o[1 - e^{-\alpha(\lambda)x}] \tag{6.6.1}$$

where $\alpha(\lambda)$ is the absorption coefficient of the material at wavelength λ.

Figure 6.4 illustrates the dependence of absorption coefficient α on wavelength for different photodiode materials. The curves show that α depends considerably on wavelength λ and therefore, a particular semiconductor material can only be used to fabricate photodetector operating over a limited range of wavelengths. The upper wavelength cutoff λ_c is determined by the band gap energy E_g of the semiconductor, when band gap energy E_g is expressed in electron volts, and λ is given in micron (µm), the relationship between them was already given as

$$\lambda_{\mu m} = \frac{1.24}{E_g(\text{eV})} \tag{6.6.2}$$

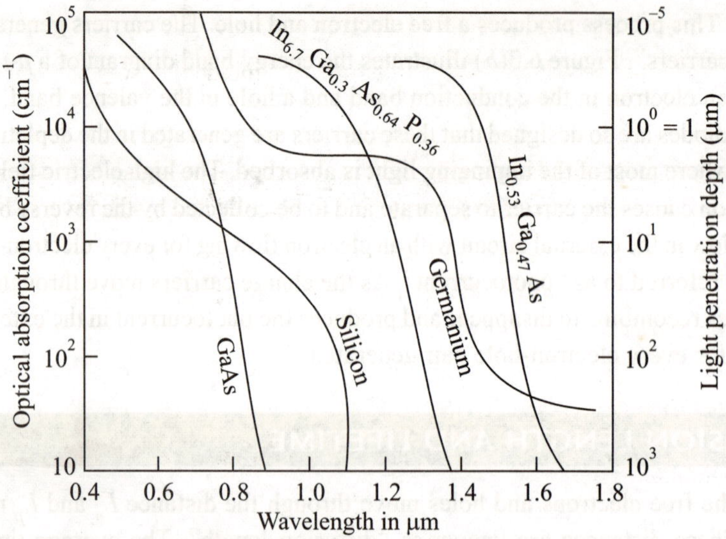

Fig. 6.4 Optical absorption and penetration for some semiconductor materials

The cutoff wavelength for silicon is 1.06 μm and that for germanium is 1.6 μm. For longer wavelengths of light, the photon energy is not sufficient to excite in electron from the valence band to the conduction band. At lower wavelength range, the value of absorption coefficient α is considerably large and the photo response cuts off. In this case photons are absorbed very close to the photodetector surface, where the recombination time for the generated carriers (electron-hole pair) is very short. Thus, the generated carriers recombine before they can be collected by the photodetector circuits.

If the width of the depletion layer is W, from Eq. (6.6.1), the total power absorbed in the depletion layer $P(W)$ is given by (putting $x = W$)

$$P(W) = P_o(1 - \varepsilon^{-\alpha(\lambda)W}) \quad (6.6.3)$$

If we consider the effect of reflectivity R_f at the entrance surface of the photo detector, $P(W)'$ which is absorbed in the photodiode is given by (using Eq. 6.6.3)

$$P(W)' = P(W)(1 - R_f) = P_o(1 - R_f)[1 - \varepsilon^{-\alpha(\lambda)W}] \quad (6.6.4)$$

Since each photon has an energy equal to hf, the number N of the incident photons absorbed will be equal to

$$N = \frac{P'(W)}{hf} = \frac{P_o}{hf}(1 - R_f)[1 - \varepsilon^{-\alpha(\lambda)W}] \quad (6.6.5)$$

If I_p is the photo current and constant e is numerically equal to electronic charge in $mks = 1.6 \times 10^{-19}$ C, the number N is equal to I_p/e. Thus the value of the photo current I_p can be obtained from Eq. (6.6.4) as

$$I_p = \frac{eP_o}{hf}[1 - R_f][1 - \varepsilon^{-\alpha(\lambda)W}] \quad (6.6.6)$$

This is an important equation in determining some parameters of the photodiode. The amount of incident power that is absorbed in the depletion depth W assuming that the layer begins d below the device surface is given by

$$P(W) = P_1\, \varepsilon^{-\alpha d} (1 - \varepsilon^{-\alpha W}) (1 - R_f) \tag{6.6.7}$$

where $P(W)$ is the power absorbed in the depletion layer, P_1 incident power, α = absorption coefficient at the operating wavelength and W = width of depletion layer.

6.7 QUANTUM EFFICIENCY

The important parameters of a photodetector are the quantum efficiency η and Responsivity R (speed of response). These parameters depend on the band gap energy of the semiconductor material, the amount of doping and thickness of the *p*-, *i*- and *n*-regions of the device.

The quantum efficiency η is defined by the number of electron-hole pairs generated per incident photon of energy hf and is given by

$$\text{Quantum efficiency } \eta = \frac{\text{number of electron-hole pairs generated}}{\text{number of incident photons}}$$

$$= \frac{r_e}{r_p} \tag{6.7.1}$$

where r_p is the incident photon rate (photons per second) and r_e the corresponding electron rate (electrons per second).

$$r_e = \frac{I_p}{e} \tag{6.7.2a}$$

and

$$r_p = \frac{P_o}{hf} \tag{6.7.2b}$$

Substituting the values from Eq. (6.7.2*a*) and Eq. (6.7.2*b*) in Eq. (6.7.1), we get

$$\eta = \frac{I_p}{e} \Big/ \frac{P_o}{hf} = \frac{I_p hf}{e P_o} \tag{6.7.3a}$$

or

$$I_p = \frac{\eta e P_o}{hf} \tag{6.7.3b}$$

From Eq. (6.6.6)

$$\eta = [1 - R_f]\,[1 - \varepsilon^{-\alpha(\lambda) W}] \tag{6.7.4}$$

Where I_p is the average photo current generated by a steady average optical power P_o incident on the photo detector.

However, in a practical photo detector diode 100 photons may create between 30 and 95 electron hole pairs and thus, the quantum efficiency ranges between 30 and 95 per cent. In order to achieve a high value of quantum efficiency, the depletion layer must be made thick enough to allow a large portion of incident light to be absorbed. But when the depletion layer is made thicker, it takes a longer time for the photogenerated carriers to drift across the reverse biased junction. Since the drift time determines the response speed of a photodetector, a compromise has to be made between the response speed and quantum efficiency.

6.8 RESPONSIVITY *R*

The performance of a photodetector is determined by the parameter responsivity, *R*. Responsivity is defined as the ratio between the average current generated in the detector and the average optical power incident on the detector.

Responsivity *R* is given by

$$R = \frac{I_p}{P_o}(AW^{-1}) \tag{6.8.1}$$

Where I_p is the output current in amperes and P_o is the incident optical power in watts. *R* is the photocurrent per unit incident optical power and gives the transfer characteristic of the detector.

Substituting the value from (6.6.6), we get

$$R = \frac{e}{hf}[1 - R_f](1 - \varepsilon^{-\alpha(\lambda)W}) \tag{6.8.2}$$

From Eq. (6.7.4) the value of *R* is given by

$$R = \eta\frac{e}{hf} = \eta\frac{e}{h}\left(\frac{\lambda}{c}\right) \tag{6.8.3}$$

The parameter *R* is quite useful, since it gives the amount of photocurrent generated per unit optical power. Figure 6.5 shows the responsivity curves of an ideal and practical photodiodes as a function of wavelength. Typical values of *R* are 0.65 μA/μW for a silicon device at 0.8 μm, 0.45 μA/μW for the germanium device at 1.3 μm and 0.6 μA/μW for indium gallium arsenide device at 1.3 μm wavelength.

In most photodiodes, the quantum efficiency is independent of the power level incident on the photodiode at given photon energy *hf*. Thus the responsivity *R* is the linear function of the incident optical power and as such, the responsivity *R* is a constant at a given wavelength.

The quantum efficiency η is not constant at all wavelengths because it varies with the photon energy. Thus, responsivity is the function of wavelength and of the photodiode material, since different materials have different values of band gap energy, E_g. Figure 6.5 shows that the responsivity decreases very rapidly beyond the cutoff wavelength.

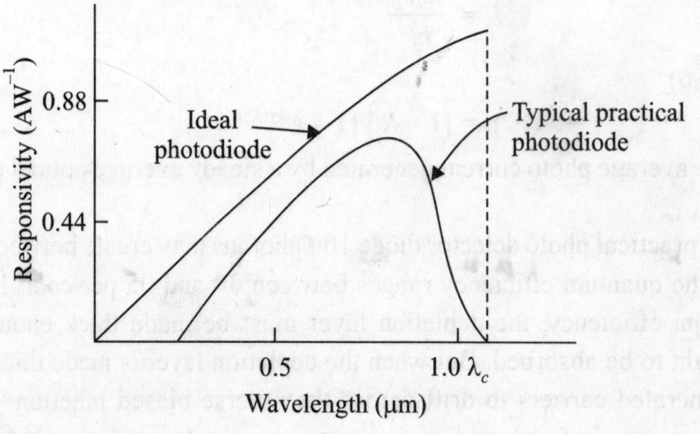

Fig. 6.5 A curve between responsivity versus wavelength for an ideal photodiode. Typical practical photodiode characteristic is also shown

Example 6.1: *A photodiode has a quantum efficiency of 70 per cent when photons with energy 2.2×10^{-19} Joule are incident on it. (a) Calculate the operating wavelength of the photodiode and (b) Calculate the incident power required to obtain a photocurrent of 2.0 mA when the diode is operating in above condition.*

Solution:

(*a*) $$E = hf = \frac{hc}{\lambda} \quad \text{or} \quad \lambda = \frac{hc}{E}$$

Data supplied:

$$\eta = 70\% \text{ photon energy} = 2.2 \times 10^{-19} \text{ joule},$$

$$I_p = 2 \text{ mA}$$

$$\lambda = \frac{hc}{E} = \frac{6.62\times10^{-34}\times3\times10^{8}}{2.2\times10^{-19}} = 0.9\ \mu\text{m}$$

$$f = \frac{c}{\lambda} = \frac{3\times10^{8}}{0.9\times10^{-6}} = 3.33\times10^{14}\text{ Hz}$$

(*b*) $$R = \left(\frac{\eta e}{hf}\right) = \left(\frac{\eta e\lambda}{hc}\right) = \frac{0.70\times1.6\times10^{-19}}{6.62\times10^{-34}\times3.33\times10^{14}} = 0.51\text{ AW}^{-1}$$

$$R = \frac{I_p}{P_o} \text{ and } P_o = \frac{I_p}{R} = \frac{2.0\times10^{-6}}{0.51} = 3.92\ \mu\text{W}$$

Example 6.2: *When 3×10^{11} photons each with a wavelength of 0.85 μm are incident on a photodiode, on average 1.5×10^{11} electrons are collected at the terminals of the device. Determine the quantum efficiency and responsivity of the photodiode at the wavelength of 0.85 μm.*

Solution: From Eq. (6.7.1), we get

Quantum efficiency $$\eta = \frac{r_e}{r_p} = \frac{\text{number of hole-pairs generated}}{\text{number of incident photon}}$$

$$= \frac{1.5\times10^{11}}{3\times10^{11}} = 0.5$$

From Eq. (6.8.3) Responsivity

$$R = \eta\frac{e\lambda}{hc} = \frac{0.5\times1.6\times10^{-19}\times0.85\times10^{-6}}{6.62\times10^{-34}\times3\times10^{8}} = 0.342\text{ AW}^{-1}$$

Example 6.3: *A photodiode has a quantum efficiency of 65 per cent when photons of energy 1.5×10^{-19} J are incident on it (a) Calculate the wavelength at which the photodiode is operating and (b) incident optical power required when the current through the diode is 3 μA.*

Solution: We know

(*a*) $$E = hf = \frac{hc}{\lambda}$$

or $$\lambda = \frac{hc}{E} = \frac{6.62\times10^{-34}\times3\times10^{8}}{1.5\times10^{-19}} = 1.324\times10^{-6}$$

$$= 1.324\ \mu\text{m}.$$

(*b*) From Eq. (6.8.3) Responsivity

$$R = \frac{\eta e}{hf} = \frac{0.65\times1.6\times10^{-19}}{6.62\times10^{-34}\times f}$$

$$= \frac{0.65\times1.6\times10^{-19}\times1.324\times10^{-6}}{6.62\times10^{-34}\times3\times10^{8}}$$

or $$R = 0.694\ \text{AW}^{-1}$$

$$R = \frac{I_p}{P_o} \quad \text{or} \quad P_o = \frac{I_p}{R} = \frac{3\times10^{-6}}{0.694} = 4.32\ \mu\text{W}$$

Example 6.4: *A photodetector is fabricated with GaAs whose band gap energy is measured to be 1.43 eV at 300K. Determine the critical frequency above which the detector will cease to operate.*

Solution: $E_g = 1.43\ \text{eV} = 1.43 \times 1.6 \times 10^{-19}$ joule

$$\text{Cut off wavelength } \lambda_c = \frac{hc}{E_g} = \frac{6.62\times10^{-34}\times3\times10^{8}}{1.43\times1.6\times10^{-19}}$$

$$= 0.867\ \mu\text{m}.$$

The GaAs photodetector will cease to operate above 0.86 μm

6.9 PRACTICAL *p-i-n* PHOTODIODE STRUCTURES

Front Illuminated *p-i-n* Photodiodes

Figure 6.6(*a*) shows the layer structure of a front illuminated silicon *p-i-n* photodiode operating in the shorter wavelength band below 1.09 μm. In order to attain high quantum efficiency (about 85%) and fast response (less than 1 ns.) together with low dark current, the front illuminated photodiode operating in the 0.8 to 0.9 μm wavelength band requires a depletion layer of width between 20 and 50 μm. In the absence of illumination, the dark current in the diode originates from surface leakage current and the generation-recombination current in the depletion region. Typical quantum efficiency may be 85%, response time 1 ns and dark current 1nA.

Side Illuminated *p-i-n* Diode

Figure 6.6(*b*) shows the side illuminated *p-i-n* structure where light is injected into the diode parallel to the junction plane. The *p-i-n* diode exhibits a large absorption width of 500 μm and hence it is sensitive at wavelengths close to the band gap limit (1.09 μm), where the absorption coefficient is comparatively smaller.

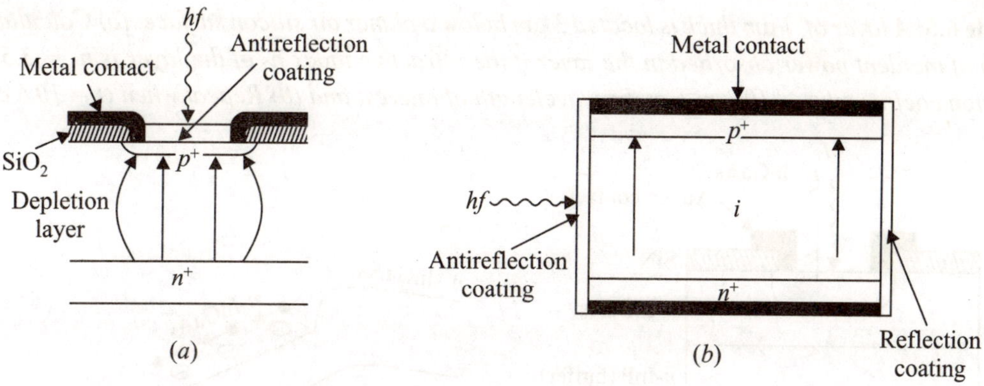

Fig. 6.6 (*a*) A structure of a front illuminated silicon photodiode (*b*) A structure of a side illuminated photodiode

Substrate Entry *p-i-n* Photodiode

Figure 6.7(*a*) shows the layer structure of a substrate entry *p-i-n* photodiode. This device uses a p^+ InGaAsP layer to provide heterostructure barrier (Schottky) and the quantum efficiency of the device is considerably improved. It is fabricated as a "mesa structure" *which reduces the parasitic capacitance.* The shortcoming of this device is that the charge trapping can occur at the $n - p^+$ (InGaAs/InGaAsP) interface. Figure 6.7(*b*) shows the energy level diagram for InGaAs type of *p-i-n* photodiode illustrating the mechanism of charge trapping. However, small area surface entry devices are capable of producing an extremely low capacitance (less than 0.1 pF), with quantum efficiency ranging between 75 and 100 per cent and dark current less than 1 nA.

Top Entry *p-i-n* Photodiode Structure

Figure 6.8(*a*) shows the structure of a top entry *p-i-n* photodiode and it has the simplest structure. In this case light enters into the device through the upper p^+ layer of the device and the incident light gets absorbed in undepleted p^+ region. The quantum efficiency is quite low.

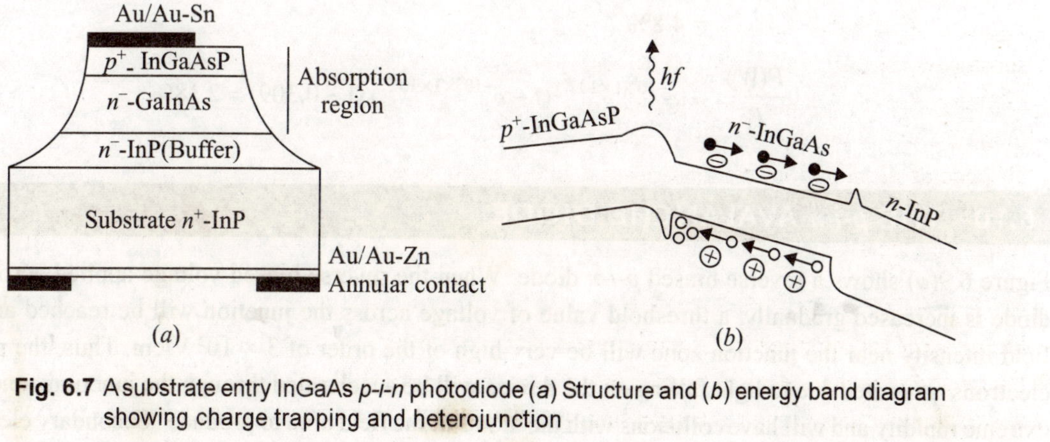

Fig. 6.7 A substrate entry InGaAs *p-i-n* photodiode (*a*) Structure and (*b*) energy band diagram showing charge trapping and heterojunction

Example 6.5: *A layer of 1 μm thick is located 3 μm below a planer air silicon surface. (a) Calculate the fraction of incident power absorbed in the layer if the refractive index n_1 of the layer is $n_1 = 3.5$. The absorption coefficient $\alpha = 10^3\ cm^{-1}$ at the wavelength of interest and (b) Repeat when $\alpha = 10^{-4}\ cm^{1}$.*

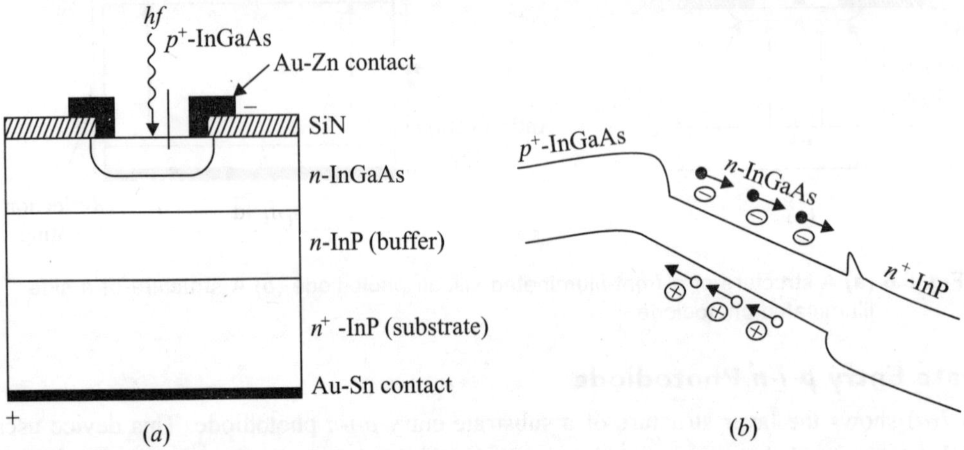

Fig. 6.8 A top entry planer InGaAs *p-i-n* photodiode showing (*a*) structure and (*b*) energy band diagram showing homojunction associated with conventional *p-i-n* structure

Solution: Data supplied: (*a*) $\alpha = 10^{-3}\ \text{cm}^{-1} = 10^{-5}\ \text{m}^{-1}$ and for (*b*) $\alpha = 10^{-6}\ \text{m}^{-1}$

The reflection coefficient $R_f = \left(\frac{n_1 - 1}{n_1 + 1}\right)^2 = \left(\frac{2.5}{4.5}\right)^2 = \left(\frac{5}{9}\right)^2 = 0.309$

The fraction of the incident power absorbed is (Eq. 6.6.7)

(*a*) $$\frac{P(W)}{P_1} = e^{-ad}(1 - e^{aW})(1 - R_f) = e^{-10^{-5}(3\times10^{-6})}(1 - 1e^{10^{-5}(3\times10^{-6})})$$

$$= e^{-10^{-5}(3\times10^{-6})}(1 - 0.309)$$

$$= 4.8\%$$

$$\frac{P(W)}{P_1} = e^{-10^{6}(1\times10^{-6})}(1 - e^{-10^{6}\times1\times10^{-6}})(1 - 0.309) = 2.18\%$$

6.10 THE *p-i-n* AVALANCHE DIODE

Figure 6.9(*a*) shows a reverse biased *p-i-n* diode. When the reverse biased voltage applied across the diode is increased gradually, a threshold value of voltage across the junction will be reached and the field intensity near the junction zone will be very high of the order of 3×10^5 V/cm. Thus, the photo-electrons generated due to light falling on the device will be accelerated through the junction zone with extreme rapidity and will have collisions with the atoms in the field area to produce "Secondary electron-hole pairs". The number of carriers generated in this way goes on multiplying exponentially with the electric field intensity. This type of carrier multiplication is known as "impact ionization" and the phenomenon is referred to as "Avalanche Effect".

It often requires a high value of reverse bias voltages so that the new carriers created by impact ionization can produce additional carriers by the same mechanism. An extremely high gains can be attained by the *p-i-n* diode and these are much greater than that would have been achieved without avalanche effect. The *p-i-n* structure of Fig. 6.3(*a*) is not suitable for the avalanche formation in the diode. In this structure, the electric field is distributed throughout the relatively thick depletion zone and the externally applied voltage requirement for obtaining avalanche threshold value of electric field is as high as 50 to 400 Volts. More recently the devices which will operate at much lower voltages (15 to 25 volts) have become available.

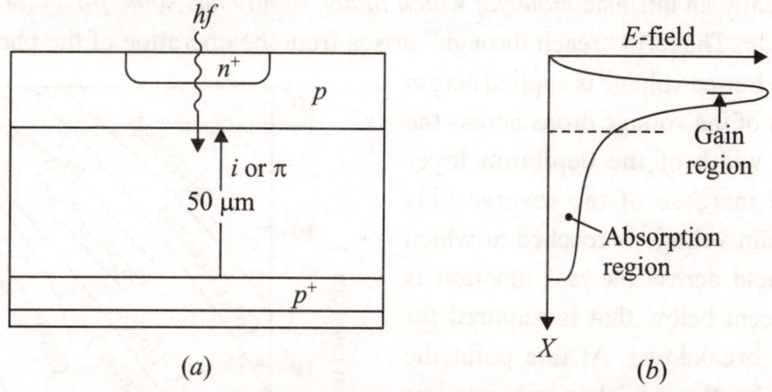

Fig. 6.9 Structure of silicon RAPD (Reach through avalanche photodiode) (*a*) Electric field distribution in RAPD illustrating gain across the *p-n* junction

6.11 REACH THROUGH AVALANCHE PHOTODIODE (RAPD)

Figure 6.10 shows the commonly used structure for achieving carrier multiplication by "impact ionization" with excess noise current and the structure is known as "reach through construction". The reach-through avalanche photodiode (RAPD) consists of a high resistivity *p*-type material deposited on an epitaxial layer on a p^+-type substrate. The p^+ type substrate is a *p*-type layer with an extremely high value of impurity concentration. A *p*-type diffusion or ion implementation is then made on the high resistivity material followed by the construction of an n^+-layer. In an n^+-layer there is an extremely high concentration

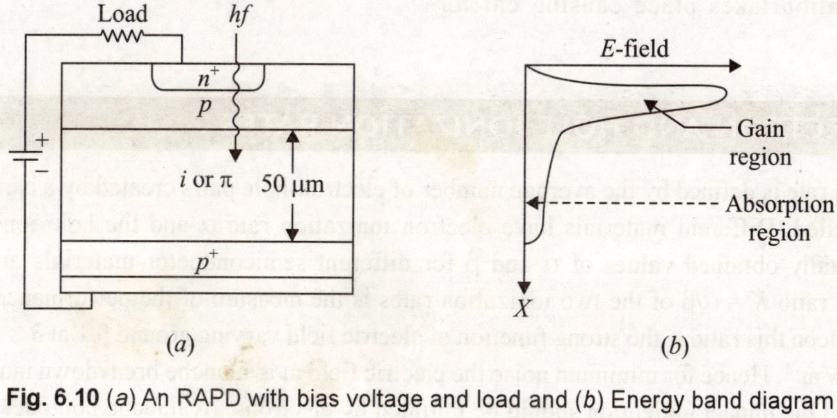

Fig. 6.10 (*a*) An RAPD with bias voltage and load and (*b*) Energy band diagram

of donor atoms. In the case of silicon device, the doping materials (impurity atoms) used to make these p^+ and n^+ layers are usually boron and phosphorous respectively. The configuration of Fig. 6.10 shows an APD with bias voltage and load and is referred to as a "$p^+ \pi p n^+$" reach through structure. Figure 6.10 shows a $p^+\pi pn^+$ avalanche photodiode structure illustrating electric field, depletion and absorption regions. i-layer is referred to π as layer.

6.12 OPERATION OF REACH THROUGH PHOTODIODE (RAPD)

The π-layer is basically an intrinsic *material which inadvertently has some p-type doping, because of imperfect purification*. The term "reach through" arises from the operation of the photodiode. When a low value of reverse biased voltage is applied across the device, the most of the voltage drops across the pn^+ junction. The width of the depletion layer increases with the increase of the reverse bias voltage until a certain voltage is reached at which the peak electric field across the pn^+ junction is about 5 to 10 per cent below that is required for causing avalanche breakdown. At this point the depletion layer "reaches through" the nearly intrinsic n-region.

In normal operation the RAPD is operated in the fully depleted mode. Light enters into the device through the n^+ region to get absorbed in the π-material that acts as the collection region for the photo generated carriers. On being absorbed photons give their energy to create electron-hole pairs. The electrons and holes are separated by their electric field in the π-region. The photo generated electrons drift through the π-region to pn^+ junction where an extremely high field exists. In this high field region impact ionization takes place causing carrier multiplication.

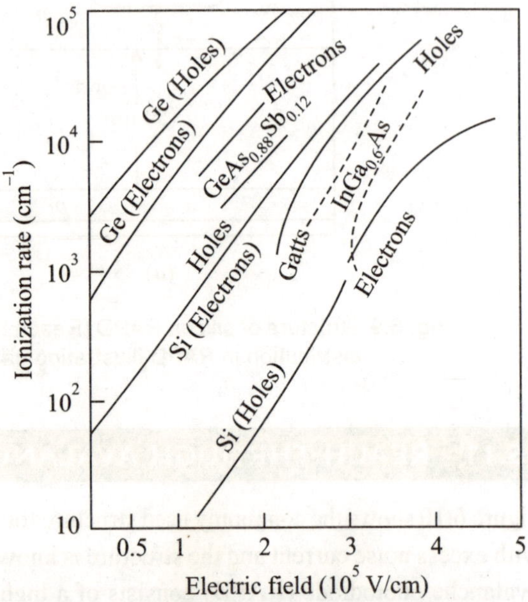

Fig. 6.11 Carrier ionization rate obtained experimentally for silicon, germanium, gallium arsenide gallium antimonide and indium gallium arsenide

6.13 ELECTRON AND HOLE IONIZATION RATES

The ionization rate is defined by the average number of electron-hole pairs created by a carrier per unit distance travelled. Different materials have electron ionization rate α and the hole ionization rate β. Experimentally obtained values of α and β for different semiconductor materials are shown in Fig. 6.11. The ratio $K = \alpha/\beta$ of the two ionization rates is the measure of the performance of a photo detector. In silicon this ratio is the strong function of electric field varying around 0.1 at 3×10^5 Vm^{-1} to 0.5 at 6×10^5 Vm^{-1}. Hence for minimum noise the electric field at avalanche breakdown must be as low as possible and the impact ionization should be initiated by electrons. Avalanche photodetectors made of materials in which one type of carriers dominates impact ionization exhibit low noise and large bandwidth product. Figure 6.11 shows that only silicon has a significant difference between electron and hole ionization rates.

The multiplication factor M (current gain) for all carriers generated in the avalanche photodiode is defined as the average value of the total multiplied output current I_M and the primary unmultiplied photo current I_p.

Current gain $$M = \frac{I_M}{I_p} \quad (6.13.1)$$

where I_p is given by Eq. (6.6.6) as $I_p = \frac{eP_o}{hf}(1 - \varepsilon^{-\alpha(\lambda)W})(1 - R_f)$

Every electron-hole pair generated in an avalanche photodiode does experience the same amount of multiplication due to avalanche effect. In practice, the avalanche mechanism is a statistical process and thus, the measured value of M is expressed as an average quantity.

Figure 6.12(*a*) and (*b*) show the curves for the variation of current gains M (or multiplication factor) for a silicon reach-through avalanche photodiode as a function of bias voltage for different wavelengths and temperature respectively. The dependence of current gain on excitation wavelength is that the mixed initiation of the avalanche process takes place by both electrons and holes when the most of the light is absorbed in the n^+p-region close to the detector surface. This is especially observed at short wavelengths where a large portion of the optical power is absorbed close to the surface as compared to the longer wavelengths. Since the ionization coefficients of holes is smaller than that for electrons in silicon, the total current gain becomes less at the short wavelengths.

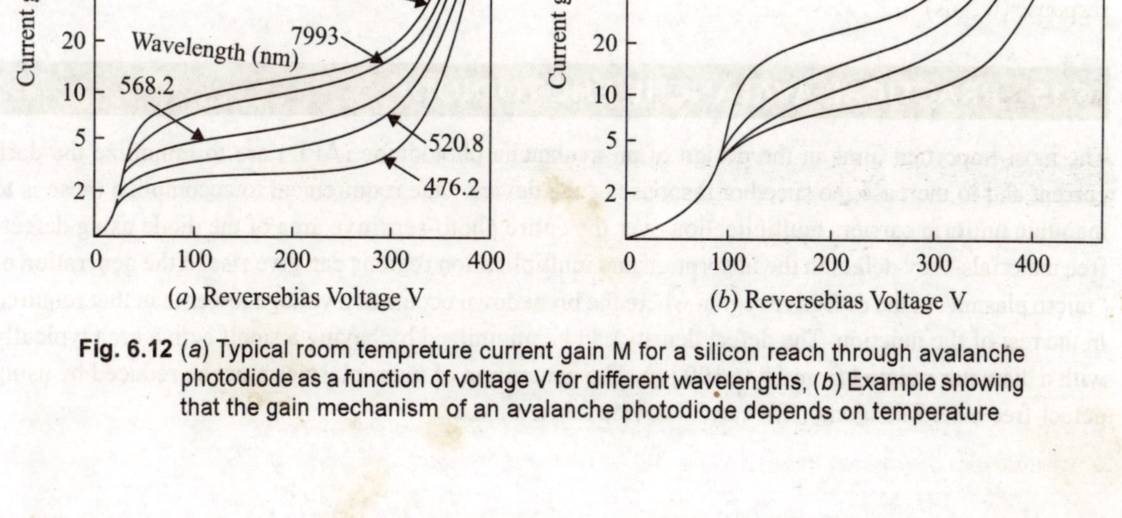

Fig. 6.12 (*a*) Typical room tempreture current gain M for a silicon reach through avalanche photodiode as a function of voltage V for different wavelengths, (*b*) Example showing that the gain mechanism of an avalanche photodiode depends on temperature

Responsivity of Avalanche Photodiode

An important characteristic of an APD is its "Responsivity" R_{APD} analogous to an *p-i-n* diode, the responsivity R_{APD} is given by

$$R_{\text{APD}} = \frac{\eta e M}{hf} = R_o M$$

where R_o is the unity gain responsivity (Eq. 6.8.3).

6.14 SILICON REACH THROUGH AVALANCHE PHOTODIODES

In order to have carrier multiplication without excess noise for a certain thickness of multiplication region within an APD, it is necessary to reduce the ratio of the ionization coefficients for electrons and that for holes ($K = \alpha/\beta$). In the case of silicon this ratio K is a strong function of electric field varying from 0.1 at 3×10^5 Vm^{-1} to 0.5 at 6×10^5 Vm^{-1}. Thus, for minimum noise in the device, the electric field at avalanche breakdown must be as low as possible and the impact ionization should be started by electrons. In order to achieve this a "reach through" structure has been fabricated with silicon avalanche photodiode.

The silicon reach through APD (or *RAPD*) consists of p^+-π-p-n^+ layer as shown in Fig. 6.10. As apparent in Fig. 6.10 the high field region where the avalanche multiplication occurs in comparatively narrow region and is situated on the p-n^+ region. Thus under low reverse bias, most of the voltage develops across the p-n^+ junction.

When the reverse bias voltage is increased the depletion layer widens across the p region until it "reaches through" to the nearly intrinsic π-region which is lightly doped. Since the π region is much wider than the p-region, the electric field in the π-region is much smaller than that at pn^+ junction (Fig. 6.10). This has the result of removing some of the excess applied voltage from the multiplication region to π-region giving a comparatively slow increase in the multiplication factor M with applied voltage. Although the field in the π-region is lower than that in the multiplication region, it is high enough (2×10^4 V/m) when the photodiode is operating to sweep the carriers through the multiplication region at their scattering lower velocity (10^7 cm sec^{-1}). This limits the transit time and ensure a fast response (0.5 ns).

6.15 PRACTICAL AVALANCHE PHOTODIODE

The most important aims in the design of an avalanche photodiode (APD) are to minimize the dark current and to increase the speed of response of the device. One requirement to accomplish those is to maintain uniform carriers multiplication over the entire photo-sensitive area of the diode using defect-free materials. Any defect in the absorption and multiplication regions can give rise to the generation of "micro plasma" in the defective region where the breakdown occurs at a voltage lower than that required in the rest of the junction. The defect density can be minimized by having a small active area typically with a diameter ranging from 10 to 100 μm. The generation of microplasmas must be reduced by using defect-free materials along with careful device fabrication.

The dark current has its origin from generation-recombination in the depletion region, leakage and tunnelling current. The excessive leakage current at the junction edges can be eliminated by the use of "guard ring structure" as shown in Fig. 6.13. The guard ring in a silicon device is an *n-p* junction created by selective diffusion around the periphery of the diode. Moreover, a SiO_2 film is deposited to minimize surface leakage and edge leakage current in Mesa-diode.

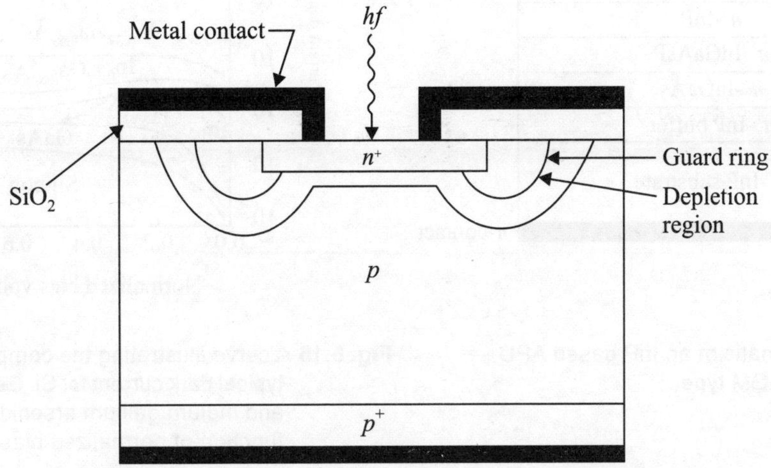

Fig. 6.13 Guard ring created in silicon APD by selective diffusion

In order to operate this device at high speed it is required to have full depletion in absorption region. When carriers are generated in undepleted region they are collected slowly by diffusion processes. This has effect of producing a long "diffusion tail" on short optical pulse. Figure 6.14 shows the configuration of recent back illuminated Mesa structure with separate absorption, grading, multiplication (SAGM) in GaAs APD. The device has gain bandwidth product of 70 GHz.

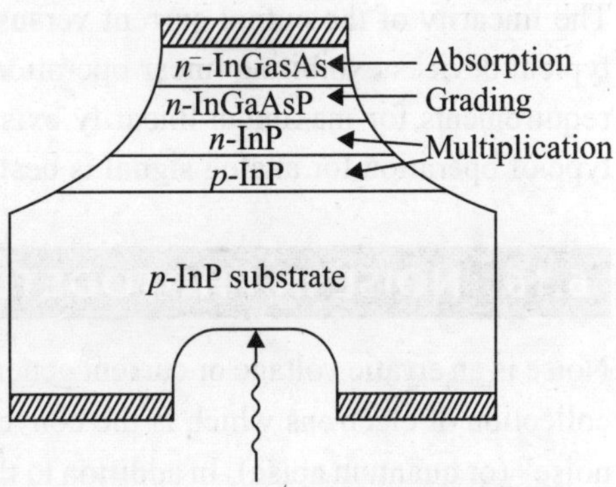

Fig. 6.14 Separate absorption, grading and multiplication (SAGM) APD with mesa structure

The choice of material is decided by the application of the device. For applications of an APD in near infrared region GaAs/AlGaAs diodes are usually used. For optical fiber communications InGaAsP with E_g lying between 0.74 and 1.35 eV is chosen. For example, $In_{0.53}Ga_{0.47}As$ has a band gap of 0.74 eV which corresponds to 1.24/0.74 = 1.66 μm. Homojunction diodes made of this low band gap materials have a high reverse leakage current at high voltage due to band to band tunnelling. The probability of tunnelling is proportional to $E_g^{-3/2}$. The high value reverse leakage current due to tunnelling is reduced considerably by using high field avalanche region of diode in high band gap material. These considerations have led to the design and realization of the "separate absorption and multiplication" (SAM) APD (or SAMAPD). A graded band gap quaternary InGaAsP layer can be inserted between the InP and InGaAs layer to alleviate the problem of large response time. The resulting structure is known as "separate absorption-graded multiplication (SAGM)-APD (or SAGM-APD) is shown in Fig. 6.15. Figure 6.16 shows a comparison of typical dark current for Si, Ge, GaAs, and InGaAs photodiodes as a function of normalized bias voltage.

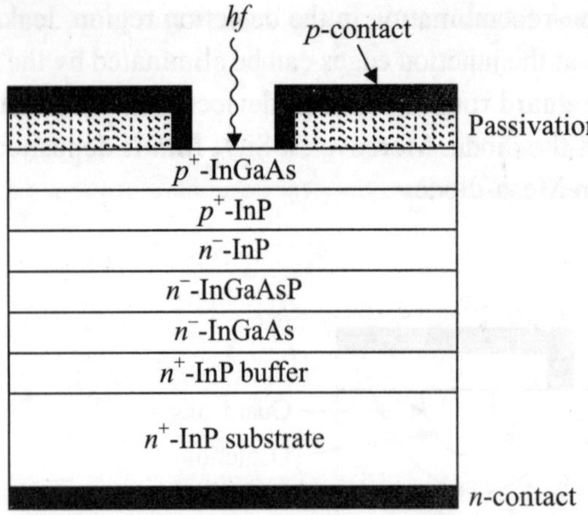

Fig. 6.15 Schematic of an InP based APD of SAGM type

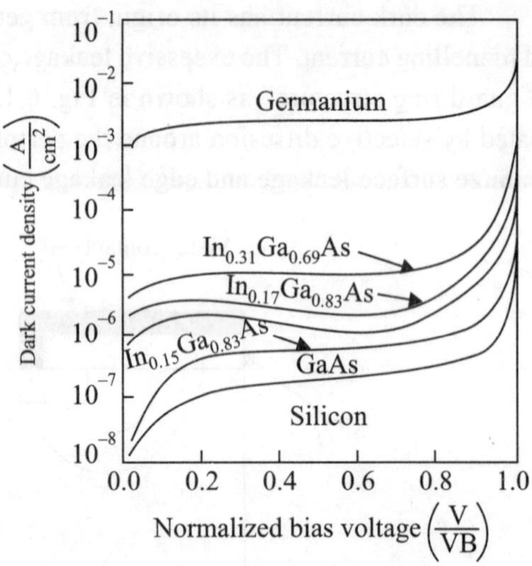

Fig. 6.16 A curve illustrating the comparison of typical dark current for Si, Ge, gallium arsenide and indium gallium arsenide photodiodes as function of normalized biased current

Linearity

The linearity of the output current versus optical input power of optical photodiode is excellent with typical devices exhibiting linear operation. The linearity of an APD is not quite good. Fortunately, the requirements for maximum linearity exist for analog signals with a high signal-to-noise output. This type of operation for analog signal is best met by *p-i-n* diodes, not APDs.

6.16 NOISE IN PHOTODETECTORS

Noise is an erratic voltage or current generated due to the statistical nature of the quantum detection and collection of electrons which is the conversion of photons to electrons. This noise is referred to "Shot noise" (or quantum noise). In addition to the shot noise, thermal noise is also produced in the resistances in the detector circuit.

Shot noise: The average current I in a detector always exhibits a random fluctuation about its mean value and this fluctuation is referred to as "shot noise" where the mean square variation of the noise current is proportional to the average value of the photo current I and the bandwidth B of the photodiode circuit.

The average value of the shot noise is

$$< i_s^2 >^{1/2} = (2e\,B\,I)^{1/2} \tag{6.16.1}$$

where B is the bandwidth, e the electronic charge and I has three parts and these are photo current I_p, background current I_B and dark current I_D so that

$$I = I_p + I_B + I_D \tag{6.16.2}$$

We shall now calculate the noise power in a *p-i-n* photodiode. We consider modulated optical power $P(W)$ expressed by

$$P(W) = P_o(1 + m\,\varepsilon^{jwt}) = P_o(1 + e^{jwt}) \tag{6.16.3}$$

where P_o is the incident power peak value and it is assumed that m (the modulation index) is unity ($m = 1$). w is the angular frequency of the modulating signal.

The r.m.s. optical power is $P_o/\sqrt{2}$ and the a.c. r.m.s. photo current will depend on the quantum efficiency η of the photodiode and r.m.s. photo current I_p is given by (6.7.3*a*)

$$I_p = \frac{e\eta P_o}{\sqrt{2}\,hf} \tag{6.16.4}$$

In addition to I_p, there is usually a background radiation of low intensity in the ambient temperature at which the photodetector is placed and this radiation has given rise to a background photo current I_B. Finally, the reverse bias current flows through the device even when no light is incident on the device and this current has its origin from the thermally generated electron hole pairs in the depletion region. This dark current I_D contributes to the total system noise and gives the random fluctuations of the output current.

The equivalent noise current can be expressed as

$$< i_s^2 > = 2e\,(I_p + I_B + I_D)B \tag{6.16.5}$$

where $I = I_p + I_B + I_D$ and B is the bandwidth of the device Eq. (6.16.1) we have a frequency term and under the condition $wT = 1$, the corresponding frequency $f_c = 1/2\pi$ is called the noise equivalent bandwidth (Fig. 6.17(*b*)).

In addition to shot noise there is a resistance or Johnson noise at the output by different resistance in the diode circuit. Figure 6.17 shows the equivalent circuit of a *p-i-n* photodiode and typical values of different components of a reverse biased (≈ 10 V) InGaAs *p-i-n* photodiode. The resistances are diode shunt resistance R_D, series resistance R_s and load resistance R_L. If the diode is followed by a pre amplifier with input resistance R_i in the receiver circuit and R_i is parallel with R_L (Fig. 6.17).

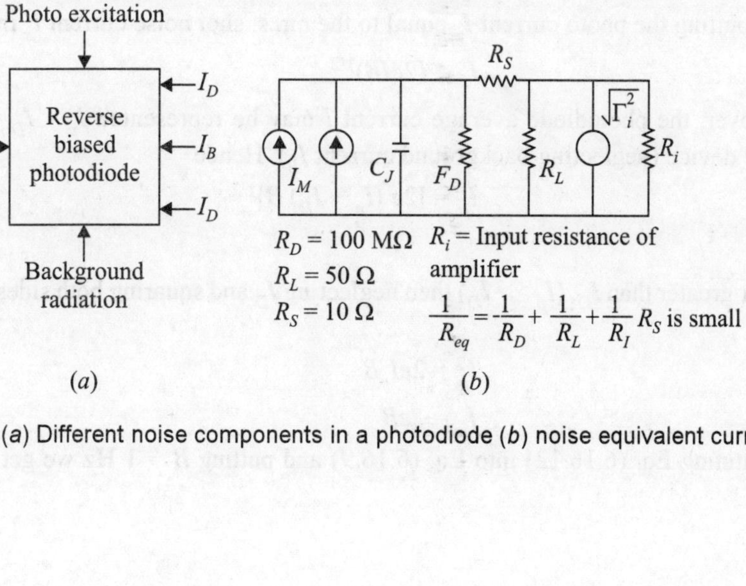

Fig. 6.17 (*a*) Different noise components in a photodiode (*b*) noise equivalent current

The mean square Johnson noise current or thermal noise is given by

$$<I_{th}^2> = \frac{4K_BTB}{R_{eq}} \tag{6.16.6}$$

where $$\frac{1}{R_{eq}} = \frac{1}{R_D} + \frac{1}{R_L} + \frac{R}{R_i}.$$

The series resistance R_s very small of order of few ohms and is neglected.

The signal power is given by $(I_0^2 R_{eq})$ and the noise power at the output is $<i_s^2> + <i_{th}^2> R_{eq}$. Therefore the signal to noise ratio is given by

$$\left(\frac{S}{N}\right)\text{power} = \frac{\frac{1}{2}\left(\frac{e\eta P_o}{hf}\right)}{2e(I_p + I_B + I_D)\,B + \frac{4K_BTB}{R_{eq}}} \tag{6.16.7}$$

The total noise current in the device can be expressed as

$$<i_N^2> = <i_s^2> + <i_{th}^2> = 2e\,[I_p + I_B + I_D] + \frac{4K_BTB}{R_{eq}} \tag{6.16.8}$$

NEP (Noise Equivalent Power)

The noise equivalent power (NEP) is defined as the incident optical power at a particular wavelength or with a specified spectral content required to produce a photo detector current equal to the r.m.s. noise current within a unit bandwidth (*i.e.*, B = 1Hz). In order to obtain an expression for NEP at a specific wavelength, Eq. (6.7.3*a*) must be arranged as follows to give

$$P_o = \frac{I_p hf}{\eta e} = \frac{I_p hc}{\eta e\lambda} \tag{6.16.9}$$

Then putting the photo current I_p equal to the r.m.s. shot noise current i_s^2 in Eq. (6.16.6) gives

$$I_p = (2eIB)^{1/2} \tag{6.16.10}$$

Moreover, the photodiode average current I may be represented $I_p + I_D$ where I_D is the dark current in the device (neglecting background current I_B). Hence

$$I_p = [2e\,(I_p + I_D)\,B]^{1/2} \tag{6.16.11}$$

Case I

When I_p is far greater than I_D, ($I_p >> I_D$) then neglecting I_D and squaring both sides of Eq. (6.16.11), we get

$$I_p^2 = 2eI_pB$$

or $$I_p = 2eB \tag{6.16.12}$$

Substituting Eq. (6.16.12) into Eq. (6.16.9) and putting B = 1 Hz we get the noise equivalent power

$$\text{NEP} = P_o = \frac{2hc}{\lambda} \tag{6.16.13}$$

it should be noted that NEP for an ideal photo detection is given by Eq. (6.16.13) when quantum efficiency $\eta = 1$.

Case II

When $I_p << I_D$ then from Eq. (6.16.11) the photo current becomes

$$I_p = (2eI_D B)^{1/2} \tag{6.16.14}$$

Hence for a photodiode in which the dark current noise is dominant, the use of Eq. (6.16.14) with $B = 1$ gives the expression for the noise equivalent power of

$$\text{NEP} = P_o = \frac{hc(2eI_D)^{1/2}}{\eta e\lambda} \tag{6.16.15}$$

Directivity D is defined as the inverse of NEP.

Thus $$D = \frac{1}{\text{NEP}} \tag{6.16.16}$$

When we consider that a photodiode receives monochromatic radiation with the dark current as its dominant noise source, then by combining Eq. (6.6.14) and (6.6.15), we get

$$D = D_\lambda = \frac{ne\lambda}{hc(2eI_D)^{1/2}} = \frac{1}{\text{NEP}} \tag{6.16.17}$$

The specific directivity D^* is a parameter which incorporates the area of the photodetector A. In order to take account of the effect of this factor on the device dark current:

$$D^* = DA^{1/2} = \frac{ne\lambda}{hc\left(\dfrac{2eI_D}{A}\right)^{1/2}} \tag{6.16.18}$$

Example 6.6: *The quantum efficiency of a silicon RAPD is 70% for the detection of radiation at a wavelength of 0.8 μm. When the incident optical power is 0.6 μW the output of the device after avalanche gain is 10 μA. Determine the multiplication factor M after this condition.*

Solution: Data given $\lambda = 0.8 \times 10^{-6}$, $n = 0.7$, $I = 10 \times 10^{-6}$, $P_o = 0.60 \times 10^{-6}$ Eq. (6.8.13) gives

Responsivity $$R = \frac{\eta e\lambda}{hc} = \frac{0.7 \times 1.6 \times 10^{-19} \times 0.8 \times 10^{6}}{6.62 \times 10^{-34} \times 3 \times 10^{8}}$$

$$= 0.45 \text{ AW}^{-1}$$

Diode current $$I_p = P_o R = 0.6 \times 0.45 \times 10^{-6} = 0.27 \times 10^{-6}$$

$$M = \frac{I}{I_p} = \frac{10 \times 10^{-6}}{0.27 \times 10^{-6}} = 37$$

Example 6.7: *A germanium p-i-n photodiode with an active area dimension of 100 × 50 μm has a quantum efficiency of 60 per cent when operating at a wavelength 1.2 μm. The measured dark current is 10 nA. Calculate the noise equivalent power and specific directivity of the device. It may be assumed that dark current is the main source of noise.*

Solution: Data given $I_D = 10$ nA, $\eta = 0.6$ and $\lambda = 1.2 \times 10^{-6}$ m. The noise equivalent power is given as

$$\text{NEP} = \frac{hc(2eI_d)^{1/2}}{\eta e \lambda}$$

$$= \frac{6.62\times10^{-34}\times3\times0^{8}(2\times1.6\times10^{-19}\times10\times10^{-9})^{1/2}}{0.6\times1.6\times10^{-19}\times1.2\times10^{-6}}$$

$$= 9.80 \times 10^{-14}\ \text{W}$$

Substituting the directivity D in Eq. (6.16.18) from Eq. (6.16.16) allows the specific directivity D^* as

$$D^* = \frac{A^{\frac{1}{2}}}{\text{NEP}} = \frac{(100\times10^{-6}\times50\times10^{-6})^{1/2}}{9.80\times10^{-14}} = 7.0\times10^{8}\ \text{m Hz}^{1/2}\ \text{W}^{-1}.$$

6.17 PHOTODIODE MATERIALS WITH BAND GAP ENERGY

Table 6.1

Materials	*Band gap energy in eV at 300K*
GaAs	1.43
GaSb	0.73
$GaAs_{0.88}Sb_{0.12}$	1.15
Ge	0.67
InAs	0.35
InP	1.35
$In_{0.53}Ga_{0.47}As$	0.75
$In_{0.14}Ga_{0.86}As$	1.15
Si	1.14

6.18 PHOTOTRANSISTORS

Like an avalanche photodiode a phototransistor is capable of providing an internal gain of the photocurrent. This gain is achieved through the "transistor action" rather than by avalanche multiplication. Figure 6.18(*a*) shows the symbolic representation of a bipolar *n-p-n* phototransistor. The difference between a phototransistor and a conventional transistor is that in a phototransistor the base is unconnected. The base collector junction is photosensitive and acts as a light gathering element. The absorbed light increases the base current and the amplification of the base current results in the increase of the primary photocurrent through the device.

Structure

Figure 6.18(*b*) shows the layer structure of an *n-p-n* InGaAsP/InP heterojunction phototransistor. The three-layer structure is grown on an InP substrate using liquid phase epitaxy (LPE). It consists of an *n*-type InP collector layer followed by a thin (0.1 μm) *p*-type InGaAsP base layer. The third layer is a wide bandgap semiconductor *n*-type In*P* emitter layer.

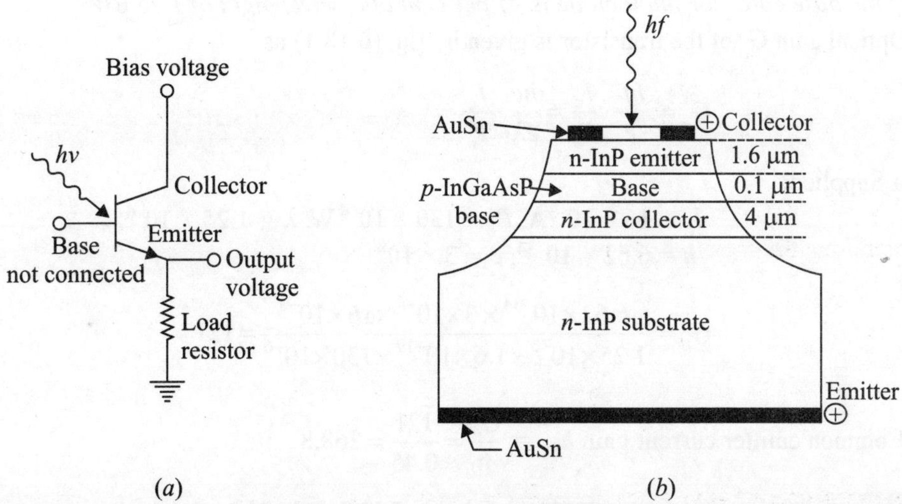

Fig. 6.18 (*a*) A symbolic representation of a bipolar *npn* phototransistor. (*b*) The layer structure of an npn InGaAsP heterojunction phototransistor

Operation

Radiation incident on the phototransistor passes unattenuated through the wide bandgap emitter and is absorbed by the base layer, base collector depletion region and the collector. A large secondary photo current flowing between the collector and emitter is obtained as the photogenerated holes are swept into the base region, increasing the forward bias on the device. The heterostructure permits a low value of capacitances between the emitter-base and the collector-base junction with a low value of base resistance. This is achieved through low emitter and collector doping levels coupled with the heavy doping of the base and these type of doping allows large current gain. Moreover, the potential barrier created by the heterostructure at the emitter base junction effectively eliminate the hole injection from the base when the junction is forward biased. This increases the emitter-base injection efficiency. The optical gain G_o of the device is given approximately by

$$\frac{I_C}{I_B} = G_o = \eta h_{FE} = \frac{hfI_c}{eP_o} \quad (6.18.1)$$

where
$$I_B = I_p = \frac{eP_o}{hf}$$

where η is the quantum efficiency of the base-collector photodiode, h_{FE} is the common emitter current gain, I_c is the collector current, P_o is the incident optical power, *e* is numerically equal to electronic charge in m.k.s = 1.6×10^{-19} C and *hf* is the photon energy.

The photo transistor of Fig. 6.18(*b*) can operate over the range of wavelengths from 0.9 to 1.3 μm and has an optical gain around 100.

Example 6.8: *The phototransistor of Fig. 6.18(b) has a collector current of 16 mA when the incident power at a wavelength of 1.25 μm is 130 μW. (a) Calculate the optical gain of the phototransistor under the above operating condition and (b) Determine common emitter current gain h_{FE} when the quantum efficiency of the base collector photodiode is 45 per cent at a wavelength of 1.25 μW.*

Solution: Optical gain G_o of the transistor is given by Eq. (6.18.1) as

$$G_o = \frac{hf}{e} \cdot \frac{I_c}{P_o} = \frac{hc}{e\lambda} \cdot \frac{I_c}{P_o}$$

Data Supplied:

$$I_c = 16 \times 10^{-3} \text{A}, P_o = 130 \times 10^{-6} \text{W}, \lambda = 1.25 \times 10^{-6},$$
$$h = 6.62 \times 10^{-34}, c = 3 \times 10^{8}$$

$$G_o = \frac{6.62\times10^{-34}\times3\times10^{-8}\times16\times10^{-3}}{1.25\times10^{-6}\times1.6\times10^{-19}\times130\times10^{-6}} = 121$$

(*b*) Common emitter current gain $h_{FE} = \frac{G_o}{\eta} = \frac{121}{0.45} = 268.8$

The common emitter current and optical gain are 268.8 and 121 respectively.

6.19 PHOTO CONDUCTING DETECTOR

The photo conductor is the simplest form of an optical detector and it is capable of exhibiting gain mechanism. The operation of a photo conducting detector is based on the fact that when light falls on a semiconductor, an electron may be raised from the valence band to the conduction band by the absorption of the photon provided the energy of the photon is greater than the band gap energy of the semiconductor (*i.e.*, $hf \geq E_g$). As long as the electron remains the conduction band, it will cause to increase the conductivity of the semiconductor. The phenomenon is referred to as "photo conductivity" Fig. 6.19 shows the structure of the photo conducting device designed for operation in the longer wavelength region. The conducting channel consists of a thin layer (1 to 2 μm) of *n*-type InGaAs which is capable of absorbing a considerable amount of incident light over the wavelengths of 1.1 to 1.6 μm. In particular, good sensitivities at reasonable high bandwidths have been achieved with a lightly doped (less than 5×10^{14} cm^{-3}) *n* type $In_{0.53}Ga_{0.47}As$ channel. The active layer is formed either by epitaxially or ion implementation on a high resistivity substrate. Low resistance contacts are made to the conducting layer through use of inter-digital anode and cathode (as shown

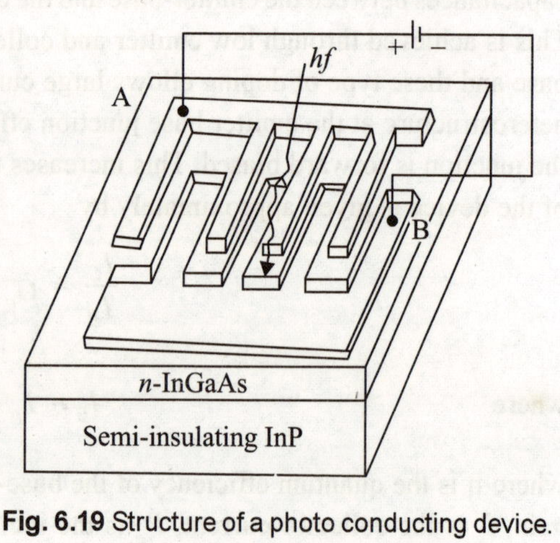

Fig. 6.19 Structure of a photo conducting device. (1.1 to 1.6 μm wavelength)

in the figure) by the evaporation of metal and subsequent alloying. A suitable bias voltage is applied across the contacts (*A* and *B*) to collect photo generated carriers.

Another important factor which affects the performance of the device is the "noise current". The large dark current in the device generates the noise current. This noise is referred to as "thermal noise" (or Johnson noise). The noise current is given by

$$\langle i_{\text{th}}^2 \rangle = \frac{4K_B TB}{R_c} \tag{6.19.1}$$

where B is the bandwidth of the device, R_c the resistance of the photo conducting channel, K_B Boltzmann constant and T the absolute temperature. The thermal noise has its origin from the random nature of the motion of the carriers which contribute to the "dark current".

Current always flows through the photo conductor device and the signal to noise ratio S/N can be expressed as

$$\frac{S}{N} = \frac{\text{Conductivity when light falls}}{\text{Conductivity when no light falls}} = \frac{\text{(Conductivity) light}}{\text{(Conductivity) dark}} \tag{6.19.2}$$

Thus in order to get a high value of S/N the dark conductivity has to be reduced. The photo generated electrons and holes move in opposite directions in the active region under the applied bias voltage. The resulting photo current will persist until both types of carriers are collected at the electrodes or until they (electrons and holes) recombine in bulk of semiconductor before reaching the respective contacts. We are interested with the recombination of minority carriers and therefore the important parameter is the minority carrier recombination time Γ. The time for detection of the photo generated current is limited by transit time between electrode of the faster carrier (usually electrons). Therefore, the shortest response time (maximum bandwidth) can be obtained by minimizing the contacts. It is also assumed that the bias voltage is such that the carrier travel with their saturation velocity. The persistence of slower holes in an *n*-type channel after the electron is collected will increase the response time.

The current response in a photo conductor remains for a time t_s (recombination time for slow carrier hole) after the end of the optical pulse and the response pulse exhibits an exponential decay with time constant equal to slow carrier transit time or lifetime), then the maximum 3dB bandwidth B_m for the device is given by

$$B_m = \frac{1}{2\pi t_s} \tag{6.19.3}$$

Photo Conductive Gain G

When the fast electrons are collected at the anode, the corresponding photo generated holes are still moving across the channel. This creates absence of electrons in the channel with a net positive charge in channel region. But the excess positive charge due to existence of holes is immediately neutralized by the injection of further electrons from the cathode to the channel. Thus the further electrons may be generated from the absorption of a single photon. This mechanism creates what is known as the photo conductive gain *G*. The photo conductive gain (*G*) may be defined as the ratio of the slow carrier transit times t_s (or lifetime) to the fast carrier transit time t_f. Hence

$$G = \frac{t_s}{t_f} \tag{6.19.4}$$

Moreover, the photo current produced by the photo conductor as given by Eq. (6.18.1) the current in the photo conductor.

$$I_p = \frac{\eta P_o e}{hf} G \tag{6.19.5}$$

where η is the device quantum efficiency, e the charge of an electron and P_o incident optical power. Combining Eq. (6.19.4) and Eq. (6.19.3), the gain-bandwidth product of the photo conductor is

$$G.B_m = \frac{t_s}{t_f} \times \frac{1}{2\pi t_s} = \frac{1}{2\pi t_f} \tag{6.19.6}$$

The conclusion from Eq. (6.19.3) is that when the fast transit time (usually of electron) is constant, the photo conductive gain can only be determined at the expense of maximum bandwidth permitted by the device. There is a compromise between gain (sensitivity) and the speed of response.

Example 6.9: *The electron transit time in an InGaAs photoconductive detector is 8 picoseconds. Calculate the maximum 3dB bandwidth when the photoconductive gain of the device is 60.*

Solution: Using Eq. (6.19.6) and substituting the values we can write

$$t_f = 8 \times 10^{-12}\ S \text{ and } G = 60$$

$$B_m = \frac{1}{2\pi t_f G} = \frac{1}{2\pi \times 8 \times 10^{-12} \times 60} = 3.28 \times 10^6 = 3.28 \text{ MHz}$$

The 3dB bandwidth is 3.28 MHz.

Example 6.10: *$GaAs_{0.88}Sb_{0.12}$ (Gallium arsenic antimonide) is a direct band gap semiconductor and has a band gap energy of 1.15 eV. Determine the critical wavelength above which an intrinsic photodetector from this material cannot be operated.*

Solution: The critical wavelength is given by

$$\frac{h_c}{\lambda_c} = E_g \quad \text{or} \quad \lambda_c = \frac{hc}{E_g}$$

Substituting the values supplied

or $$\lambda_c = \frac{h_c}{E_g} = \frac{6.62 \times 10^{-34} \times 3 \times 10^8}{1.15 \times 1.6 \times 10^{-19}} = 1.06\ \mu\text{m}$$

$$= 1060 \text{ nm}$$

The critical wavelength is 1.06 μm (= 1060 nm).

6.20 SOLAR CELL

Introduction

A solar cell is a *p-n* junction in which the generation of carriers by an external source of energy, usually sunlight, is utilized to generate the electrical power. In other words a solar cell directly converts solar energy into electrical power. The design of most solar cells involves efficiency of energy conversion to be maximized. In this section, however, we will exemplify the operation of a solar cell using a simple *p-n* junction structure. Solar cell operation is based on the generation of electron-hole pairs in the transition region, and the separation of both types of carriers by the junction electric field. A *p-n* junction solar cell operates with minority carrier and, hence, are commonly classified as "minority carrier devices."

Working Principle

Consider the *p-n* junction shown with a resistive load. Even with zero bias applied to the junction, an electric field exists in the space charge region. Incident photon illumination can create EHP (electron hole pair) in the space charge region that will be swept out to produce photocurrent I_L in the reverse bias direction as shown.

The photocurrent produces a voltage drop across the resistive load which forward biases the *p-n* junction. The forward bias voltage produces a forward bias current I_D as indicated in the Fig. 6.20 below.

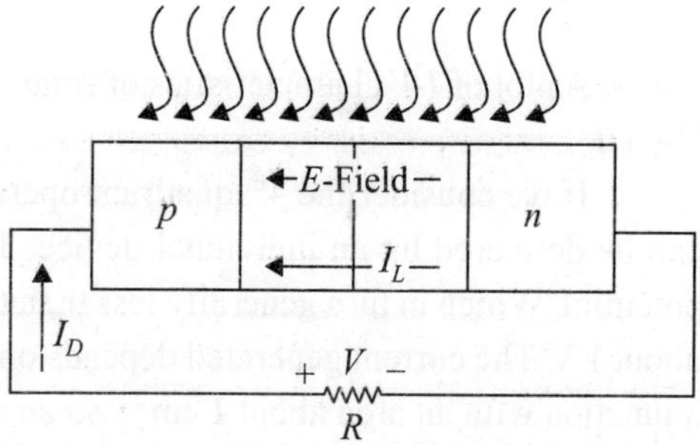

Fig. 6.20

$$I = I_L - I_D = I_L - I_o(e^{qV/KT} - 1) \tag{6.20.1}$$

The photocurrent is always in the reverse biased direction and the net solar cell current is always in the reverse bias direction.

When $V = 0$, $R = 0$ is referred as the short circuit condition and the current is referred as the short circuit current $I_{SC} = I_L$. This is the reason why the net current is not zero in an illuminated solar cell.

Figure 6.21 below shows the equivalent circuit of a solar cell.

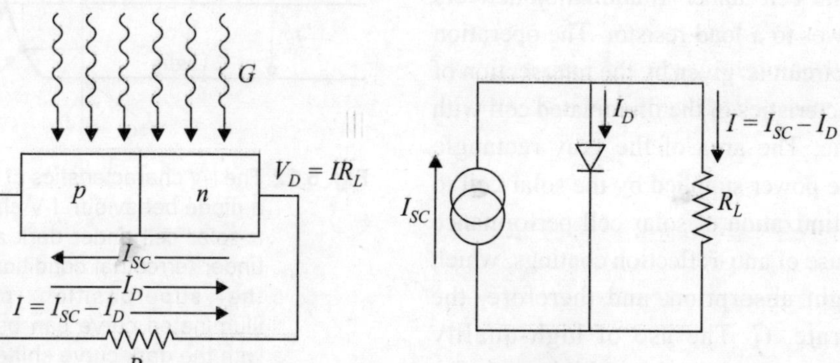

Fig. 6.21 Equivalent circuit of a solar cell

In a solar cell the forward bias is not explicitly applied across the cell. It is generated by the flow of the current across the load (which may be the resistance of a light bulb for instance). This is shown schematically in Fig. 6.21 along with the equivalent circuit of the solar cell. The total cell current goes to zero at a voltage, V_{oc}, termed the open circuit voltage, when the forward diode current is equal and opposite to the generated current.

When $R \to \infty$ the net current is zero and the voltage produced is open circuit voltage. Here $I = 0$ so photo current is just balanced by the forward biased junction current.

$$I = 0 = I_L - I_o(e^{qV_{oc}/KT} - 1)$$

$$V_{OC} = V_t \ln\left(1 + \frac{I_L}{I_S}\right) \tag{6.20.2}$$

A plot of *I-V* characteristics of solar cell is shown in Fig. 6.22. The power delivered to the load is $P = VI$.

If we consider the 4th quadrant operation of the last figure it appears doubtful that much power can be delivered by an individual device. The voltage is restricted to the values less than the contact potential. Which in turn generally less than the band gap voltage E_g/q. For Si, the voltage V_{oc} is less than about 1 V. The current generated depends on contact area, but typically I_{op} is in the 10-100 mA range for a junction with an area about 1 cm^2. So an array of such devices is used to deliver a significant amount of power. Commercial solar cells are basically *p-n* junction of silicon and have large surface area. Top contact is a metal grid attached to the surface *n*-region where bottom *p*-type substrate has a normal metal contact. Further, load resistor is connected between the top and bottom contacts of the solar cell.

Figure 6.22 shows the current-voltage characteristics of a solar cell in the dark and under illumination. The inset shows a simple circuit where the solar cell under illumination delivers electrical power to a load resistor. The operation point of the circuit is given by the intersection of the *I-V* characteristics of the illuminated cell with the load line. The area of the gray rectangle represents the power supplied by the solar cell to the load. Optimization of solar cell performance involves the use of anti-reflection coatings, which increases light absorption, and therefore, the generation rate, *G*. The use of high-quality semiconductor material with a high minority carrier lifetime, and the choice of a load resistance value, maximizes the power transferred to the load (which maximizes the area of the gray rectangle in Fig. 6.22).

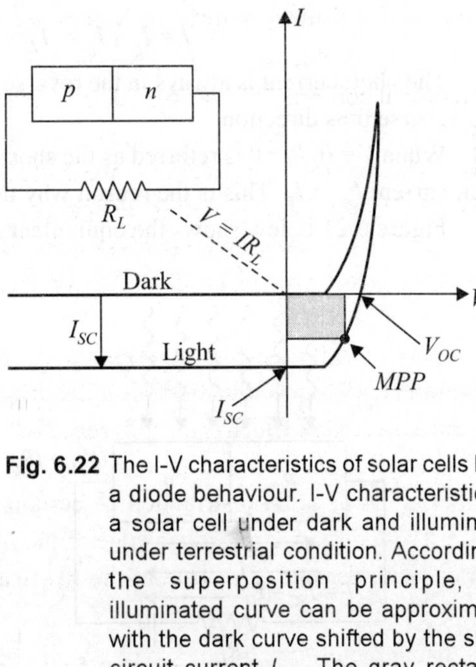

Fig. 6.22 The I-V characteristics of solar cells have a diode behaviour. I-V characteristics of a solar cell under dark and illuminated under terrestrial condition. According to the superposition principle, the illuminated curve can be approximated with the dark curve shifted by the short-circuit current I_{SC}. The gray rectangle represents the power supplied to a load resistance R_L. The operating point with maximum power output (MPP) is where $P = I \cdot V$ is maximal

The short-circuit current and the open-circuit voltage of an illuminated solar cell are noted as I_{SC} and V_{OC} respectively. Assume the gray-coloured rectangle in Figure 6.22 above and 6.23 is the largest possible rectangle that can be inscribed between the axes and the *I-V* characteristics of the device, *i.e.*, a rectangle that represents the maximum power that the solar cell can deliver for a given level of illumination. Let its area be noted S. One can then define a fill factor (FF) by the following relationship:

Fill Factor = $S/I_{SC} \cdot V_{OC}$

Conversion efficiency in watt/cm^2/incident light intensity in watt/cm^2 × 100%

Conversion efficiency (η) $\eta = V_m I_m$

Fill factor = $V_m I_m / V_{oc} I_{sc}$ (6.20.3)

It is expected to be in the range of 0.8 for good cell is a figure of merit for solar cell. The fill factor depends on the design and the fabrication parameters of a solar cell and is optimized to increase the energy conversion efficiency of the device.

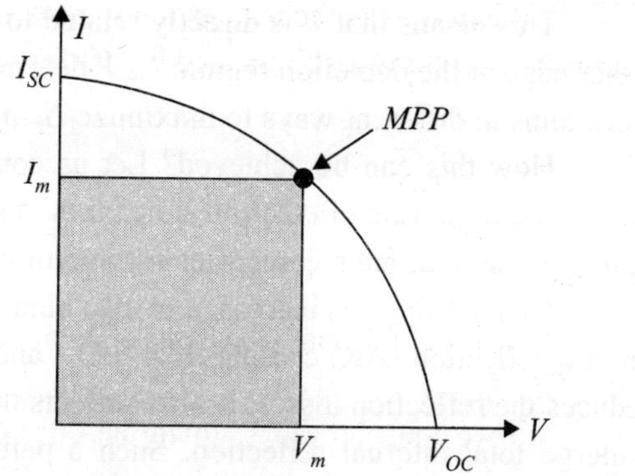

Fig. 6.23 I-V characteristics of an illuminated solar cell. The maximum power rectangle is shaded. The operating point with maximum power output (MPP) is where $P = I \cdot V$ is maximal

How to Optimize Performance of Solar Cells?

A solar cell can deliver electrical power $P = I \cdot V$ because it generates both a current I and a voltage V at the terminal contacts. The cell can do so because the photogenerated electrons and holes thermalize in the separate Fermi–Dirac distributions with their quasi-Fermi energies E_{Fn} and E_{Fp}, respectively.

The electrons are injected at one contact and extracted at other contact, because to deliver current I, the electrons and holes need to be separated. This is achieved by having a highly doped n-type region in front of one contact and having a highly doped p-type region in front of the other. The source of current I is not the p-n junction, but the gradients ∇E_{Fn} and ΔE_{Fp} that drive the free carriers toward the contacts.

Only small gradients are required, as the resistivity of the silicon material is selected to be low. Therefore, the quasi-Fermi energies of the majority carriers are relatively constant in the whole device, and the delivered V, which is the difference between the photo-induced E_{Fn} and E_{Fp} near the p-n junction.

The free carriers distribute themselves throughout the device in the most probable way, which is the Fermi–Dirac (FD) distribution. A certain carrier type (either electrons or holes) is the minority carrier (with density $c_{\min}$) at one edge of the p-n junction, whereas at the other edge, it is the majority carrier (with density c_{maj}). Due to the FD distribution, $c_{\min}$ and c_{maj} are related to each other by the Boltzmann factor, *i.e.*, by: $c_{\min} = c_{\text{maj}}\, e^{-\frac{q\psi}{kT}}$ because the potential barrier ψ across the p-n junction is larger than the thermal voltage kT/q. In thermal equilibrium, ψ is the built-in voltage ψ_0, and equation becomes $c_{\min,0} = c_{\text{maj},0}\, e^{-\frac{q\psi_0}{kT}}$, where $c_{\min,0}$ and $c_{\text{maj},0}$ denote the minority and majority carrier densities in the thermal equilibrium, respectively. If additional carriers are injected, ψ_0 is lowered by the separation of the quasi-Fermi levels, *i.e.*, by $E_{Fn} - E_{Fp} \approx V$, and we can obtain $c_{\min} = c_{\min,0}\, e^{\frac{qV}{kT}}$ after inserting the value of $c_{\min,0}$ from the above equation.

This means that V is directly related to the excess minority carrier density $\Delta c = c_{min} - c_{min,0}$ at either edge of the depletion region. As V determines the cell's power output $P = I \cdot V$, most optimization work aims in different ways to maximize Δc near the *p-n* junction depletion region.

How this can be achieved? Let us consider the general expression $\Delta c = G\tau$, where G is the photogeneration rate of electron-hole pairs, and τ is the excess carrier lifetime. This expression shows that there are two main categories for maximizing Δc, via either G or τ.

Commonly, G is increased by two kinds of surface treatments: first, by covering the surface with an anti-reflection (AR) coating, like SiO_2; and second by texturing it. The second procedure not only reduces the reflection losses, it also deflects the incoming sun rays sideways, so that they are likely to undergo total internal reflection. Such a path-length enhancement leads to better absorption and is called light-trapping. In the other category of optimization, the excess-carrier lifetime τ is increased, and this procedure acts mainly on V. τ is the mean time that elapses between the generation of an excess electron-hole pair and its subsequent recombination. If there were only one excess electron-hole pair (per cm^3), then the rate R of recombination events (per cm^3 per second) would be $1/\tau$; but as there are Δc pairs in excess, we have $R = \Delta c/\tau$. Hence, τ is a measure of the driving force that restores thermal equilibrium in the solar cell, and we want to minimize this driving force so the photogenerated carriers can reach the terminal contacts and deliver power.

Equilibrium is attained in the different device parts in different ways. In lowly doped device regions, R is usually dominated by recombination via defects, which is quantified using the Shockley–Read–Hall (SRH) formalism. In highly doped or injected regions, R is dominated by Auger recombination. Mainly because the excess carriers can move to various device parts, they have a range of recombination options. In this case, their lifetime is called effective lifetime τ_{eff}. This is the quantity that is normally measurable. Hence, numerical modelling is a helpful tool to quantify the contributions to τ_{eff} from various device parts and from various recombination mechanisms. Such investigations provide concrete strategies of how to minimize recombination losses and of how to optimize the device.

This kind of optimization is not a straightforward task, quite intricate, mainly as because the movement of the carriers, their transport properties are involved and other factors.

Types of Solar Cells-materials

The most commonly used types of solar cells are

1. Silicon solar cells: crystalline silicon is used as main material.
2. Homojunction solar cells: most of the solar cells are homojunction cells.
3. Heterojunction solar cells: As the name suggests these are made from two different types of materials. One example is cadmium sulphide-copper sulphide cells, the two metals form a heterojunction.
4. GaAs solar cells: they are most efficient but also most expensive solar cells.
5. Amorphous solar cells: amorphous thin film silicon is used. They are popular as, amorphous Si has greater absorption coefficient than mono-crystalline silicon and amorphous Si can be formed on any surface

Applications of Solar Cells

1. Independent power system
2. Solar house

3. Rural electrification
4. Military applications
5. Replacement of expensive dry cells
6. Low power application like weather monitoring
7. Marine navigation systems positioned in rural areas.

Example 6.11: *Determine the conversion efficiency of a solar cell of input power of 900 mW, V_{OC} = 600 mV, I_{SC} = 240 mA, FF = 75%.*

Solution:

Maximum power $P_{max} = V_{OC} \times I_{SC} \times FF = 600 \times 10^{-3} \times 240 \times 10^{-3} \times 0.75 = 108$ mW.

$$\text{Conversion efficiency } \eta = \frac{P_{max}}{P_{in}} = \frac{108 \text{ mW}}{900 \text{ mW}} = 12\%.$$

Example 6.12: *The area of a solar cell is 4 cm² with a conversion efficiency of 12% producing 0.5 V. To generate 12 W output power of the solar panel, determine the number of cells, total area of the panel and current capacity of each cell. Assume solar constant on earth or input radiation as 100 mW/cm².*

Solution:

$$\text{Conversion efficiency } \eta = \frac{P_{max}}{P_{in}} = \frac{P_{total}}{\text{input radiation} \times \text{active area of the panel}}$$

$$\text{Active area of the panel} = \frac{P_{total}}{\text{input radiation} \times \eta} = \frac{12}{100 \times 10^{-3} \times 0.12} = 1000 \text{ cm}^2$$

Number of cells = active area of panel/active are of each cell = 1000/4 = 250.

$$\eta = \frac{P_{max}}{P_{in}} = \frac{V \times I}{\text{input radiation} \times \text{active area of the panel}}$$

$$\text{Current capacity } I = \frac{\eta \times \text{input radiation} \times \text{active area of the panel}}{V}$$

$$= \frac{0.12 \times 100 \times 10^{-3} \times 4}{0.5} = 96 \text{ mA}$$

REVIEW QUESTIONS

6.1 Outline the reasons for the adoption of the materials for the devices used for photo detection in optical fiber communications.

6.2 A *p-i-n* photodiode on average generates one electron-hole pair per two incident photons at a wavelength of 0.8 mm. Assume that all the electrons are collected, calculate (*a*) the quantum efficiency of the device (*b*) its maximum possible band gap energy of the device material and (c) the mean output photo current when the received optical power is 10^{-7} Watt.

6.3 Explain the detection process in a *p-n* photodiode. Compare the device with *p-i-n* photodiode.

6.4 A *p-n* photodiode has a quantum efficiency of 50 per cent at a wavelength of 0.85 mm. Calculate (*a*) its responsivity at 0.85 mm, (*b*) the received optical power if the mean photo current is 1mA and (*c*) the corresponding received photons at this wavelength.

6.5 A silicon RAPD has a multiplication power factor of 10^3 when operating at a wavelength of 0.82 mm. At the operating point the quantum efficiency of the device is 90 per cent and the dark current is 1 nA. Determine the number of photons per second at the wavelength of 0.82 mm required to register a light input corresponding to the output current after avalanche gain which is greater than the level of the dark current (*i.e.*, $I >$ 1nA).

6.6 Draw the diagram and the operation of a *p-i-n* diode. Give the diagrams for three practical *p-i-n* diodes show that the detector current.

$$I_p = \frac{ep_o}{\eta f}[1 - \exp(-\alpha(\lambda)W)](1 - R_f)$$

where R_f is the reflectivity, W is the width of the depletion layer. η = efficiency, P_o = optical power.

6.7 Define the quantum efficiency h of the photodiode and show that

$$h = \frac{I_p}{P_o}\frac{hf}{e}[1 - \exp\{(-\alpha(W))\}](1 - R_f)$$

Define responsivity R and show that

$$R = \eta\frac{e}{hf} = \frac{e}{hf}(1 - R_f)[1 - \exp\{(-\alpha(\lambda))W\}].$$

6.8 Describe with relevant diagram the operation of a silicon RAPD describe how it differs from a *p-i-n* photodiode. Outline the advantages and drawbacks with the use of RAPD as detector in optical fiber communications.

6.9 A photodiode has a quantum efficiency of 70% when photons of energy 2.3×10^{-9} Joule are incident on it (*a*) Calculate the operating wavelength of the photodiode and (*b*) Calculate the incident optical power required to obtain a photo current of 2 mA. When the diode is operating in that condition.

6.10 Describe with necessary diagram the working principle of a phototransistor and give an expression for the gain G.

The electron transit time for an InGaAs photo conductive detector is 8 picoseconds. Calculate the maximum 3 dB bandwidth when the photo conductive gain of the device is 60.

6.11 Describe the different types of noise encountered in a photo detector. A germanium *p-i-n* photodiode with an active area dimensions of 120 × 60 mm has a quantum efficiency of 50 per cent when operating at a wavelength of 1.3 mm. The measured current is 8 nA. Calculate the noise equivalent power (NEP) and specific directivity of the device D^*. Assume that the dark current is the main source of noise.

6.12 Determine the relative power loss due to series connection of two solar cells with their peak voltages and currents of 0.21 V, 0.31 V and 45mA, 55m A respectively. The peak values are 0.9 V and 55 mA, when connceted in series.

6.13 The voltage and current output of a solar cell is 0.4 V and 90 mA respectively with a conversion efficiency of 25%. Determine the area required by the solar panel so that a power of 4 W can be produced on the earth assuming the solar constant of the earth as 100m W/cm^2.

6.14 Calculate the fill-factor of a solar cell having an input power of 800m W, open-circuit voltage of 500m V, short-circuit current density of 30m A/cm^2, area of 5 cm^2 and conversion efficiency of 10%

CHAPTER 7

Manufacture of Fibers and Cables

CHAPTER FOCUS

- *Introduction: Fiber production method*
- *Vapour deposition method*
- *Vapour axial deposition method (VAD)*
- *Improved drying method*
- *Preform production scheme*
- *Modified chemical vapour deposition (MCVD)*
- *Pulling machine for fiber drawing*
- *Control mechanism for fiber diameter*
- *Polymer coating on fiber*
- *Plasma activated chemical vapour deposition (PCVD)*
- *Improved PCVD technique*
- *Optical fiber cables*
- *Typical structure of multimode and singlemode fibers*
- *Single mode fibers*
- *Plastic clad silica fibers*
- *All plastic fiber*
- *The Splices and connectors*
- *Mechanical misalignments*
- *Power loss in joining technique*
- *Fiber end face preparation*
- *The Score and break technique*

- *Splicing technique*
- *V-groove splice technique*
- *Loose tube splice*
- *Elastic tube splice*
- *Optical fiber connectors*
- *The precision-bore in a jewel bearing system*
- *Power launching and coupling*
- *Review Questions*

7.1 INTRODUCTION: FIBER PRODUCTION METHOD

The techniques used to prepare highly transparent glass materials for the optical fiber may be divided into two broad categories:

These are (1) Crucible or liquid Phase Method and (2) Vapour deposition Method.

In both cases the success of achieving low attenuation requires that there should be strict control over the purity of the starting materials and the contamination of any type must be avoided during the whole manufacturing processes. Out of the vapour deposition methods only one (Vapour axial deposition) can readily be used for the continuous production of fibers, but it is not normally done. Out of the crucible methods the double crucible arrangement (Fig. 7.1) has become common in the production of inexpensive fiber by a continuous process.

Crucible or Liquid Phase Method

(a) Furnace Heating: Figure 7.1 shows the double crucible arrangement for the production of inexpensive fibers using glass with low melting points by a continuous process. The highly purified powdered components of glass are heated together in a platinum or silica crucible. When an electric furnace is used for heating, the glass components are heated by radiation from the furnace walls which themselves have to be maintained free from contaminants.

In the case of double crucible method, the starting material may be fed in powdered form or by means of high purity preformed rods. The double crucible is mounted inside a vertical silica-lined muffled furnace, which is capable of raising the temperature of the melt to 1000 to 1200°C. An inert gas atmosphere must be maintained inside the furnace. When a dopant such as thallium with a high rate of diffusion in silica is used to maintain the refractive index difference of the core and cladding, an index grading at the interface takes place during the pulling process. The refractive index difference is maintained by varying the concentration of different components such as SiO_2, B_2O_3, Na_2O and CaO and the diffusion across the core-cladding boundary within the melt produces sufficient grading so as to reduce the multipath dispersion to the range between 1 and 5 ns/km.

(b) Radio Frequency Heating: Sometime radio frequency heating method may also be used as an alternative to the furnace heating. With a metal crucible, the *rf* coupling may be made to the crucible and heat transfer occurs by conduction. But in order to make use of the silica crucible, direct coupling to the melt has to be obtained so the powdered components have to be preheated and a higher radio frequency is used. The melt is then at the higher temperature than the crucible and so the melt is less

susceptible to contamination from it. Silica crucibles cannot normally be used more than once, since they do not withstand thermal cycling. Traditionally a rod of core glass and a tube of cladding glass are produced. The rod is assembled inside the tube and the two are drawn down together to produce the clad fiber.

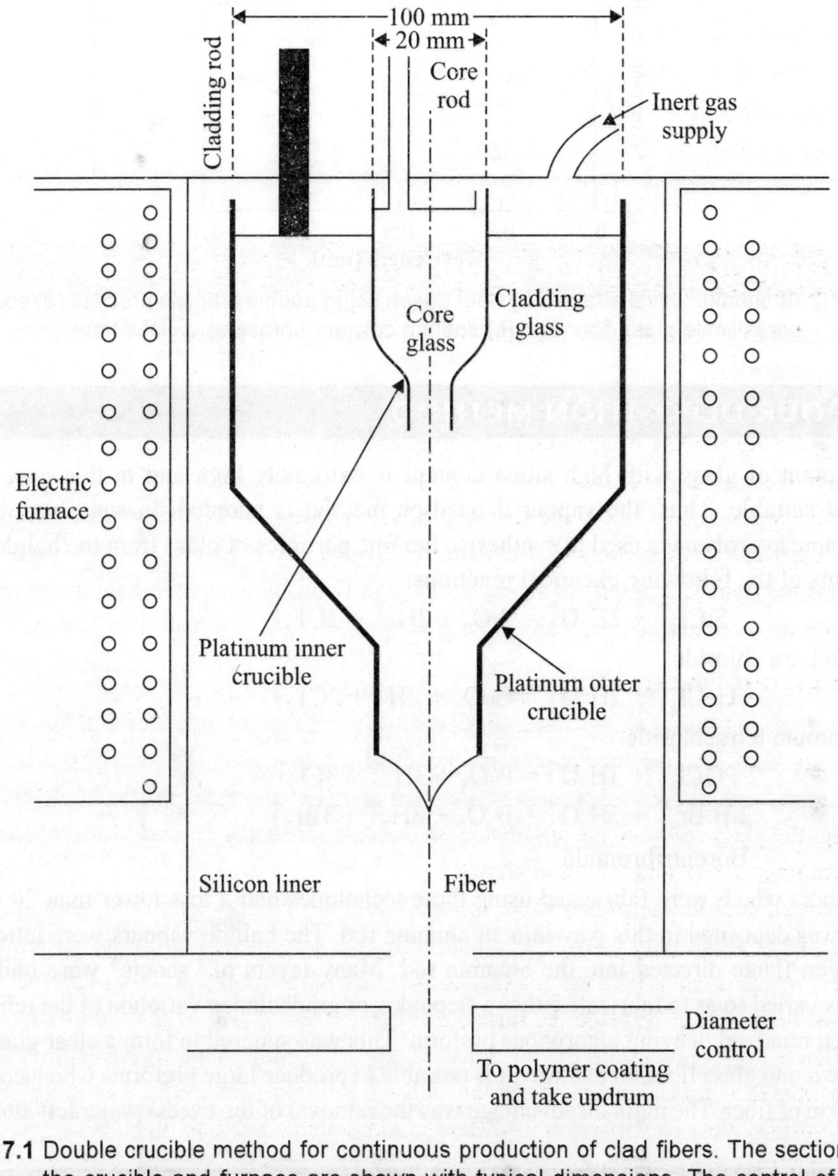

Fig. 7.1 Double crucible method for continuous production of clad fibers. The sections of the crucible and furnace are shown with typical dimensions. The control of core and cladding diameters is maintained by fiber pulling speed and the head of the molten glass in each crucible. With a suitable choice of material and control of the melt temperature at the nozzle, index grading can be obtained by ionic defusion

The levels of attenuation that is achieved is shown in Fig. 7.2.

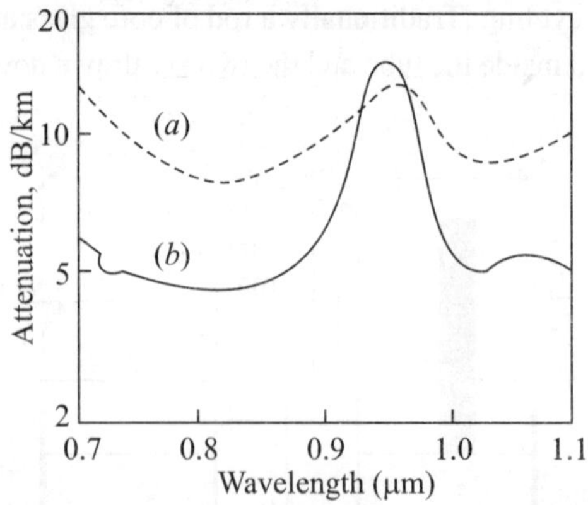

Fig. 7.2 Attenuation level attained by fiber drawn using double crucible process (*a*) sodium borosilicate glass fiber and (*b*) sodium calcium borosilicate glass fiber

7.2 VAPOUR DEPOSITION METHOD

The melting point of glass with high silica content is extremely high and in that case the crucible method is not suitable. Thus, the vapour deposition method is adopted. In some vapour deposition method the flame hydrolysis is used to synthesise the fine particles of glass from the halides vapours of the constituents of the following chemical reactions:

$$SiCl_4\uparrow + 2H_2O\uparrow = SiO_2 + 2H_2\uparrow + 2Cl_2\uparrow$$

Silicon tetra chloride

$$GeCl_4\uparrow + 2H_2O\uparrow = GeO_2 + 2H\uparrow + 2Cl_2\uparrow$$

Germanium tetrachloride

$$2\ POCl_3\uparrow + 3H_2O\uparrow = P_2O_5 + 3H_2\uparrow + 3Cl_2\uparrow$$

$$2\ B\ Br_3\uparrow + 3H_2O\uparrow = B_2O_3 + 3H_2\uparrow + 3Br_2\uparrow$$

Borontribromide

The fibers which were fabricated using these techniques had a loss lower than 20 dB/km and a glass "soot" was deposited in this way into an alumina rod. The hallide vapours were introduced into a methane-oxygen flame directed into the alumina rod. Many layers of "shoots" were built up and the composition is varied so as to fabricate either a stepindex or gradedindex variation of the refractive index. The rod is then removed, leaving amorphous preform. This was sintered to form a clear glass and this, in turn, was drawn into glass fiber. In this way it is possible to produce large preforms which could be drawn into 40 to 50 km of fiber. The main disadvantage was the removal of the excess water left after hydrolysis.

7.3 VAPOUR AXIAL DEPOSITION METHOD (VAD)

Large preform can be prepared by vapour axial deposition (VAD) method which is illustrated in Fig. 7.3. Both core and cladding glasses are deposited simultaneously on the end of the seed rod. The seed rod is rotated to ensure the azimuthal homogeneity and is drawn up in an electric furnace at about 2.5 mm per minute. There it is heated to about 1500°C in an atmosphere of oxygen and thionyl chloride ($SOCl_2$) vapour. This serves two important functions:

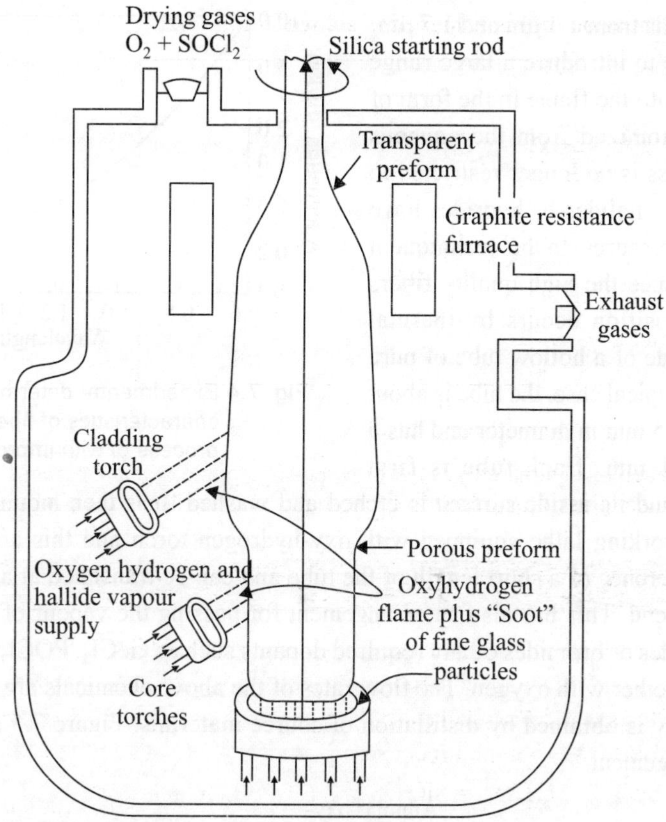

Fig. 7.3 A schematic diagram of preform manufacture using vapour phase axial deposition (VAD) process

(*a*) Any water vapour present is removed by the following chemical reactions:

$$SOCl_2\uparrow + H_2O \rightarrow SO_2\uparrow + Cl_2\uparrow + H_2\uparrow$$

$$2SOCl_2 + 2OH^- \rightarrow 2SO_2\uparrow + 2Cl_2\uparrow + H_2\uparrow$$

(*b*) At the same time the rod which is porous and initially may be about 60 mm in diameter and 200 mm in length, is consolidated into a transparent glassy preform of some 20 mm diameter.

By carefully maintaining the temperature across the porous preform during the time of deposition, it is possible to maintain a reasonably well controlled index grading with germanium doped core. The concentration of germanium in the synthesized glass particle is found to increase steadily with the deposition temperature over the range of 300 to 800°C. Long graded index fibers having attenuation less than 0.5 dB/km at the wavelength of 1.55 μm and a dispersion less than 1 ns/km have been fabricated by this process.

7.4 IMPROVED DRYING METHOD

Recently improved drying methods have been adopted and by this method it has been possible to produce fibers with water vapour contents less than 1 part in 10^9 parts. Figure 7.4 shows a typical attenuation curve with wavelength for such a fiber, and the OH^- absorption peak at the wavelength of 1.39 μm is the commonly one which is detectable. The range of wavelength over which the attenuation remains below 1 dB/km extends from 1.1 μm and 1.7 μm. It have been possible to introduce a large range of doping materials into the flame in the form of a "mist" which is atomized from the aqueous solutions. This process is no longer restricted to those materials whose halides or hydrides have the suitable vapour pressures. In this technique it was possible to produce the high quality fibers and the vapour deposition occurs by thermal oxidation on the inside of a hollow tube of pure synthetic silica. In a typical case, the tube is about 1 metre long, about 15 mm in diameter and has a wall thickness of 1 mm. Each tube is first thoroughly cleaned and its inside surface is etched and washed. It is then mounted horizontally and rotated on a glass working lathe equipped with oxy-hydrogen torch and this arrangement heats the whole of the circumference of a short length of the tube and can be translated at a controlled rate from one end to the other end. This has also an arrangement for passing the vapour of silicon tetrachloride ($SiCl_3$) and of chlorides or bromides of any required dopants such as $GeCl_4$, $POCl_3$, and $B\,Br_3$ along the inside of the tube together with oxygen. The flow rates of the above chemicals are carefully controlled, and the highest purity is obtained by distillation of source materials. Figure 7.5 shows the schematic diagram of the arrangement.

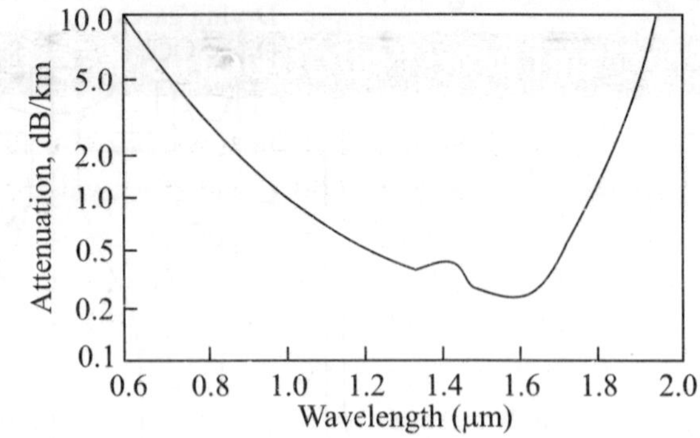

Fig. 7.4 Experimental determination of attenuation characteristics of fibers prepared by (VAD) process of with improved drying

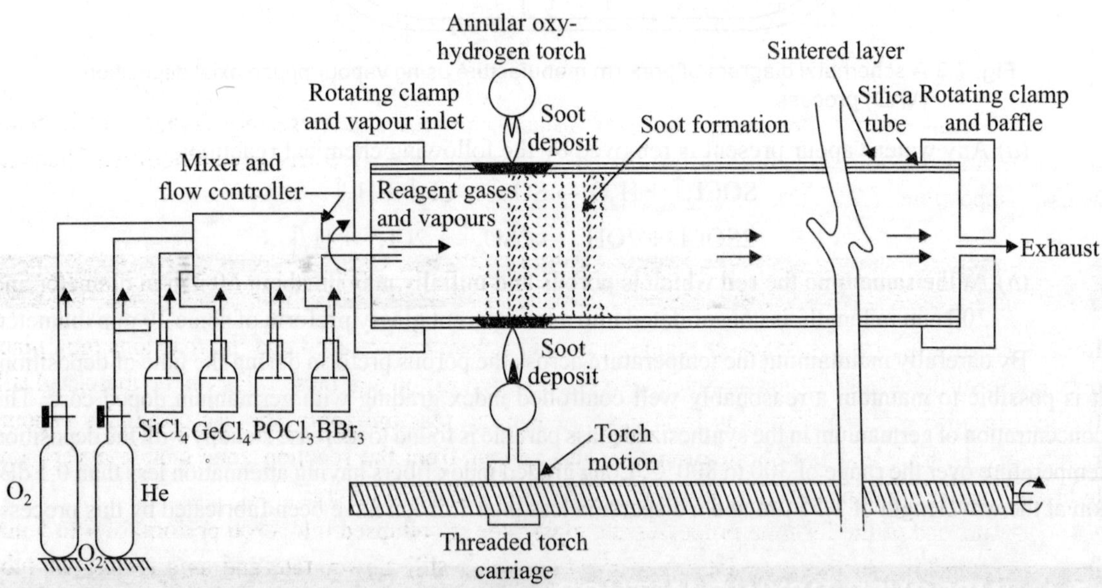

Fig. 7.5 A schematic diagram of preform manufacture using internal chemical vapour deposition. The oxy-hydrogen torch mounted on a carriage is moved along the tube at a speed which is servocontrolled to maintain a desired temperature at a heated band

7.5 PREFORM PRODUCTION SCHEME

The normal production scheme starts with several passes of the flame with only oxygen flowing through. This heats the tube up to 1500°C and effectively polishes the inside surface. Then a number of passes are made with oxygen and the vapour of dopant used in cladding such as $SiCl_4$ and $B\ Br_3$ flowing. These are followed by more passes of flame with core dopants such as silicon tetrachloride $SiCl_4$ and germanium tetrachloride passing through. When the preforms for graded index fibers are being made, some 50 to 100 such passes may be made in forming the core with concentration of the core dopants are being gradually increased from pass to pass. Each of these processes have taken 4 to 5 minutes.

An interaction between the vapours and the oxygen occurs on the inside surface of silica in the heated band. The result is that a layer of silicon dioxide SiO_2 plus dopants (GeO_2, B_2O_3) are deposited and the liberated halogen gas is carried off in the vapour stream. The form of the deposit may be a "soot" or more or less a glassy layer depending on the wall temperature. If sooty layers are formed and a heat treatment is subsequently needed to fuse them into glassy layers of varying composition. The total layer thickness varies from 200 to 300 μm.

Typical chemical reactions are:

$$\left.\begin{array}{lll} SiCl_4\uparrow + O_2\uparrow \rightarrow & SiO_2 + & 2Cl_2\uparrow \\ GeCl_4\uparrow + O_2\uparrow \rightarrow & GeO_2 + & 2Cl_2\uparrow \\ 4POCl_3\uparrow + 3O_2\uparrow \rightarrow & 2P_2O_5 + & 6Cl_2\uparrow \end{array}\right] \text{Core}$$

$$\left.\begin{array}{lll} SiCl_4\uparrow + O_2\uparrow \rightarrow & SiO_2 + & 2Cl_2\uparrow \\ 4B\ Br_3\uparrow + 3O_2\uparrow \rightarrow & 2B_2O_3 + & 6Br_2\uparrow \end{array}\right] \text{Cladding}$$

The temperature and vapour pressure are usually chosen so that these above reactions take place in vapour phase in the volume of the tube. The technique used here is referred to as "Modified Chemical Vapour Deposition" (MCVD).

7.6 MODIFIED CHEMICAL VAPOUR DEPOSITION (MCVD)

In the case of CVD processes lower temperature and pressure were used and the reactions took place only at the surface. MCVD allows much faster rate of deposition and these rates can be increased still further provided a microwave plasma is generated in the reaction zone. The use of halides vapours rather than hydrate and the elimination of water vapour from the reaction zone enable a very low hydroxyl (OH^-) ion concentration to be achieved in the deposited layers.

At the end of the forming processes the glass tube is collapsed into a rod preform of 4 to 5 mm in diameter making further translations of the flame at a still slower rate and as a result, the tube temperatures is raised to around 1770°C. The torch speed can be controlled by means of a servo which notes down the tube temperature pyrometrically. The final pass during which the rod is formed may take as much as thirty minutes. In this pass a great care is necessary, since in this step the circular circumference of the preform and the concentric location of the core and the cladding are determined.

7.7 PULLING MACHINE FOR FIBER DRAWING

Figure 7.6 shows the schematic diagram of a pulling machine which is used for drawing fiber and the continuous production methods this is integrated with the rest of the process. But when the preforms are made the pulling machine may be separated. The temperature of the hot zone has to be maintained at 1900° to 2000°C. This temperatures are generally obtained from a small, Zirconia-lined, electric furnace but other methods including one which uses carbon dioxide laser radiation to heat the rod have been found to be successful.

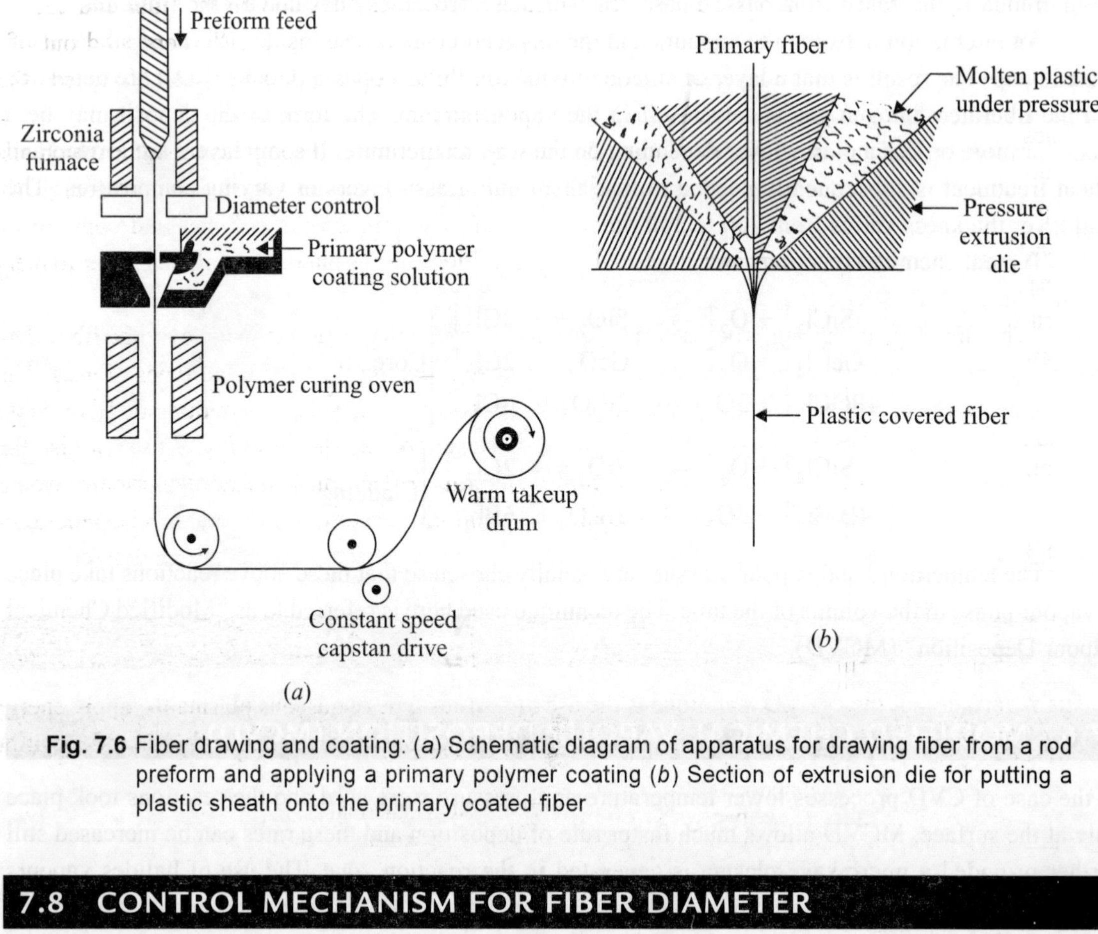

Fig. 7.6 Fiber drawing and coating: (*a*) Schematic diagram of apparatus for drawing fiber from a rod preform and applying a primary polymer coating (*b*) Section of extrusion die for putting a plastic sheath onto the primary coated fiber

7.8 CONTROL MECHANISM FOR FIBER DIAMETER

It is essential to maintain the outside diameter of the fiber to sub-micron tolerance. Because when fibers are joined together, it is the outside surfaces which are aligned. Any variation in the outside diameter will result in the off-set of the cores of the fibers. The most common system to monitor the diameter of the drawn fiber consists of a laser beam that illuminates the fiber and a photodetector placed within so that the diffraction pattern is produced. This changes with any variation of the fiber diameter and causes the current in the photodiode to change. This, in turn, acts as a signal that controls the servomechanisms determining the winding rate of the fiber and the rate at which the preform is fed into the puller. A stability better than 0.1 per cent can be obtained. The winding rate is best controlled by a precision capstan drive. Then the lengths of the fiber can be wound into rests without necessity of stopping and restarting the drawing process at each changeover.

7.9 POLYMER COATING ON FIBER

The important obstruction to the higher production rates is not the pulling process itself, but the subsequent polymer coating operation. It is essential to give the drawn fiber a primary coating of polymer immediately after it leaves the diameter monitoring cell. This step is taken to protect the fiber silica surface. Otherwise, even under the normal atmospheric conditions, surface microcracks develop on the fiber and cause a dramatic loss of fiber breaking strength. The primary polymer coating is usually deposited out of a solution as shown in Fig. 7.6(*a*). A layer of about 40 μm thickness of a polymer such "Sylgard" that adheres well to the fiber surface is found to stop the development of microcracks.

A further degeneration the fibers may face during its lifetime is the ingress of water which also weaken the fiber surface. In order to maintain fiber strength during service it is necessary to prevent the fiber from coming in contact with water which would diffuse through the polymer and come to the silica surface. Some groups have experimented with metallic coatings put on the polymer layer to make a water-tight seal.

The final step in pulling processes is to form a high strength plastic extrusion over the fiber. This protects against longitudinal and lateral mechanical forces acting on the fiber in all types of uses. This is true whether the fiber is used as a singlemode fiber or is to be made up into a multimode fiber cable. The pulled fiber may be taken straight into a plastic extrusion die as shown in Fig. 7.6(*b*) or else this process may become the part of a subsequent cabling operation. The finished fiber is normally wound on a warmed aluminium drum. When it is cooled to room temperature, this decrease in temperature shrinks and releases any tension on the fiber. Thereby the micro-bending effects are minimized.

7.10 PLASMA-ACTIVATED CHEMICAL VAPOUR DEPOSITION (PCVD)

A slight variation of the MCVD technique is the use of various type of gaseous plasma to supply energy for vapour phase oxidation of the halides. This method uses plasma induced chemical vapour deposition

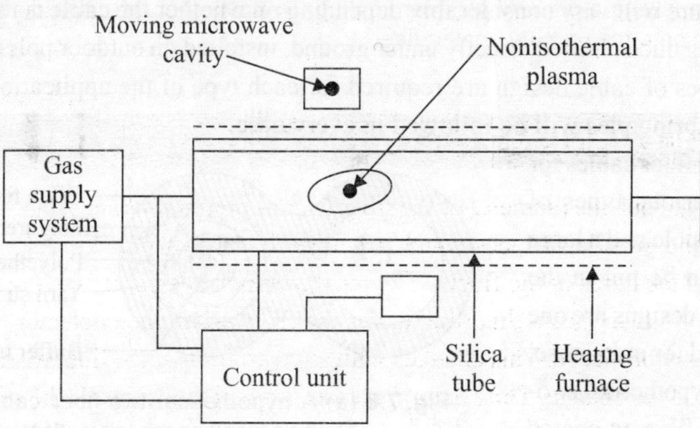

Fig. 7.7 Schematic diagram of the apparatus utilized in plasma activated vapour deposition process

inside a silica tube as shown in Fig. 7.7. The main difference between this technique and MCVD process is the stimulation of oxide formation by means of non-isothermal plasma maintained at low pressure in a microwave cavity at the frequency of 2.45 GHz and the cavity surrounds the tube. Volatile reactants are introduced into the tube where they react heterogeneously within the microwave cavity and no particulate matter is formed in the vapour phase.

The reaction zone is moved backwards and forwards along the tube by the control of the microwave and circularly symmetric layer growth is formed. The rotation of the tube is unnecessary and the efficiency of the deposition is 100 per cent. Film deposition can occur at a temperature as low as 500°C but a high chlorine content may cause expansivity and the cracking of the film. Hence the tube is heated to near about 1000°C during deposition using a stationary furnace.

The high deposition efficiency allows the composition of the layers to be accurately varied by the control of vapour-phase reactants. When the plasma zone is moved rapidly backwards and forwards along the tube, a very thin layer deposition may be achieved with the result of formation of up to 2000 individual layers. This enables a very good graded index profile to be obtained and it is approximately near the parabolic profile. Thus, low pulse dispersion of less than 0.8 ns/km for fibers with attenuations of between 3 and 5 dB/km at the wavelength of 0.85 μm have been reported.

7.11 IMPROVED PCVD TECHNIQUE

An improved PCVD technique uses an inductively coupled radio frequency argon plasma which is excited by the r.f. of 3.4 MHz. The deposition takes place in one atmospheric pressure and is predominantly homogeneous vapour phase reaction. The reaction causes a fusion of the deposited material in glass through the high temperature discharge. This technique has proved to have a reaction rate five times faster than the conventional MCVD process. But the attenuation in the fiber was somewhat higher with the losses 6 dB/km at a wavelength of 1.06 μm.

7.12 OPTICAL FIBER CABLES

For any practical use of the optical fiber waveguide, the fiber should be put inside some type of cable structure. The cable structure will vary considerably depending on whether the cable is to be pulled into underground, intra-building ducts, buried directly under ground, installed on outdoor poles or submerged under water. Different types of cable design are required for each type of the applications, but certain fundamental cable design principles will be followed in every case.

In designing optical fiber cables for different applications, various types of fiber arrangements are possible and a large variety of components can be put in the construction. The simplest designs are one or two fiber cables intended for indoor use. Figure 7.8(*a*) shows four hypothetical two-fiber design system. The fiber is coated with a buffer material and placed loosely

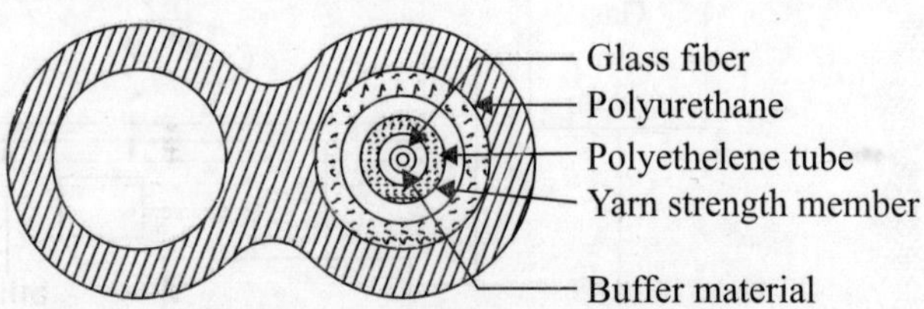

Fig. 7.8 (*a*) A hypothetical two-fiber cable design. The building block on the left is identical to that shown in the right band fiber

in tough oriented polymer tube such as polyethylene. For adding strength the tube is surrounded by strands of aramid yarn, which in turn, is encapsulated in a jacket of polyurethane. A final outer jacket of polyurethane, polyethylene or nylon binds the two encapsulated fiber units together.

When it is desired to fabricate larger cables, these are done by stranding several fiber building blocks around a central strength member. Figure 7.8(*b*) shows the arrangement for a cable with six fibers. The fiber units are bound to the strength member with plastic or paper-binding tapes and then surrounded by an outer jacket. When the repeaters are required along the route where the cable is to be installed, it is advantageous to put copper wires within the cable structure and the copper wire is used to boost the power of the signal in the repeaters. The copper wires are used for fault isolation or as an engineering order wire for voice communications during the cable installation.

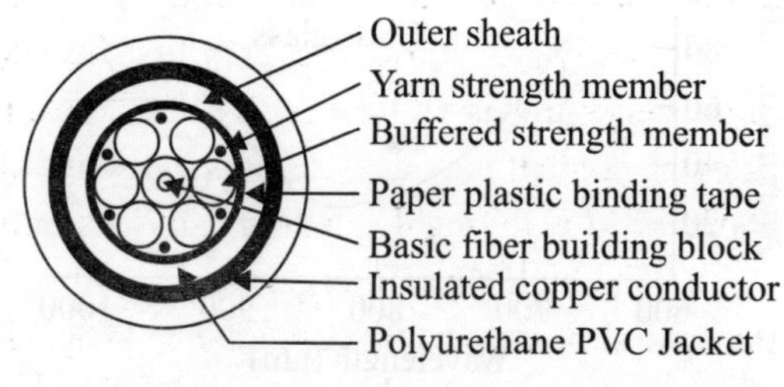

Fig. 7.8 (*b*) Arrangement for a cable with six fibers

7.13 TYPICAL STRUCTURE OF MULTIMODE AND SINGLEMODE FIBERS

Multimode step-index fibers may be prepared from either multi-component glass compounds or doped silica. The fibers can have considerably large core diameters and the large value of numerical aperture to facilitate efficient coupling to the incoherent light sources such as light emitting diodes (LEDs). Multicomponent glass and doped silica fibers are usually referred to as multicomponent glass/glass (or glass clad glass) and silica/silica (or silica clad silica) respectively. Figure 7.9 shows the structure of a typical glass multimode stepindex fiber.

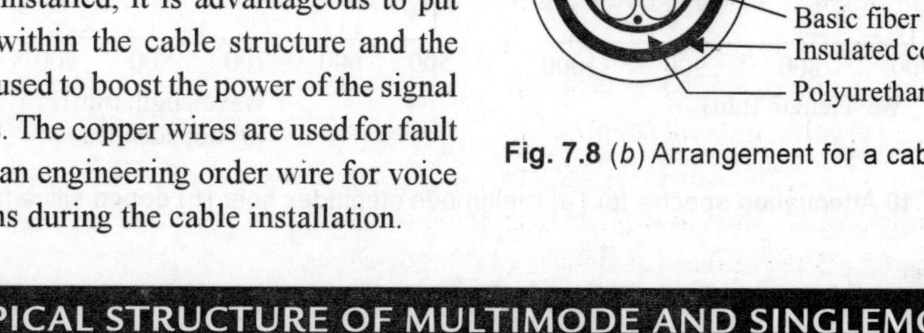

Fig. 7.9 Typical structure of a glass multimode stepindex fiber

Structure

Core diameter	:	50 to 400 μm
Cladding diameter	:	125 to 500 μm
Buffer Jacket diameter	:	250 to 1000 μm
Numerical aperture	:	0.16 to 0.5

Bandwidth 6 to 50 MHz. km.

Attenuation : 2.6 to 50 dB/km at a wavelength of 0.85 μm.

Applications : This type of fibers are suitable for short haul, limited bandwidth and low cost applications.

Figure 7.10(*a*) shows the attenuation curves for multicomponent glass fiber and 7.10(*b*) doped silica fiber (or silica clad silica).

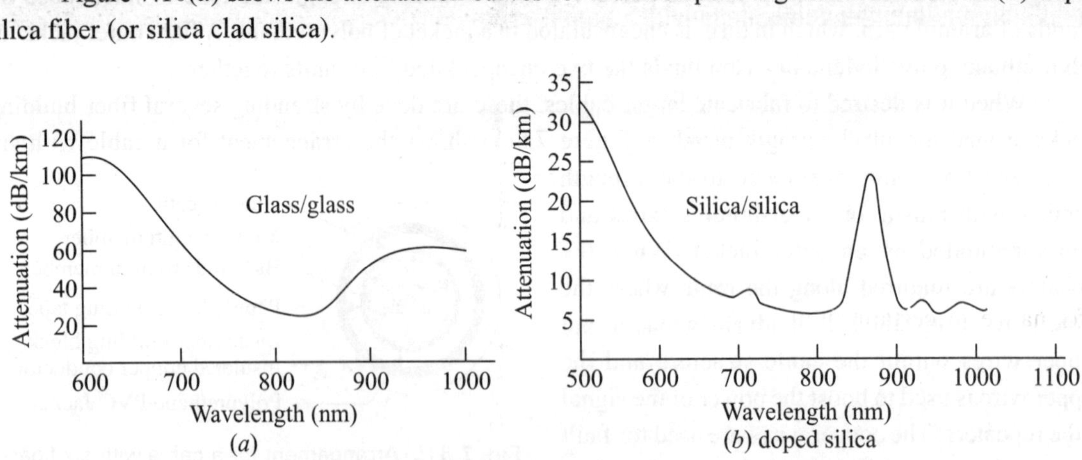

Fig. 7.10 Attenuation spectra for (*a*) multimode stepindex fiber (*b*) doped silica fiber

7.14 SINGLE MODE FIBERS

At present commercially available singlemode fibers are usually stepindex fibers. In order to reduce attenuation they are generally prepared from doped silica (silica/clad silica). These fibers are of high quality, suitable for wide band, long haul transmission.

Although a singlemode fiber has a small core diameter to allow singlemode propagation, the cladding diameter must be at least ten times the core diameter so that optical loss from the evanscent field is avoided. In order to provide protection strength by buffer jacket, a singlemode fiber must have a similar overall diameter as a multimode fiber. A typical example of a singlemode stepindex fiber is shown in Fig. 7.11.

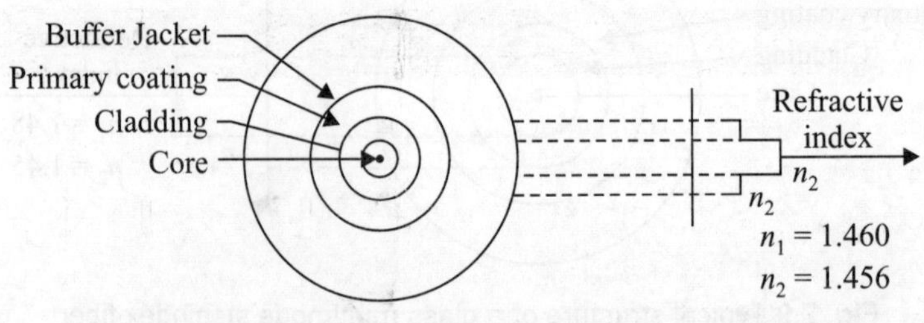

Fig. 7.11 Typical structure for a silica singlemode stepindex fiber

Structure

Core diameter	:	5 to 10 μm typically around 8.5 μm
Cladding diameter	:	Generally around 125 μm
Buffer Jacket	:	250 to 1000 μm
Numerical aperture	:	0.08 to 0.15 usually around 0.10
Bandwidth	:	Greater than 50 MHz km.

These types of fibers are suitable for high bandwidth, long haul applications using singlemode injection laser sources.

7.15 PLASTIC CLAD SILICA FIBERS

The plastic fibers are multimode ones having either a stepindex or graded index profile. They usually have a plastic cladding (often a silicons rubber) and a glass core which is mostly silica (Plastic Code Silica (PCS). The PCS fibers exhibit lower radiation losses than the silica clad silica fibers and have more satisfactory performance in certain environments. The plastic clad fibers are comparatively cheaper. Figure 7.12 shows the typical structure of a common type of a stepindex plastic clad fiber.

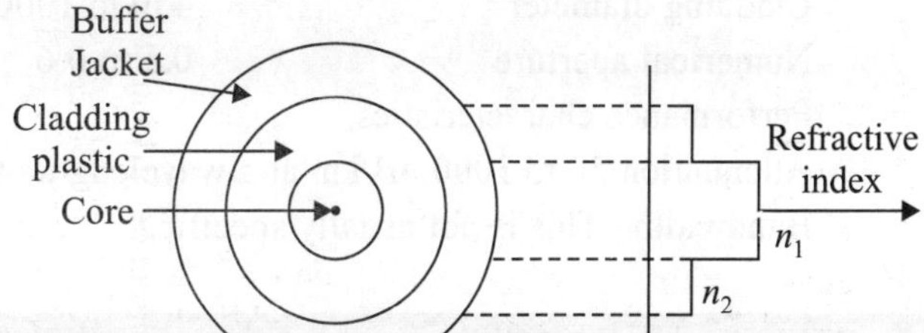

Fig. 7.12 Typical structure for plastic clad silica multimode fiber

Structure

Core diameter	:	Stepindex 100 to 500 μm Graded index 50 to 100 μm
Cladding diameter	:	Stepindex 300 to 800 μm Graded index 125 to 150 μm
Buffer Jacket diameter	:	Stepindex 500 to 1000 μm Graded index 250 to 1000 μm
Numerical aperture	:	Stepindex 0.2 to 0.5 Graded index 0.2 to 0.3
Performance Characteristics	:	
Attenuation	:	Stepindex : 5 to 50 dB/km Graded index : 4 to 15 dB/km.

7.16 ALL PLASTIC FIBER

All plastic (polymeric) fibers are of multimode stepindex type with large core and cladding diameters. Thus, they do not require a buffer jacket for the protection and strengthening of the fiber. They are usually cheaper and easier to handle than the corresponding silica based glass variety. Their performance is restricted especially for optical transmission in infrared and they have limited use in communications. All plastic fibers have large values of numerical aperture and allow easier coupling of light into the fiber from a multimode sources.

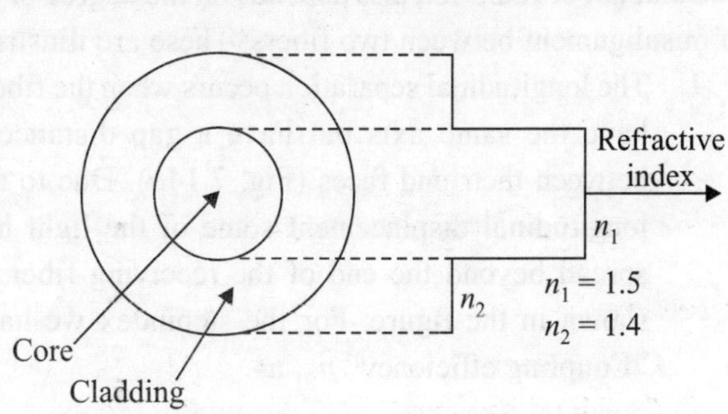

Fig. 7.13 A typical all plastic fiber structure

A typical all plastic fiber structure is shown in Fig. 7.13.

Structure

Core diameter	:	200 to 600 μm
Cladding diameter	:	450 to 1000 μm
Numerical aperture	:	0.5 to 0.6
Performance characteristics	:	

Attenuation 50 to 1000 dB/km at a wavelength of 0.65 μm

Bandwidth : This is not usually specified.

7.17 SPLICES AND CONNECTORS

The most important factor in the installation of any fiber optic system is the interconnection of fibers with a minimum possible power loss. The interconnections are needed at the optical sources in the transmitter, at the photodetector, in the receiver and at intermediate points within a cable where two fibers are joined together. The particular technique for joining two fibers depends on whether a permanent bond or easily remountable connection is desired.

The permanent bond is known as a "Splice" whereas the demountable joint is referred to as a "Connector". Every joining is subjected to certain conditions which become responsible for causing various types of optical power loss. The amount of loss depends on different parameters such as the input power distribution at the joint, the geometrical and waveguide characteristics of two fibers which are coupled, various types of misalignments between the fiber ends and joints and the qualities of the fiber end faces.

The optical power can be coupled from one fiber to another which is controlled by number of modes that can propagate in each fiber. For example, a fiber in which 500 modes can propagate is connected to another fiber in which 400 modes can propagate, and then at best 80 per cent of the optical power from the first fiber can be coupled to the second one provided all the modes are equally excited.

7.18 MECHANICAL MISALIGNMENTS

The amount of radiation loss depends on the degree of misalignment. There are three fundamental types of misalignment between two fibers. These are illustrated in Fig. 7.14.

1. The longitudinal separation occurs when the fibers have the same axis but have a gap distance S between their end faces (Fig. 7.14*a*). Due to the longitudinal displacement some of the light has spread beyond the end of the receiving fiber as shown in the figure. For the stepindex we have "Coupling efficiency" n_{S1} as

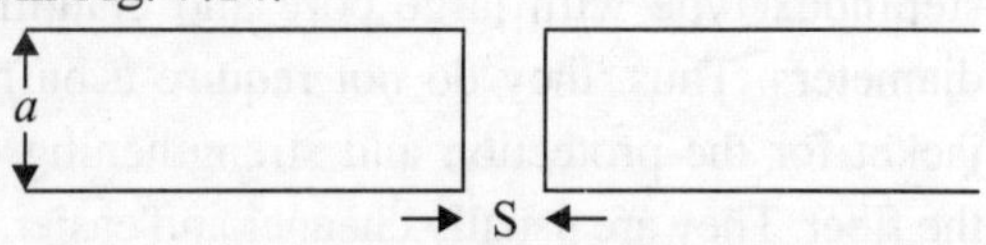

Fig. 7.14 (*a*) The longitudinal separation occurs when the fibers have the same axis but have a gap distance "*S*" between their ends

$$n_{S1} = \left(\frac{1}{1 + S/a \tan \alpha_m} \right)^2 \tag{7.18.1}$$

where S is the separation distance of the fiber ends and α_m is the critical acceptance angle $= (\sin^{-1} \text{NA})$ and a is the radius of the core.

2. Lateral displacement which is the axial displacement occurs when the axis of the two fibers are separated by a distance d as shown in Fig. 7.14(*b*). Mechanical misalignments of the fibers cause losses because the areas of the fiber core do not overlap perfectly. In the analysis of the misalignments, the usual assumptions are that the fibers have equal radii, index profile and numerical aperture to isolate the effect of the misalignment. The fiber is assumed to have uniform power distribution across its cross sectional area. For a stepindex fiber the coupling efficiency depends on the lateral misalignment distance d as

$$\eta_{S1} = \frac{2}{\pi}\cos^{-1}\left(\frac{d}{a}\right) - \frac{d}{\pi a}\sqrt{1-\left(\frac{d}{2a}\right)^2} \tag{7.18.2}$$

The equation is derived from a calculation of the overlapping areas of two equal circles with centre separated by distance d.

3. Angular misalignment is encountered when the two fiber axes form an angle θ so that the fibers are no longer parallel to each other (Fig. 7.14*c*). The common misalignment encountered in practice is the axial displacement and it is responsible for the greatest power loss. The axial offset reduces the overlap area of the two fiber end faces and consequently reduces the amount of optical power than can be coupled from one end fiber to the other one.

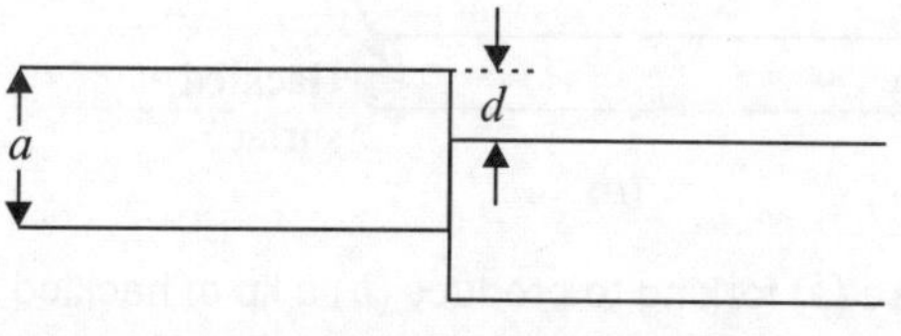

Fig. 7.14 (*b*) The lateral displacement occurs when the axis of the two fibers are separated by distance *"d"* and this misalignment causes losses because the core areas of the two fibers are not overlapping completely

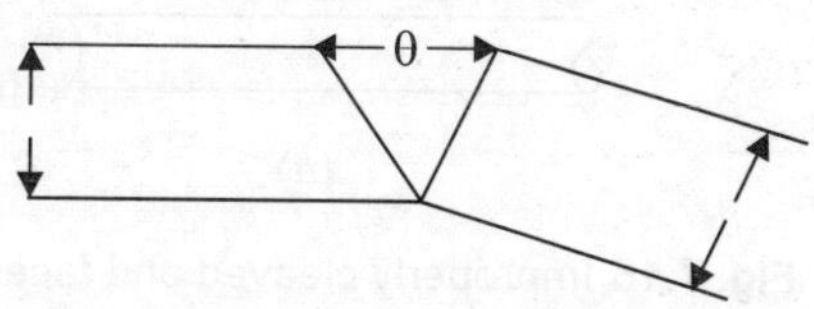

Fig. 7.14 (*c*) In angular misalignment the fiber axis form an angle θ so that the fibers are no longer parallel to each other

4. Core size Mismatch: If the core size of the outgoing fiber is smaller than that of the incoming core, then only a part of the light from the incoming core will reach the outgoing fiber and the remaining part will be lost adding total power loss of the fiber (Fig. 7.14*d*) which shows a cone of light escaping around the outgoing core.

7.19 POWER LOSS IN JOINING TECHNIQUE

Each joining technique is subjected to certain conditions which cause different amount of optical power loss at each joint. These losses depend on parameters such as:

1. The geometrical characteristics of the fiber: For example, if the emitting fiber has a larger core diameter than the receiving fiber, then the optical power will be lost because the end face area mismatches (Fig. 7.14*d*).

2. The waveguide characteristics of the fiber: For example if an emitting fiber has larger numerical aperture than the receiving fiber, all optical power falling outside the accepting core of the receiving fiber will be lost.

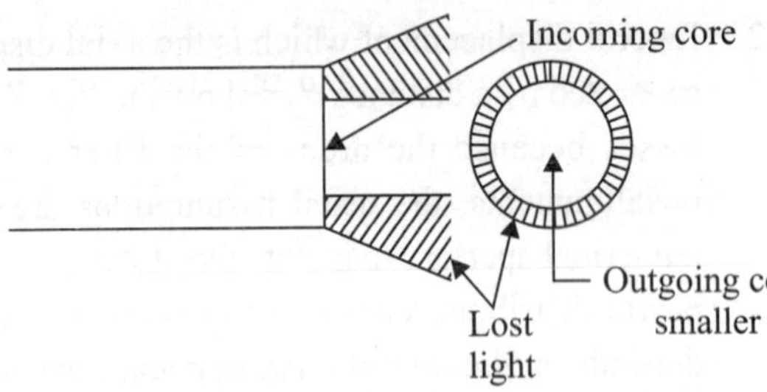

Fig. 7.14 (*d*) In core mismatch the core size of the outgoing is smaller than that of the emitting fiber a cone of light is escaping from the outgoing core

3. Fiber end-face quality: One criterion for low loss joints is that the fiber end faces must be clean, smooth and perpendicular to the axis. End preparation techniques include sawing, grinding polishing and control fracture. Figure 7.15(*a*) shows a fiber which was cleaved at an angle to the axis such that when light from the matching sloping surface, some portion of light will be lost by reflection and more by coupling into the leaky modes. Figures 7.15(*b*) shows a fiber which was not cleaved properly due to improper application of cleaving stress. The end of the fiber has a scalloped irregular surface which scatters light. These defects arise due to improper fiber end preparation.

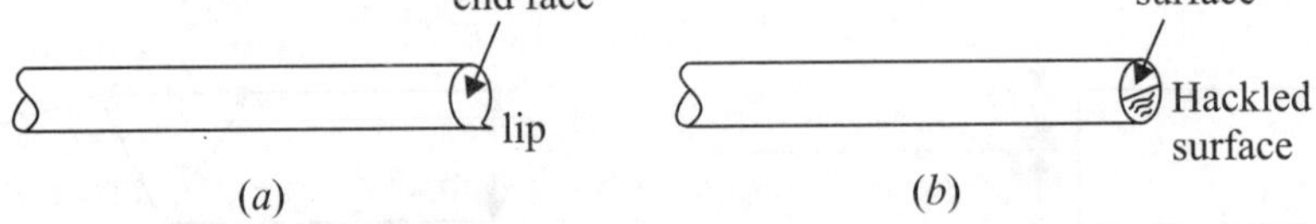

Fig. 7.15 Improperly cleaved and faces can cause (*a*) forking to produce (*b*) a lip of hackled surface at the end of the fiber

4. Power loss due to dirt: Any dirt or foreign substance which gets into a connector or splice during or after the assembly may increase its losses or even completely block it. Extreme care should be taken during installation to prevent the inclusion of dirt.

7.20 FIBER END FACE PREPARATION

It is essential to properly prepare the end faces of a fiber before the fibers are connected together or spliced with each other. In order to eliminate the deflection or scattering of light at the joint, the fiber end must be flat, smooth and perpendicular to the fiber axis. The commonly followed the end preparation technique consists of sawing, grinding, polishing and controlled fracture. Conventional grinding and polishing techniques can produce a smooth surface which is perpendicular to the fiber axis. However, this method is quite time-consuming and requires a good amount of technical skill of the operator.

7.21 THE SCORE AND BREAK TECHNIQUE (CONTROLLED FRACTURE TECHNIQUE)

This technique uses traditional glass cutting methods to get a uniform surface. A fiber is placed under a mild tension as shown in Fig. 7.16 and scratched with a scoring blade. The scratch produces a stress concentration at the tip of the scratch. The tension is then increased and the crack tip propagates across the fiber. If the curvature of the fiber and the application of tension are carefully controlled, then the crack will propagate perpendicular to the fiber axis and produce a clean smooth break. Improper control of the crack propagation can cause a formation of fork into many cracks shown in Fig. 7.15(*a*). The forking produces either a lip or a hackled portion at the fiber end as shown in Fig. 7.15(*b*).

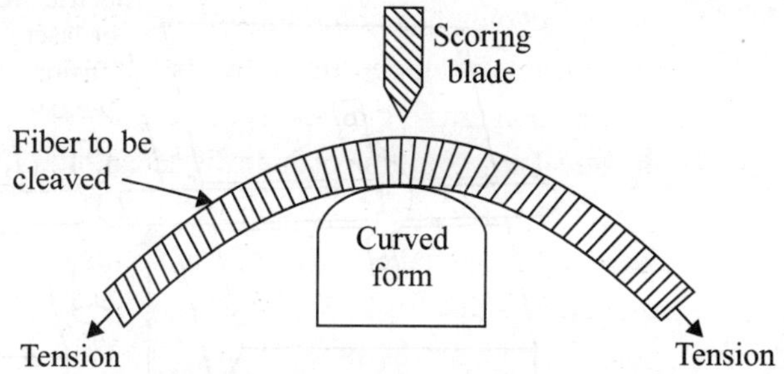

Fig. 7.16 A fiber placed under a mild tension and scratched with a scoring blade

Example 7.1: A silicon fiber has a core radius of 25 mm and the longitudinal displacement $S = 2.5$ mm and the numerical aperture NA = 0.2.

Calculate the coupling efficiency n_{S1} (Fig. 7.14(*a*) and Eq. (7.18.1)

The critical angle is found by

$$\alpha_m = \sin^{-1}(\text{NA}) = \sin^{-1}(0.2) = 11.54°$$

The coupling efficiency is found as

$$n_{S1} = \left(\frac{a}{a + S\tan\alpha_m}\right)^2 = \left(\frac{25\times10^{-6}}{25\times10^{-6} + 2.5\times10^{-6}\tan 11.54}\right)^2$$

$$= (0.980)^2 = 0.96 = 96\%.$$

7.22 SPLICING TECHNIQUE

There are different methods of splicing the fibers. The popular methods are:

1. Fusion splice method
2. V-groove Butt splice
3. Elastic tube splice and
4. Loose tube splice.

Fusion Splicing Technique

Fusion splice may be made by thermally bonding together prepared fiber ends as shown in Fig. 7.17(*a*). In this method the fiber ends are pre aligned and butt together. This is accomplished either in a grooved fiber holder or under a microscope with micro manupulators. The butt-joint is then heated with an electric arc or a laser pulse so that the fiber ends are momentarily melted and bonded together. This technique can produce an extremely small splice loss as low as 0.1 to 0.2 dB for identical fiber end. However, an extreme care must be taken in this technique. If it is not followed surface damages due to handling and surface defect growth during heating may develop. As a result of change in chemical composition can produce weak splice. The change in chemical composition arises from the melting of the material.

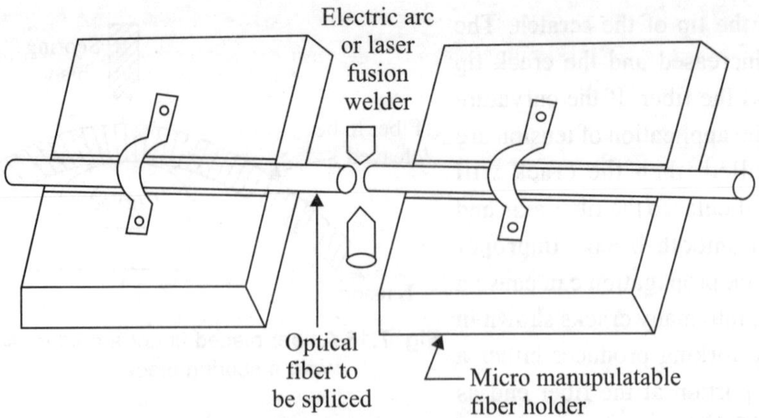

Fig. 7.17 (*a*) A fusion splice may be made by bonding together prepared fiber ends and the fiber ends are prealigned and butt together

7.23 V-GROOVE SPLICE TECHNIQUE

This method uses a V-shaped groove as an alignment aid to bring the two fibers into mechanical alignment Fig. 7.17(*b*) before applying as adhesive, such as epoxy or being fastened in place with a cover plate. The V-shaped channel could be either a grooved silicon, plastic ceramic or metal substrate. The loss in the splice in this method depends largely on the fiber size and eccentricity. The outside dimension and core diameter variations determine the fiber size, while the eccentricity depends on the position of the core relative to the centre of the fiber.

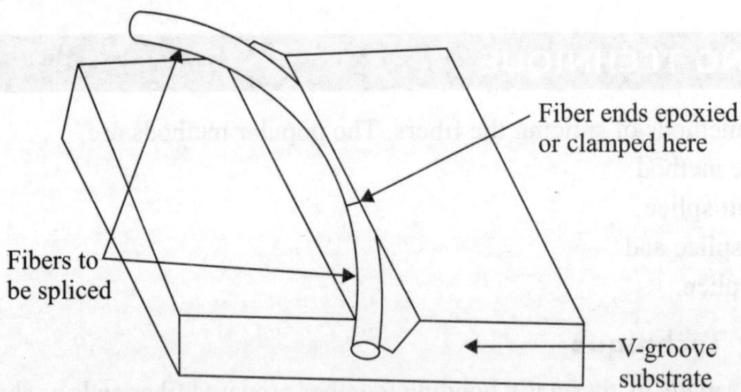

Fig. 7.17 (*b*) A V-grooved splicing as an alignment aid to bring two fibers into mechanical arrangement

7.24 LOOSE TUBE SPLICE

Figure 7.17(*c*) shows the diagram for loose tube splice system which is slightly different from V-groove splice. In this system the corner of a rectangular tube is used for the alignment aid as shown in Fig. 7.17(*c*). Since this technique uses the outside surface of the fiber as the reference surface, it is susceptible to losses due to variations in the core elipticity, the core concentricity and core size. The fiber ends require preparation before splicing. Losses as low as several hundredth of a dB have been reported with these techniques. But splicing fibers of unequal diameters by this technique will give result in unacceptable misalignments.

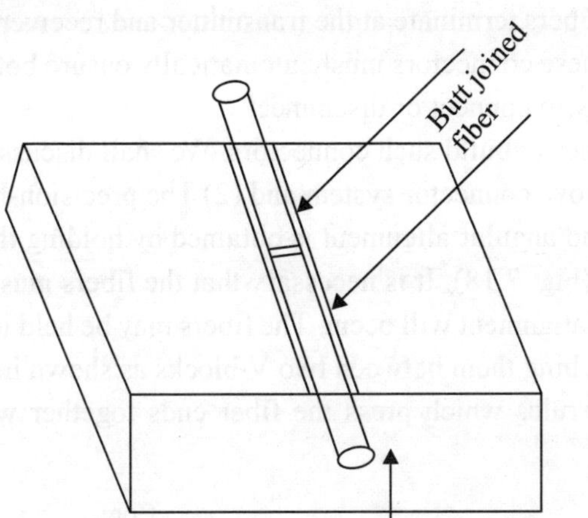

Fig. 7.17 (*c*) A loose tube splicing system

7.25 ELASTIC TUBE SPLICE

Figure 7.17(*d*) shows the elastic tube splice system cross-sectionally. It is a unique device which can automatically perform lateral, longitudinal and angular alignment. It splices multimode fiber with loss having the same order of values as in commercial fusion splices but less equipments and technical skill are required. The applied mechanism basically consists of a tube made of elastic material. The central hole diameter of the tube is slightly smaller than that of the fibers to be spliced and is tapered in each end of the smooth insertion of the fiber. When the fiber is inserted into the elastic tube, it expands the hole diameter so that the elastic material exerts a symmetrical force on the fiber. The symmetry feature allows an accurate and automatic alignment of the axes of the two joined fibers. A wide range of fiber diameters can be inserted into the elastic tube. Thus, the fibers which will be spliced need not to be equal in diameter, since each fiber moves into position independently relative to the tube axis. This aspect is a great advantage over the V-groove splice in which fibers of unequal diameter are automatically misaligned.

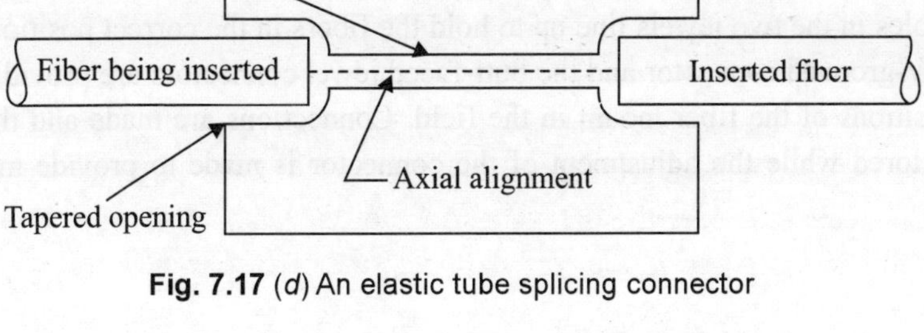

Fig. 7.17 (*d*) An elastic tube splicing connector

7.26 OPTICAL FIBER CONNECTORS

The fiber connectors must have demountable connectors so that two fibers can be separated. They are used at the points where the fibers terminate at the transmitter and receivers either at the line ends or at the intermediate repeaters. These connectors must automatically ensure both a minimum loss connector and at the same time it is easy to connect or disconnect.

There are many schemes to build such connectors. We shall discuss two important and common methods. These are (1) V-groove connector system and (2) The precision-bore hole in a jewel bearing. In V-groove system lateral and angular alignment is obtained by holding the fiber ends into the bottom of a V-groove cut in a block (Fig. 7.18). It is necessary that the fibers must have the same dimensions, otherwise the lateral core misalignment will occur. The fibers may be held in position either by a spring-loading system or by sandwiching them between two V-blocks as shown in Fig. 7.18. The V-blocks are made parts of the mating ferrules which press the fiber ends together when they are made to slide together.

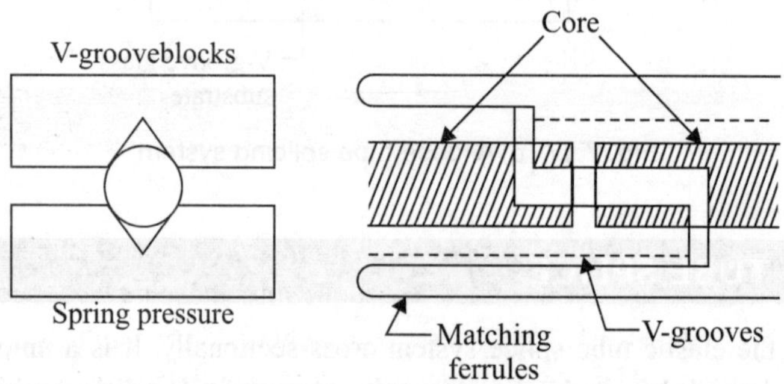

Fig. 7.18 A V-groove system connector

7.27 THE PRECISION-BORE IN A JEWEL BEARING SYSTEM

This alignment method utilizes a precision bored hole in a jewel bearing similar to the type used in watch making as shown in Fig. 7.19(*a*) and (*b*). The figure shows a self-aligning jewel connectors in which one fiber is permanently clamped part-way through the bore of a jewel mount in one ferrule, while the second is held in a flexible rubber mount in a second ferrule. When the two connectors are pushed together, the flexibly mounted fiber is guided into the conical bore of the jewel to butt against the other fiber, while the rubber mount is pushed back and deformed to clamp the rubber in place.

In another version of this type of connector, both the fibers are held in place by jewels mounted in both ferrule so that the cleavage faces of the fiber are flush with the jewel surfaces. When the connectors are mated, the holes in the two jewels line up to hold the fibers in the correct position to mate.

Both the V-grooved connector and the butt-faced jewel connectors are provided with means for adjusting the positions of the fiber mount in the field. Connections are made and the received signal strength is monitored while the adjustment of the connector is made to provide maximum received signal.

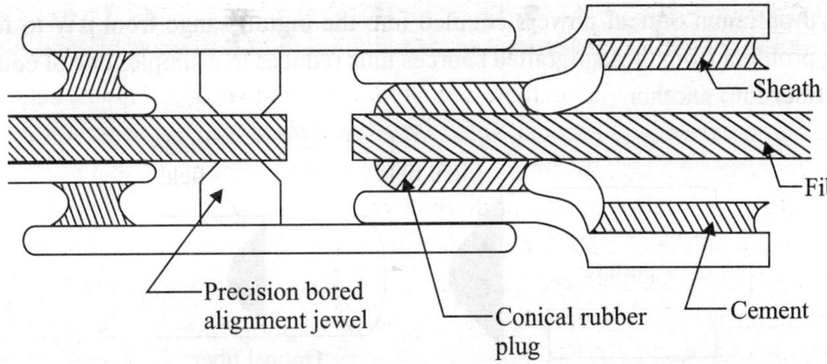

Fig. 7.19 Coupling arrangement methods (*a*) V-groove. (*b*) Alignment (*c*) Precision bored jewel alignment

Losses in this type of connector typically range between 1 and 2.5 dB per connector when the fiber gap is not filled within an optical matching material and the loss in self-aligning connectors is near about 1 dB per connector.

7.28 POWER LAUNCHING AND COUPLING

Introduction

In any fiber-optic system, the coupling of light into an optical fiber is of fundamental importance. The source of the light can be either another fiber or a light source such as a semiconductor laser or light emitting diode (LED). Regardless of what the source of the light might be, the recipe for efficient coupling is to match the light field from the source with that of the mode field in the fiber, as shown in Fig. 7.20 below.

The design objective for any transmitter is to couple as much light as possible into the optical fiber to increase the coupling efficiency. In practice, the coupling efficiency depends on the type of optical source (LED versus laser) as well as on the type of fiber (multimode versus singlemode). The coupling can be very inefficient when light from an LED is coupled into a single-mode fiber. The coupling efficiency for an LED changes with the numerical aperture, and can become < 1% in the case of single-mode fibers. In contrast, the coupling efficiency for edge-emitting lasers is typically 40–50% and can exceed 80% for VCSELs (vertical-cavity surface-emitting LASERs) because of their circular spot size.

Source coupling into an optical fiber

The source coupling into an optical fiber is considered an important one in the design of optical communication system because of the substantial power less at the interface. Coupling efficiency is defined as a measure of the amount of optical power emitted from the source that is coupled into an optical fiber. Most of the sources available in the market offer fiber optic sources with a short length of optical fiber (1 m or less) already attached in an optimum power coupling configuration. This section of fiber commonly is referred to as a flylead or a pigtail.

A small piece of fiber (also known as a pigtail) is included with the transmitter so that the coupling efficiency can be maximized during packaging; a splice or connector is used to join the pigtail with the fiber cable.

The typical optimum optical powers coupled into the pigtail range from μW to few mW. The power-launching problem from these pigtailed sources thus reduces to a simpler one of coupling optical power from one fiber into another.

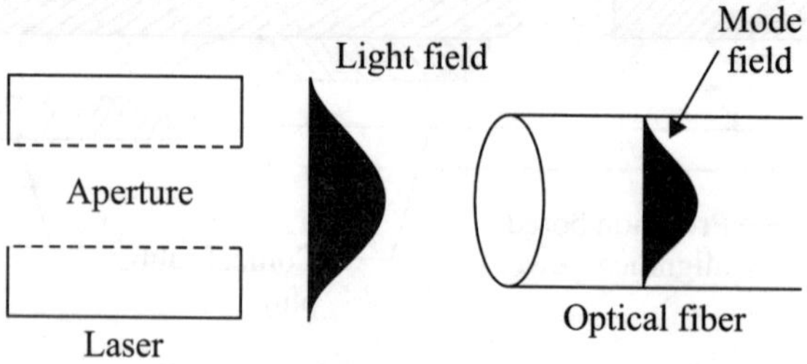

Fig. 7.20 Coupling of light into an optical fiber

We define the coupling efficiency η as the fraction of the source power that is coupled into the fiber

$$\eta = \frac{P_{LM}}{P_{in}} \tag{7.28.1}$$

where P_{LM} is the power in a particular mode in the fiber and P_{in} is the power of the incoming light from the source. Generally coupling loss is described in terms of units of decibels a, rather than in terms of percentage. Where $a = -10 \log \eta$.

Source emission pattern

The light accepting capability of a fiber depends on the radiation pattern of the source. The output pattern of the surface emitting sources are circularly symmetric cone in contrast to asymmetric cone delivered by edge emitting sources. Figure 7.21 below shows a circular core attached to a source is suffering to power loss.

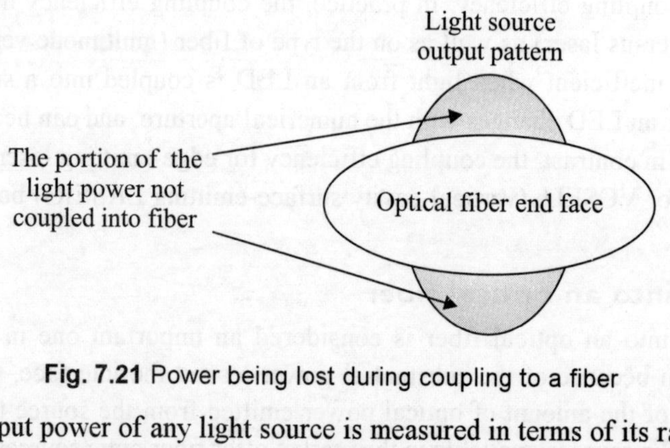

Fig. 7.21 Power being lost during coupling to a fiber

The optical output power of any light source is measured in terms of its radiance or brightness. Radiance is the optical power radiated into a unit solid angle per unit of emitting surface area. In other words, the radiance is the spatial distribution of optical power. Radiance is an important parameter for the source to the fiber coupling.

A source with a well behaved symmetric emission pattern has a symmetrical spherical emission pattern and is known as "Lambertian Source". In case of a surface emitting source, the output pattern is a circularly symmetric cone, where the power delivered at an angle Φ, measured relative to a line perpendicular to the emitting surface, varies as $\cos^n \theta$. The brightness of the emission pattern of such a circular source is given by $B(\theta, \phi) = B_0 \cos^n \theta$ where B_0 is the radiance along the line perpendicular to the emitting surface. Figure 7.22 below shows a spherical coordinate system defined with R, θ and Φ. The radiance is a function of both θ and Φ.

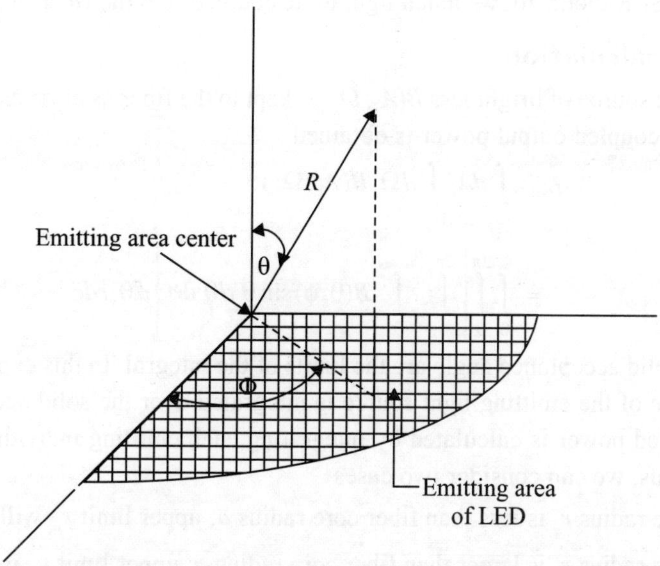

Fig. 7.22 Emission pattern of an optical source

If $n = 1$ then source is treated as a "Lambertian source" *i.e.*, it is equally bright when viewed from any direction. Surface emitting LEDs are known to have Lambertian output pattern.

Figure 7.23 below compares the output of a Lambertian source and a highly directed LASER diode.

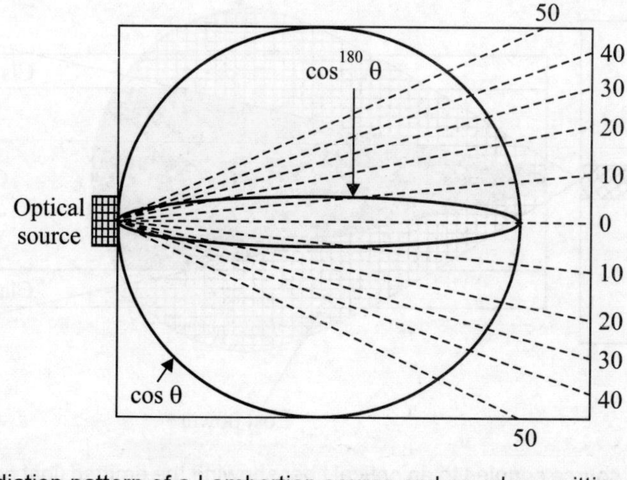

Fig. 7.23 Radiation pattern of a Lambertian source and an edge emitting LASER diode

Edge emitting LEDs and LASER sources with their more complex asymmetric emission distribution is modelled as "Lambertian source" approximated by

$$\frac{1}{B(\theta,\phi)} = \frac{\sin^2\phi}{B_0\cos^T\theta} + \frac{\cos^2\phi}{B_0\cos^L\theta} \quad (7.28.2)$$

where T, L are transverse and lateral power distribution coefficients. For edge emitting LASER diodes $T = 1$ and L can be approximated to 100 or larger due to its larger directionability. The much narrower output beam of a LASER diode allows much light to be coupled into the fiber than a surface LED.

Power coupling calculation

Consider a symmetric source of brightness $B(A_s, \Omega_s)$ is kept to the fiber as close as possible as shown in Fig. 7.24 below. The coupled output power is obtained

$$P = \int_{A_f} dA_s \int_{\Omega_f} d\Omega_s B(A_s, \Omega_s)$$

$$= \int_0^{r_m}\int_0^{2\pi}\left[\int_0^{2\pi}\int_0^{\theta_{0,\max}} B(\theta,\phi)\sin\theta\, d\theta\, d\phi\right] d\theta_s r dr \quad (7.28.3)$$

where the area and solid acceptance angle are the limits of the integral. In this expression an individual radiating point source of the emitting light source is integrated over the solid acceptance angle of the fiber. The total coupled power is calculated by integrating each emitting individual point source over the emitting area. Thus, we can consider two cases

Case 1: source radius r_s is less than fiber core radius a; upper limit r_m will be equal to r_s.

Case 2: source radius r_s is larger than fiber core radius a; upper limit r_m will be equal to a.

Figure 7.24 below shows an example of a surface LED connected to a fiber with r_s less than fiber core radius a.

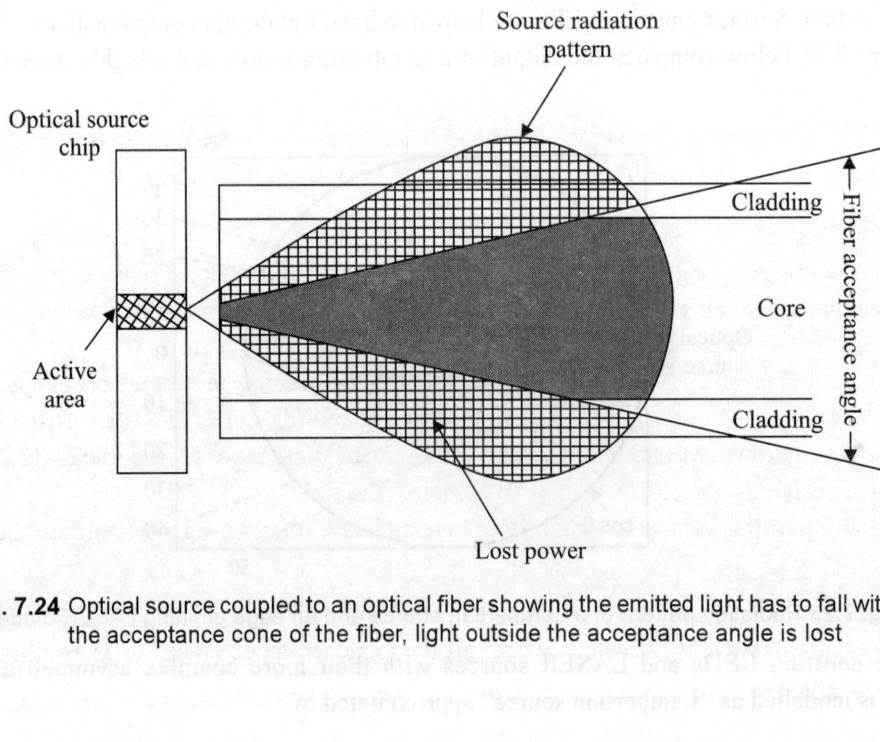

Fig. 7.24 Optical source coupled to an optical fiber showing the emitted light has to fall within the acceptance cone of the fiber, light outside the acceptance angle is lost

$$P = \int_0^{r_s}\int_0^{2\pi}\left(2\pi B_0 \int_0^{\theta_{0,\max}} \cos\theta \sin\theta \, d\theta\right) d\phi_s \, r dr$$

$$= \pi B_0 \int_0^{r_s}\int_0^{2\pi} \sin^2\theta_{0,\max} \, d\theta_s r \, dr$$

$$= \pi B_0 \int_0^{r_s}\int_0^{2\pi} NA^2 d\theta_s r \, dr$$

We know that for stepindexed fiber NA does not depend on θ_s and r, therefore above equation becomes

$$P_{\text{LED, step}} = \pi^2 r_s^2 B_0 (NA)^2 \cong 2\pi^2 r_s^2 B_0 n_1^2 \Delta \tag{7.28.4}$$

Now P_s is the total power emitted from the source area A_s given by

$$P_s = A_s \int_0^{2\pi}\int_0^{\pi/2} B(\theta,\phi) \sin\theta \, d\theta \, d\phi$$

$$= \pi r_s^2 2\pi B_0 \int_0^{\pi/2} \cos\theta \, \sin\theta \, d\theta = \pi r_s^2 B_0$$

Therefore above equation becomes

$$P_{\text{LED, step}} = P_s (NA)^2 \quad \text{for } r_s \le a \tag{7.28.5}$$

Similarly one can derive if $r_s > a$ then

$$P_{\text{LED, step}} = \left(\frac{a}{r_s}\right)^2 P_s (NA)^2 \quad \text{for } r_s > a \tag{7.28.6}$$

Thus if the source is larger than the fiber end face, the power coupled decreases by the ratio of the fiber-to-source areas. The ratio P_F/P_S is defined as the coupling efficiency.

Equilibrium numerical aperture

In order to incur a low coupling loss, the pigtail end of the source is connected to the system fiber with almost identical numerical aperture and core diameter. Depending upon connection technique a certain amount of optical power loss is incurred.

In addition to this, an excess power loss will occur in the first few tens of meters of system fiber. This power loss is generated by non-propagating modes scattering out of the fiber as the launched modes come to an equilibrium condition. Surface emitting LEDs are more susceptible to this loss than LASERs as surface LEDs launch power into all modes of the fiber.

If the light-emitting area is less than fiber core radius a, then power coupled into the fiber is given by

$$P_{\text{LED, step}} = P_s (NA)^2 \text{ for } r_s \le a \text{ where } NA = NA_{\text{in}}.$$

However, after few tens of meter, when all the launched mode comes to an equilibrium, the optical power is given by

$$P_{eq} = P_{s0}\left(\frac{NA_{eq}}{NA_{in}}\right)^2 \tag{7.28.7}$$

Here the optical power scales down suffering to a power loss. Here P_{S0} is the expected power after attaining equilibrium. Since most optical fibers attain equilibrium after 50 m or around, then it is important to consider NA_{eq} for calculation of launching optical power.

Lensing schemes

Two approaches have been used for source–fiber coupling. In one approach, known as direct or butt coupling, the fiber is brought close to the source and held in place by epoxy. In the other, known as lens coupling, a lens is used to maximize the coupling efficiency. The use of a lensed fiber can improve the coupling efficiency, and values close to 100% have been realized with an optimum design. Each approach has its own merits, and the choice generally depends on the design objectives. An important criterion is that the coupling efficiency should not change with time; mechanical stability of the coupling scheme is therefore a necessary requirement.

Law of brightness

If the source-emitting area is larger than the fiber core radius, then maximum optical power can be coupled into the fiber. This is called the "law of brightness" resulting from the principle of conservation of energy and radiance.

However, if the source-emitting area is smaller than the core area of the fiber, a miniature lens can be placed between the source and the fiber to improve the coupling efficiency. The function of this lens is to magnify the source emitting area to match exactly the core area of the fiber end face. If the emitting area is increased by a magnification factor M, the solid angle within which optical power is coupled to the fiber is increased by the same factor.

If the micro-lens is properly aligned, then the improvement in coupling efficiency is the magnification factor M. The maximum magnification factor is given by

$$M_{max} = \frac{\text{diameter of the fiber}}{\text{diameter of the source}} \tag{7.28.8}$$

Figure 7.25 below shows a lens-coupling approach for transmitter design. The coupling efficiency can exceed 70% for such a confocal design in which a sphere is used to collimate the laser light and focus it onto the fiber core. The alignment of the fiber core is less critical for the confocal design because the spot size is magnied to match the fiber's mode size. The mechanical stability of the package is ensured by soldering.

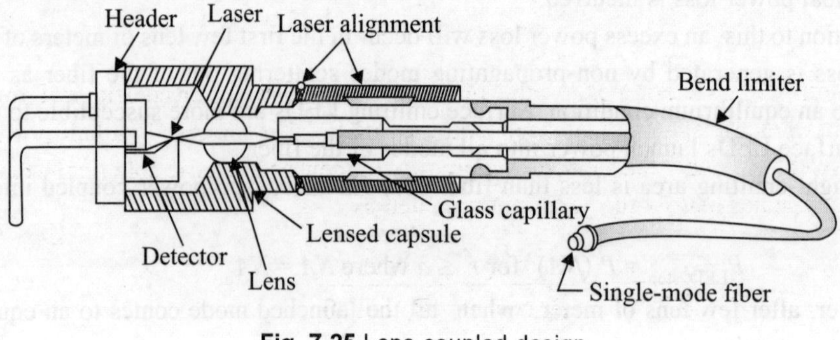

Fig. 7.25 Lens coupled design

Different lensing schemes are shown in Fig. 7.26 below. The goal remains the same, *i.e.*, to magnify the effective emitter area and to match the fiber core area.

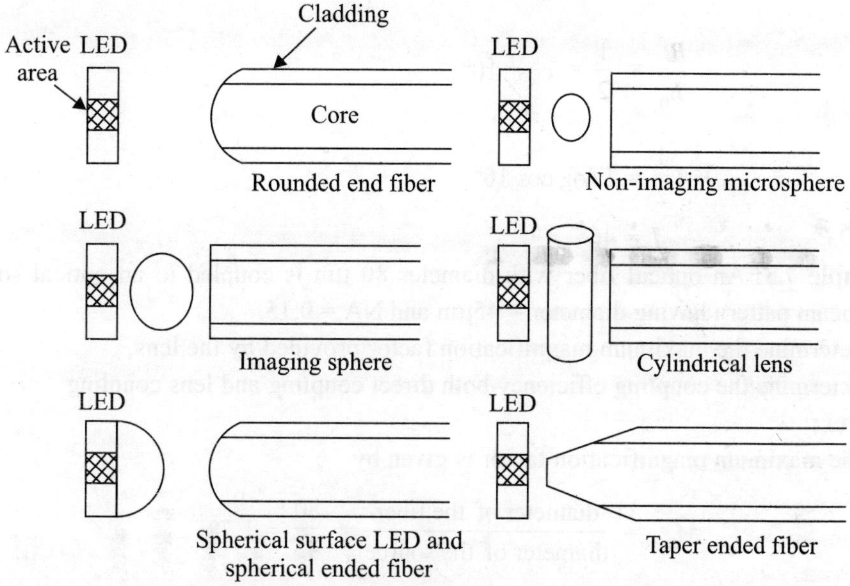

Fig. 7.26 Various lensing schemes

Sloved examples

Example 7.1: Calculate the value of n at -3dB (half power points) that allows a spatial power distribution with variation $(\cos\theta)^n$ for a surface emitting sources where the power delivered at an angle $\theta = 30°$.

Answer: Given : $\theta = 30°$

$$\therefore \quad \frac{1}{2} = (\cos\theta)^n$$

$$\therefore \quad n = \frac{\log\cos 30°}{\log\frac{1}{2}}$$

$$\therefore \quad n = \frac{0.301}{0.062} = 4.85$$

Example 7.2: Determine the value of lateral power distribution coefficient L for a LASER diode having an asymmetric spherical emission pattern with $\theta = 10°$ and $\varphi = 0°$ at half-power points.

Answer: We know that, LASER sources with their asymmetric spherical emission distribution modeled as "Lambertian source" are approximated by

$$\frac{1}{B(\theta,\phi)} = \frac{\sin^2\phi}{B_0\cos^T\theta} + \frac{\cos^2\phi}{B_0\cos^T\theta}$$

$$\frac{1}{B} = 0 + \frac{1}{B_0 \cos^L 10^\circ}$$

$$\frac{B}{B_0} = \frac{1}{2} = \cos^L 10^\circ$$

$$\log \frac{1}{2} = L \log \cos 10^\circ$$

$$L = 46$$

Example 7.3: An optical fiber with diameter 80 μm is coupled to an optical source with a Lambertian beam pattern having diameter = 45μm and NA = 0.15.

(*a*) Determine the maximum magnification factor provided by the lens.

(*b*) Determine the coupling efficiency both direct coupling and lens coupling.

Answer:

(*a*) The maximum magnification factor is given by

$$M_{max} = \frac{\text{diameter of the fiber}}{\text{diameter of the source}} = \frac{80}{45} = 1.78$$

(*b*) Considering direct coupling, coupling efficiency = $(NA)^2 = (0.15)^2 = 2.25\%$
Considering lens coupling, coupling efficiency = $M_{max} (NA)^2 = 1.78\ (0.15)^2 = 4.005\%$

REVIEW QUESTIONS

7.1 Describe with relevant diagrams the double crucible method of manufacturing optical fiber.

7.2 In the following reactions SiO_2, GeO_2, P_2O_5 and B_2O_3 in are produced complete the following chemical reactions:

$$SiCl_4\uparrow + 2H_2O\uparrow =$$
$$GeCl_4\uparrow + 2H_2O =$$
$$2POCl_3\uparrow + 3H_2O\uparrow =$$
$$4BBr_3\uparrow + 3H_2O\uparrow =$$

7.3 Give the diagram for vapour axial deposition (VAD) method for fabricating large preform. Discuss the attenuation characteristics of fibers fabricated by the VAD process with improved drying.

7.4 Explain with the schematic diagram of preform manufacturing using the technique of internal chemical vapour deposition (ICVD). In the following chemical reactions oxides of Si, Ge, P are produced in the core of a fiber and oxides of silicon and Boron are produced in the cladding. Complete the chemical reactions

$$\left.\begin{array}{l} SiCl_4\uparrow + O_2\uparrow \rightarrow \\ GeCl_4\uparrow + O_2\uparrow \rightarrow \\ 4POCl_3\uparrow + 3O_2\uparrow \rightarrow \end{array}\right] \text{Core}$$

$$\left.\begin{array}{l} SiCl_4\uparrow + O_2\uparrow \rightarrow \\ 4BBr_3\uparrow + 3O_2\uparrow \rightarrow \end{array}\right] \text{Cladding}$$

7.5 (*a*) Describe with diagram the apparatus used for drawing fiber from a rod preform and applying a primary coating (*b*) Give the diagrams of the section of extrusion die for putting plastic sheath on the primary coated fiber.

7.6 A fiber has a core radius of 20 mm and the longitudinal displacement $S = 2.0$ mm and the numerical aperture NA = 0.22 calculate the coupling efficiency n_{S1}.

7.7 Discuss in brief two methods for preparing highly transparent glass material for the preparation of optical fiber.

7.8 Describe with necessary diagram the controlled fracture procedure for the end preparation of a fiber. If proper precautions in the procedure are not taken, what will be the defects? Describe the defects with necessary diagram.

7.9 State the expression for the coupling efficiency for a step-index fiber when there is a longitudinal separation. Describe with diagram the different types of misalignment.

7.10 Describe with relavant diagrams different splicing techniques. What are the different types of connectors? Describe them with necessary diagram.

7.11 The total power emitted from an optical source is 50 μW. Calculate the coupled output power with NA of the fiber equals to 0.15 for both

(*a*) Optical source radius r_s =50 μm and fiber core radius a = 100 μm.

(*b*) Optical source radius r_s = 60 μm and fiber core radius a = 40 μm.

7.12 Calculate the value of transverse power distribution coefficient T for an optical source with complex asymmetric emission pattern having $\phi = 10°$ and $\theta = 0°$ at half-power points.

7.13 Calculate the value of n of a circular source that allows a spatial power distribution with variation $(\cos \theta)^n$ with θ the angle at which power delivered equals to 30° and the ratio of the radiance of the source with radiance along normal to the radiating surface equals to 0.25.

CHAPTER 8

Optical Fiber Communication Receiver

CHAPTER FOCUS

- *Introduction: Signal path through optical data link*
- *Receiver configuration with noises*
- *Receiver noises*
- *Noise at the input of the amplifier*
- *Receiver capacitance and bandwidth*
- *Noise in avalanche photo diode (APD) receiver*
- *Front ends of a receiver*
- *Low impedance front end*
- *High impedance (Integrating front end)*
- *Transimpedance front end*
- *Block diagram of an optical receiver*
- *Optical receiver circuits using BJT and FET*
- *Transimpedance Front end circuits*
- *Automatic gain control (AGC) circuit*
- *Constant current source to provide simple AGC*
- *Equilizer circuit*
- *Digital receiver performance*
- *Receiver sensitivity and bit error rate*
- *Simplified model*
- *Optoelectronic integrated circuit (OEIC)*
- *Review Questions*

8.1 INTRODUCTION: SIGNAL PATH THROUGH OPTICAL DATA LINK

The design of an optical receiver is more complicated than that of an optical transmitter, because the function of a receiver is to detect weak and distorted signals and to determine the nature of a signal whether it is analog or digital. The source generally used in optical fiber communication system is a two-level binary digital system.

Figure 8.1 shows a schematic diagram of an optical transmission system using digital signal input. The transmitted signal is the two-level data stream consisting of a "0" and a "1". Each has a time slot duration T_b. This time is known as the "bit period". One of the simplest method of such data is an amplitude shift keying. In amplitude shift keying method the voltage level is switched between two values which are usually "on" and "off ". The resultant voltage signal waveform consists of a voltage pulse of amplitude V relative to zero voltage level. When the amplitude is V, then a binary "1" occurs and a "1" is transmitted, so that a voltage pulse of amplitude V and of duration T_b occurs at the output. When a "0" is sent, the signal voltage remains at zero for the period T_b at the output. The main function of optical transmitter to convert electrical signal to an optical signal. Any time varying electric current $i(t)$ can be used to modulate an optical source such as an LED or laser diode to produce an optical power output $P(t)$. The optical signal coming out of the transmitter a "1" is represented by a pulse of optical power $p(t)$ of duration T_b, whereas a "0" is the absence of any light pulse for the time T_b.

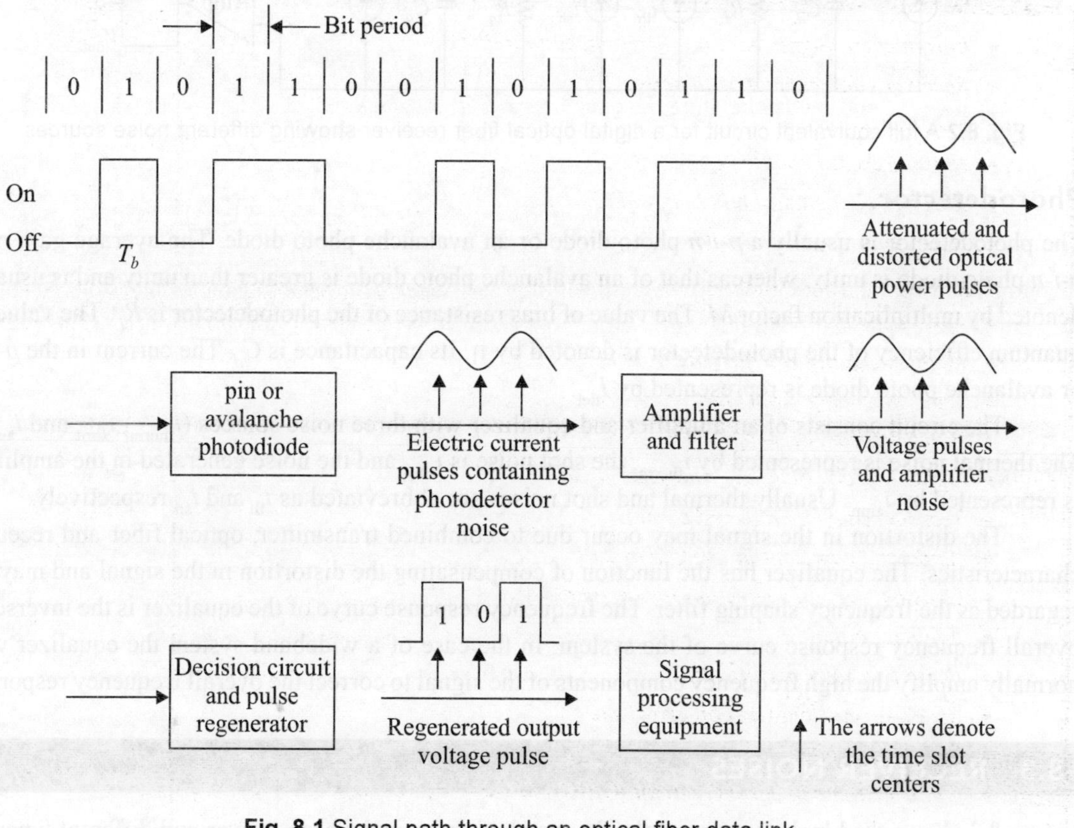

Fig. 8.1 Signal path through an optical fiber data link

The optical signal emitted from the light source is coupled to one end of a fiber, and it gets attenuated and distorted as the light pulse reaches the other end of the fiber. Then the signal is coupled to the input of a receiver. The optical signal is detected in the receiver with the help of either a *p-i-n* photo diode or an avalanche photodiode which converts the optical signal back to the original electrical signal. Of course, the original signal is further attenuated and distorted in the detector. Then the electrical signal coming out of the detector is amplified and filtered.

A decision circuit compares the signal in each time slot T_b to a certain reference signal level, which is known as "threshold level". If the received signal level is greater than the threshold level a "1" is said to have been received. If the received signal is less than the threshold level, a "0" is assumed to have been received.

8.2 RECEIVER CONFIGURATION WITH NOISES

Figure 8.2 shows the schematic diagram of a typical optical receiver with different noise sources. In the receiver there are three main stages. These are (*a*) photodetector (*b*) amplifier and (*c*) an equalizer.

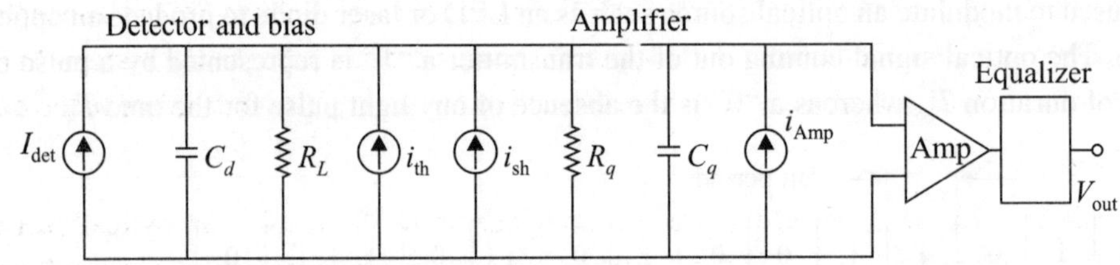

Fig. 8.2 A full equivalent circuit for a digital optical fiber receiver showing different noise sources

Photodetector

The photodetector is usually a *p-i-n* photo diode or an avalanche photo diode. The average gain of a *p-i-n* photo diode is unity, whereas that of an avalanche photo diode is greater than unity and is usually denoted by multiplication factor M. The value of bias resistance of the photodetector is R_L. The value of quantum efficiency of the photodetector is denoted by η. Its capacitance is C_d. The current in the *p-i-n* or avalanche photo diode is represented by I_{det}.

The circuit consists of an amplifier and equalizer with three noise sources (i_{therm}, i_{shot} and i_{amp}). The thermal noise is represented by i_{therm}, the shot noise is i_{shot} and the noise generated in the amplifier is represented by i_{amp}. Usually thermal and shot noises are abbreviated as i_{th} and i_{sn} respectively.

The distortion in the signal may occur due to combined transmitter, optical fiber and receiver characteristics. The equalizer has the function of compensating the distortion in the signal and may be regarded as the frequency shaping filter. The frequency response curve of the equalizer is the inverse of overall frequency response curve of the system. In the case of a wideband system the equalizer will normally amplify the high frequency components of the signal to correct the overall frequency response.

8.3 RECEIVER NOISES

Figure 8.3 shows the block diagram of the front end of an optical fiber receiver and different types of noise associated with each block. The main optical detector in a receiver is either a *p-i-n* photodetector or an avalanche photo diode. The different noises shown in Fig. 8.3 apply to both types of receiver (analog/digital).

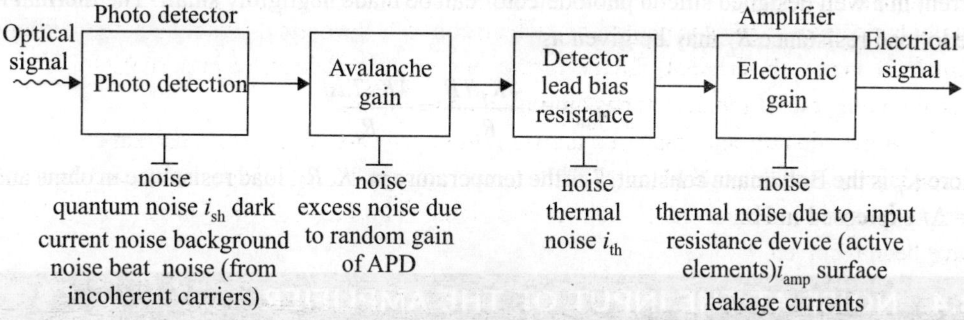

Fig. 8.3 A schematic block diagram of the front end of an optical receiver showing various noise sources

Noise is used to describe the unwanted erratic components of electrical current or voltage that has the tendency to disturb the transmission and processing of the desired signal in a physical system. We have an incomplete control over this noise. The external noise sources are the atmospheric noise, noise generated in external equipments. The internal noises are generated in the system. In this section we shall discuss noises which are produced in the active or passive elements in the receiver.

Beat Frequency Noise

The beat noise is generated from the various components of the incoherent optical carriers having slightly different frequencies, and it can be shown to have insignificant effects in multimode optical propagation. Hence the beat noise will not be considered here.

We shall consider the other types of noises as shown in Fig. 8.3. As shown in Fig. 8.3 these are:

1. Quantum noise (or shot noise) is produced due to the photo current I_p, dark current I_d and background noise due to I_b.
2. The excess noise (*shot noise*) is generated due to random gain mechanism in an avalanche photo diode.
3. The thermal noise is produced in the bias resistance of the detector and at the input of *p-n* and *p-i-n* photo diodes in the receiver.

Shot Noise (Quantum Noise)

The *p-n* and *p-i-n* photo diode do not have any internal gain (multiplication factor $M = 1$). The important noises associated with a photo diode without any internal gain are the quantum noise (shot noise) due to photo current I_p, dark current I_d and background current I_b. These noises are also referred to as the shot noise on the photo current. The total average shot noise $\langle i^2_{Tsh} \rangle$ be given by

$$\langle i^2_{Tsh} \rangle = 2eB\,(I_p + I_d + I_b) \tag{8.3.1}$$

where B is the bandwidth, I_p and I_d are the photo current and dark current respectively, I_b is the photo current due to background radiation and $e = 1.6 \times 10^{-19}$ C.

When the photo current I_b due to background radiation is negligible, the expression for the shot noise is

$$\langle i^2_{Tsh} \rangle = 2eB(I_p + I_d) \tag{8.3.2a}$$

When the photo diode does not have any internal gain ($M = 1$), the thermal noise from the load resistance R_L of the detector and the active element in the amplifier tends to predominate. The dark current in a well designed silicon photodetector can be made negligibly small. The thermal noise $<i^2_{ther}>$ due to load resistance R_L may be given as

$$<i^2_{th}> = \frac{4K_BTB}{R_L} = \frac{4K_BT\Delta f}{R_L} \tag{8.3.2b}$$

where K_B is the Boltzmann constant, T is the temperature in °K, R_L, load resistance in ohms and bandwidth $B = \Delta f$ expressed in Hz.

8.4 NOISE AT THE INPUT OF THE AMPLIFIER

The noise sources in the amplifier are associated with both active and passive elements in the amplifier. This noise is represented by a series noise voltage $<v^2_a>$ and a shunt noise current $<i^2_a>$.

The total noise current $<i^2_{amp}>$ associated the amplifier is given by

$$<i^2_{amp}> = \int_o^B <i_a^2> + <V_a^2> |Y|^2 \, df \tag{8.4.1}$$

where $|Y|$ is the shunt admittance formed by the capacitance C_a and resistance R_a of the amplifier and f is the frequency.

Figure 8.4 shows the equivalent circuit for the front end of the receiver along with the effective capacitance C_a and input resistance R_a of the amplifier. The capacitance C_d of the detector is also included in the equivalent circuit and as such, the noise generated in C_d is also included in the expression for $<i^2_{amp}>$ in Eq. (8.4.1).

Here $\quad Y = \frac{1}{R_{TL}} + jwC_{sh}$; where $C_{sh} = C_d + C_a$ and $\frac{1}{R_{TL}} = \frac{1}{R_L} + \frac{1}{R_a}$

The signal to noise ratio S/N for the *p-n* or *p-i-n* photo diode receiver is modified by including the noise contribution from Eq. (8.4.1), Eq. (8.3.2*a*), (8.3.2*b*)

$$\frac{S}{N} = \frac{I_p^2}{2eB(I_p + I_d) + \frac{4K_BTB}{R_L} + <i^2_{amp}>} \tag{8.4.2a}$$

It is possible to reduce the thermal noise by increasing the load resistance R_L. This reduction has effect of reducing bandwidth, since the bandwidth is

$$B = \frac{1}{2\pi C_{sh} R_L} \tag{8.4.2b}$$

Noise Figure F_n

The noise associated with an amplifier is usually referred to load resistance R_L, and noise figure F_n may be obtained. This enables us to combine $<i^2_{amp}>$ with the thermal noise across the load resistor $<i^2_{th}>$ to give

$$<i_{\text{th}}^2> + <i_{\text{amp}}^2> = \frac{4K_BTB}{R_L} F_n \tag{8.4.3}$$

where F_n is known as noise figure.

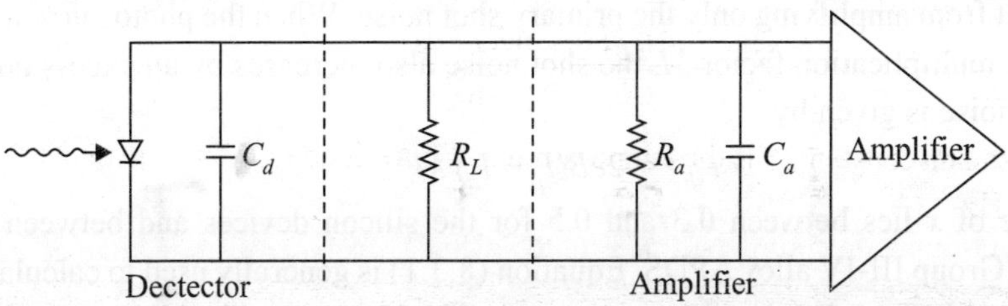

Fig. 8.4 A schematic equivalent circuit for the front end of an optical fiber receiver showing effective capacitance C_a and input resistance R_a of the amplifier

The expression for the signal-to-noise ratio *S/N* of a *p-n* or *p-i-n* photo diode may be written in the modified form as

$$\frac{S}{N} = \frac{I_p^2}{2eB\,(I_p + I_d) + \dfrac{4K_BTB}{R_L} F_n} \tag{8.4.4}$$

Thus if the noise figure F_n for the amplifier is known, *S/N* can be determined.

8.5 RECEIVER CAPACITANCE AND BANDWIDTH

Combining two capacitances C_d and C_a in the equivalent circuit of Fig. 8.4, the equivalent total capacitance C_T is given by

$$C_T = C_d + C_a \tag{8.5.1}$$

where C_d and C_a are the detector and amplifier input capacitance respectively. The total capacitance should be made as small as possible for improvement of bandwidth and the reduction of noise.

The reciprocal of time constant $2\pi R_L C_T$ must be greater than or equal to post-detection bandwidth B so that

$$\frac{1}{2\pi R_L C_T} \geq B \tag{8.5.2}$$

when $\dfrac{1}{2\pi R_L C_T} = B$, it expresses the maximum possible bandwidth B for straightforward termination.

8.6 NOISE IN AVALANCHE PHOTO DIODE (APD) RECEIVER

The internal gain due to carrier multiplication mechanism in an avalanche photo diode causes an increase in the signal current into the amplifier and also the signal-to-noise ratio *S/N*. This happens due to the fact that the load resistance and the amplifier noise remain the same. However, the dark current and quantum noise are increased by the multiplication process of the avalanche photo diode. The multiplication factor becomes an important factor in determining the signal to noise ratio, *S/N*. The random gain mechanism of current produces an excess noise in the receiver in terms of the shot noise above the level that would result from amplifying only the primary shot noise. When the photo current increases by the mean avalanche multiplication factor *M*, the shot noise also increases by an excess noise factor M^x so that the excess noise is given by

$$<i_{sh}^2> = 2eB(I_p + I_d)\, M^{2+x} \tag{8.6.1}$$

where the value of *x* lies between 0.3 and 0.5 for the silicon devices and between 0.7 and 1.0 for germanium and Group III-IV alloy APDS. Equation (8.3.1) is generally used to calculate the total noise term in order to calculate the signal to noise ratio *S/N*. Hence the signal to noise ratio for an APD receiver is given by

$$\frac{S}{N} = \frac{M^2 I_p^2}{2eB(I_p + I_d)M^{2+x} + \dfrac{K_B TBF_n}{R_L}} \tag{8.6.2}$$

Dividing the right hand side of denominator and numerator by M^2, we get

$$\frac{S}{N} = \frac{I_p^2}{2eB(I_p + I_d)M^x + \dfrac{4K_B TBF_n M^{-2}}{R_L}} \tag{8.6.3}$$

Optimum Value of Multiplication Factor *M*

It may be noticed that the first term in denominator of Eq. (8.6.3) increases with the increase of *M* and the second term decreases with the increase of *M*. For the low values of *M*, combined amplifier and thermal noise term predominates and the total noise power is usually unaffected. When the signal voltage is increased so that the signal to noise ratio *S/N* improves. However, when *M* is large the combined thermal and amplifier noise term becomes negligible and *S/N* decreases with the increasing *M* at the rate of M^x. Thus, there exists an optimum value of multiplication factor *M* and this optimum value maximizes the term *S/N*.

The optimum value is obtained by differentiating *S/N* in Eq. (8.6.3) with respect to *x* and putting the result equal to zero. It is given by

$$\frac{2eB(I_p + I_d)M_{\text{opt}}^x}{\dfrac{4K_B TB\, F_n}{R_L} M_{\text{opt}}^{-2}} = \frac{2}{x} \tag{8.6.4}$$

Therefore,

$$M_{\text{opt}}^{2+x} = \frac{4K_B TF_n}{ex\, R_L (I_p + I_d)} \tag{8.6.5}$$

Example 8.1: *An optical fiber receiver using silicon p-i-n photo diode has a quantum efficiency of 50 per cent operating at an wavelength of 0.85 μm. The dark current in the photo diode at this wavelength is 4 nano amperes. The load resistance is 6 K ohms. The incident optical power at this wavelength is 250 nW and the post-detection bandwidth of the receiver is 8 MHz. Calculate*

(a) *the photo current in the photo diode*

(b) *the shot noise generated in the photo diode and*

(c) *the thermal noise generated in the load resistance at 27°C.*

Solution: (*a*) The photo current I_p is given by Eq. (6.7.3*b*) as

$$I_p = \frac{\eta e P_o}{hf} = \frac{\eta e P_o \lambda}{hc}$$

Data supplied: $\eta = 0.50$, $e = 1.6 \times 10^{-19}$, $P_o = 250 \times 10^{-9}$ W, $B = 8 \times 10^6$, $\lambda = 0.85 \times 10^{-6}$, $h = 6.63 \times 10^{-34}$, $I_D = 4 \times 10^{-9}$ $K_B = 1.38 \times 10^{-23}$, $c = 3 \times 10^8$ C and $t = 273 + 27°$ C $= 300°$ K.

Substituting the values in the above equation

$$I_p = \frac{0.5 \times 1.6 \times 10^{-19} \times 250 \times 10^{-9} \times 0.85 \times 10^{-6}}{6.63 \times 10^{-34} \times 3 \times 10^8} = 86.4 \text{ nA}$$

(*b*) Total shot noise is given by Eq. (8.3.2*a*)

$$<i_{sh}^2> = 2eB\,(I_p + I_d)$$

$$= 2 \times 1.6 \times 10^{-19} \times 8 \times 10^6\,[(86.4 + 4) \times 10^{-9}]$$

$$= 25.6 \times 10^{-13} \times 90.4 \times 10^{-9} = 2.31 \times 10^{-19} \text{A}^2$$

or $$<i_{sh}^2>^{1/2} = 4.8 \times 10^{-10} \text{A}$$

(*c*) The thermal noise in the load resistor is given by Eq. (8.3.2*b*)

$$<i_{th}^2> = \frac{4K_BTB}{R_L} = \frac{4 \times 1.38 \times 10^{-23} \times 300 \times 8 \times 10^6}{6 \times 10^3} = 22 \times 20^{-18} \text{ A}^2$$

or $$<i_{th}^2>^{1/2} = 4.7 \times 10^{-9} \text{ A}$$

Example 8.2: *A p-i-n photo diode has capacitance of 5 pf. Calculate the maximum value of load resistance R_L which will make the post-detection bandwidth of 10 MHz and estimate the decrease in bandwidth with the same load resistance when the following amplifier has an input capacitance of 5 pf.*

Solution: Data: $C_d = 5$ pf, $B = 10 \times 10^6$

The maximum bandwidth is given by Eq. (8.4.2*b*)

$$B = \frac{1}{2\pi R_L C_d}$$

or $$R_L = \frac{1}{2\pi B C_d} = \frac{1}{2\pi \times 10 \times 10^6 \times 5 \times 10^{-12}}$$

$$R_L = 3.18 \times 10^3 \text{ ohms}$$

But when the system is connected to an amplifier whose input capacitance is 5 pf and load resistance is 3.18 × 10 ohms, the bandwidth is

$$B' = \frac{1}{2\pi \times 3.18 \times 10^3 \times (5+5) \times 10^{-12}} = 5 \times 10^6$$

As soon as the detector is connected to the amplifier the bandwidth decreases to 5 MHz.

Example 8.3: *A silicon avalanche photo diode (APD) with x = 0.3 has a capacitance of 6 pf and negligible dark current. The APD is operating with a post-detection bandwidth of 40 MHz. When the photo current before gain is 2×10^{-7} A and the temperature is 27°C.*

Calculate

(a) the signal to noise ratio S/N when the multiplication factor M = 1.

(b) optimum value of the multiplication factor M_{opt} when S/N is maximum and

(c) the signal to noise ratio S/N assuming that all operating conditions are maintained and $M = M_{opt.}$

Solution: Data given: $C_d = 6$ pf $I_d = 0$, $B = 40$ MHz, $I_p = 2 \times 10^{-7}$A $T = 273 + 27 = 300$ °K and $F_n = 1$

(*a*) The maximum value of load resistance R_L is given by

$$R_L = \frac{1}{2\pi C_d B} = \frac{1}{2\pi \times 6 \times 10^{-12} \times 40 \times 10^6} = 666 \text{ ohms.}$$

(*b*) When the multiplication factor $M = 1$, the signal to noise ratio

$$= \frac{I_p^2}{2eBI_p + \dfrac{4K_BTB}{R_L}}$$

Shot noise is given by $\langle i_{sh}^2 \rangle = 2eB\, I_p = 2 \times 1.6 \times 10^{-19} \times 40 \times 10^6 \times 2 \times 10^{-7} = 2.56 \times 10^{-18}$

$$\langle i_{sh}^2 \rangle = 2.56 \times 10^{-18}\, A^2$$

Thermal noise is givne by

$$\langle i_{th}^2 \rangle = \frac{4KTB}{R_L} = \frac{4 \times 1.38 \times 10^{-23} \times 300 \times 40 \times 10^6}{666}$$

$$= 9.95 \times 10^{-16}\, A^2$$

$\langle i_{th}^2 \rangle$ is greater than $\langle i_{sh}^2 \rangle$

$$\frac{S}{N} = \frac{(2 \times 10^{-7})^2}{9.95 \times 10^{-16}} = 40.2$$

When $M = M_{\text{opt}}$ and $x = 0.3$

$$M_{\text{opt}}^{2.3} = \frac{4K_BT}{exR_LI_p} = \frac{4 \times 1.38 \times 10^{-23} \times 300}{1.6 \times 10^{-19} \times 0.3 \times 666 \times 2 \times 10^{-7}}$$

$$M_{\text{opt}} = 29.8$$

The signal-to-noise ratio with $M_{\text{opt}} = 29.8$ and $R_L = 666.5$ may be written as

$$\frac{S}{N} = \frac{M_{\text{opt}}^2 I_p^2}{2eBI_pM^{2.3} + \dfrac{4K_BTB}{R_L}}$$

or

$$\frac{S}{N} = \frac{(29.8)^2 \times (2 \times 10^{-7})^2}{2 \times 1.6 \times 10^{-19} \times 40 \times 10^6 \times 2 \times 10^{-7} \times (29.8)^{2.3} + \dfrac{4 \times 1.38 \times 10^{-23} \times 300 \times 40 \times 10^6}{666}} = 53.12$$

8.7 FRONT ENDS OF A RECEIVER

Three basic amplifier configurations are frequently used in optical fiber communication receivers. These are:

(*i*) Low impedance Front end

(*ii*) High impedance Front end

(*iii*) Transimpedance Front end.

8.8 LOW IMPEDANCE FRONT END

The low impedance front end type of amplifier in a receiver is the simplest and most common type and it consists of a voltage amplifier with an effective resistance R_a as shown in Fig. 8.5(*a*).

In order to make proper design choices, it is essential to consider both bandwidth and noise. When the detector has a load resistance R_L, the bandwidth B is given by

$$B = \frac{1}{2\pi R_L C_T} \tag{8.8.1}$$

where C_T is the equivalent parallel capacitance $C_T = C_a + C_d$ where C_a and C_d are the input capacitance of the amplifier and the detector capacitance respectively.

However, in most practical receivers, the detector is loaded with resistance R_b and the input resistance of the amplifier as shown in Fig. 8.5(*a*). In determining bandwidth, R_L will be the parallel combination of all passive resistances which appears across the detector terminal.

However, R_L may be modified to include the parallel resistance of the detector bias and the input resistance R_a of the amplifier. The modified total load resistance is given by

$$R_{TL} = \frac{R_b R_a}{R_b + R_a} \tag{8.8.2}$$

considering Eq. (8.8.1) and Eq. (8.8.2) we replace R_L by R_{TL} to achieve optimum bandwidth and thus R_a and R_b must be minimized. This leads to a low impedance front end design for the optical receiver amplifier. But the inherent defect of the design is that the thermal noise will be predominant and the noise factor may severely limit the sensitivity of the receiver. Thus a compromise between the bandwidth and sensitivity must be made.

$$B' = \frac{1}{2\pi R_{TL} C_T} \tag{8.8.3}$$

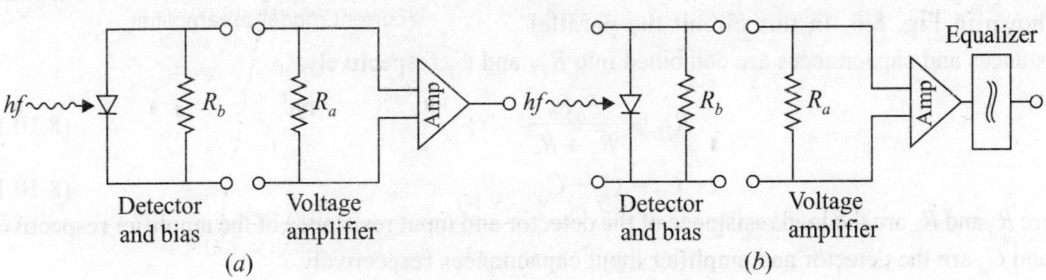

Fig. 8.5 (*a*) A low impedance front end type of an optical receiver, (*b*) A high impedance type of integrating front end optical fiber receiver with an equalizer

8.9 HIGH IMPEDANCE (INTEGRATING FRONT END)

The second configuration consists of a high impedance amplifier together with a large detector bias resistance so that the effective shunt resistance R_{TL} is high. The thermal noise is reduced considerably by the large value of the load resistance. However, this configuration tends to produce inferior response characteristics with a small value of bandwidth. The bandwidth relationship $\frac{1}{2\pi R_{TL} C_T} \geq B$ is not maintained for wide band operation.

The detector output is effectively integrated over a long time constant ($R_{TL}C_T$) and this shortcoming must be restored by differentiation. This may be accomplished by correct equalization at a later stage after the amplifier as shown in Fig. 8.5(*b*).

Therefore the high impedance front end structure provides a considerable improvement in the sensitivity of the receiver over the low-impedance structure, but it is essential to make equalization for the frequency response. The shortcoming of the high-impedance structure is that it has a small dynamic range which means that the ratio between the maximum and minimum input signal is small. The problem of low dynamic range is the result of attenuation of low frequency signal component by equalization process. As a result, the amplifier saturates at high value of input signal. When the amplifier reaches saturation before the equalization has occurred, the signal becomes highly distorted in its wave shape. The reduction in dynamic range depends on the amount of integration and subsequent equalization applied.

8.10 TRANSIMPEDANCE FRONT END

Low impedance and high impedance front end types of receivers are important theoretically, but out of all three types transimpedance type with feedback amplifier is preferred. The advantage of transimpedance structure lies in the fact that the need of equalization is avoided and the noise in the amplifier is low due to negative feedback. An equivalent circuit for an optical fiber receiver incorporating a transimpedance front end structure is shown in Fig. 8.6. In this circuit the parallel resistances and capacitances are combined into R_{TL} and C_T respectively.

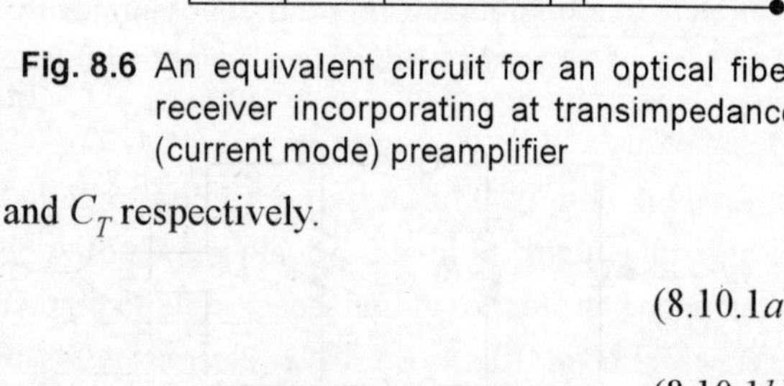

Fig. 8.6 An equivalent circuit for an optical fiber receiver incorporating at transimpedance (current mode) preamplifier

$$R_{TL} = \frac{R_a R_b}{R_a + R_b} \tag{8.10.1a}$$

$$C_T = C_d + C_a \tag{8.10.1b}$$

where R_a and R_b are the load resistance of the detector and input resistance of the amplifier respectively. C_d and C_a are the detector and amplifier input capacitances respectively.

At node *A*,
$$I_{det} + I_F = I_R + I_c \tag{8.10.2}$$

where $I_{det} = I_p(f)$ for *p-i-n* diode and $I_{det} = M\,I_p(f)$ for avalanche diode (8.10.3)

$$I_F = \left(\frac{V_{in} - V_o}{R_F}\right) \quad (8.10.4)$$

$$I_R = \frac{V_{in}}{R_{TL}} \quad (8.10.5)$$

$$I_C = \frac{V_{in}}{\dfrac{1}{jwC_T}} = V_{in} j2\pi f C_T \quad (8.10.6)$$

where the detector current I_{det} is equal to $MI_p(f)$ in the case of an avalanche photodiode.

From Eqs. (8.10.2), (8.10.3), (8.10.4), (8.10.5) and (8.10.6)

$$MI_p(f) - \frac{V_o - V_{in}}{R_F} = V_{in}\left(\frac{1}{R_{TL}} + j2\pi f C_T\right) \quad (8.10.7a)$$

$$MI_p(f) = V_{in}\left(\frac{1}{R_{TL}} + j2\pi f C_T\right) + \frac{V_{in}}{R_F} - \frac{V_o}{R_F} \quad (8.10.7b)$$

when the gain of the amplifier is $-A$, then $\dfrac{V_o}{V_{in}} = -A$ (8.10.8)

Substituting the value of A in terms of V_{in} Eq. (8.10.7*b*) becomes

$$I_{det} = MI_p(f) = V_{in}\left(\frac{1}{R_{TL}} + j2\pi f C_T + \frac{1}{R_F} + \frac{A}{R_F}\right) \quad (8.10.9)$$

or

$$MI_p(f) = -V_o\left(\frac{1}{R_{TL}} + \frac{1}{R_F} + \frac{1}{AR_F} + \frac{j2\pi f C_T}{A}\right) \quad (8.10.10)$$

or

$$V_o = \frac{-MI_p(f)R_F}{1 + \dfrac{1}{A} + \dfrac{R_F}{AR_{TL}} + \dfrac{j2\pi_f C_T R_F}{A}} \quad (8.10.11)$$

Dividing both numerator and denominator by $\dfrac{1}{1 + \dfrac{1}{A} + \dfrac{R_F}{AR_{TL}}}$, we get

$$V_o = \frac{-\dfrac{MI_p(f)\,R_F}{1 + \dfrac{1}{A} + \dfrac{R_F}{AR_{TL}}}}{1 + \dfrac{j2\pi f C_T R_F}{1 + \dfrac{1}{A} + \dfrac{R_F}{AR_{TL}}}} \quad (8.10.12)$$

When the gain $A >> \dfrac{R_F}{R_{TL}}$, Eq. (8.10.12) becomes

$$V_o = -\frac{R_F MI_P(f)}{1+\dfrac{j2\pi fC_T R_F}{A}} \quad (8.10.13)$$

When A >> 2 $\pi f C_T R_f$, no further equalization of the signal is required then

$$V_o = -R_F MI_p(f) = -R_F I_{det} \quad (8.10.14)$$

The voltage transfer function

$$\frac{V_o}{I_{det}} = \frac{V_o}{MI_p(f)} = \frac{-R_f}{1+\dfrac{j2\pi fC_T R_F}{A}} \quad (8.10.15)$$

The permitted bandwidth may be written

$$B \le \frac{A}{2\pi fC_T R_F} \quad (8.10.16)$$

Hence it may be noted that the transimpedance amplifier (Feedback amp.) has a much greater bandwidth than the amplifiers without feedback. This is pronounced when the gain A is large.

Example 8.4: *A high input impedance amplifier is used in an optical fiber receiver. The effective input resistance of the amplifier is 5MΩ, which is matched to the bias resistance of the detector of the same value. Determine the following:*

(a) The maximum bandwidth that may be obtained without equilization if the total capacitance C_T is 5 pf.

(b) The mean square thermal noise current per unit bandwidth generated by this high input impedance configuration when operating at the temperature of 27°C.

(c) Compare the values calculated in (a) and (b) with those obtained when the high input amplifier is replaced by a transimpedance amplifier with the feedback resistor R_F = 100 K and the open loop gain of 400. It may be assumed that $R_F << R_{TL}$ and the total capacitance remain 5 pf.

Solution: (*a*) Using Eq. (8.8.2) the total effective load resistance

$$R_{TL} = \frac{5\times10^6 \times 5\times10^6}{5\times10^6 + 5\times10^{-6}} = 2.5\ \text{M}\Omega$$

Hence from Eq. (8.8.1) the maximum bandwidth is given by

$$B = \frac{1}{2\pi\times R_{TL}C_T} = \frac{1}{2\pi\times 2.5\times10^6\times5\times10^{-12}} = 1.27\times10^4\ \text{Hz}$$

The maximum bandwidth that may be obtained without equalization is 1.27 × 10^4 Hz.

(*b*) The mean thermal energy noise current per unit bandwidth for the high impedance configuration following Eq. (8.3.2(*b*)), putting B = 1 Hz and T = 273 + 27 = 300

$$<i_{th}^2> = \frac{4K_BT}{R_{TL}} = \frac{4\times1.38\times10^{-23}\times300}{2.5\times10^6} = 6.62\times10^{-27}\ \text{A}^2\ \text{Hz}^{-1}$$

(*c*) The maximum bandwidth without equalization for the transimpedance configuration

$$B = \frac{A}{2\pi R_F C_F} = \frac{400}{2\pi \times 10^5 \times 5 \times 10^{-12}} = 1.24 \times 10^8 \text{ Hz}$$

Bandwidth with transimpedance design is 124 MHz.

Assuming that $R_F << R_{TL}$ the mean square thermal noise current per unit bandwidth for the transimpedance configuration is given by

$$<i_{th}^2> = \frac{4K_B T}{R_F} = \frac{4 \times 1.38 \times 10^{-23} \times 300}{10^5} = 16.56 \times 10^{-25} A^2 \text{ Hz}.$$

8.11 BLOCK DIAGRAM OF AN OPTICAL RECEIVER

Figure 8.7 shows the block schematic showing the main elements of an optical fiber receiver. The blocks are optical detector, preamplifier, main amplifier, equalizer and filter.

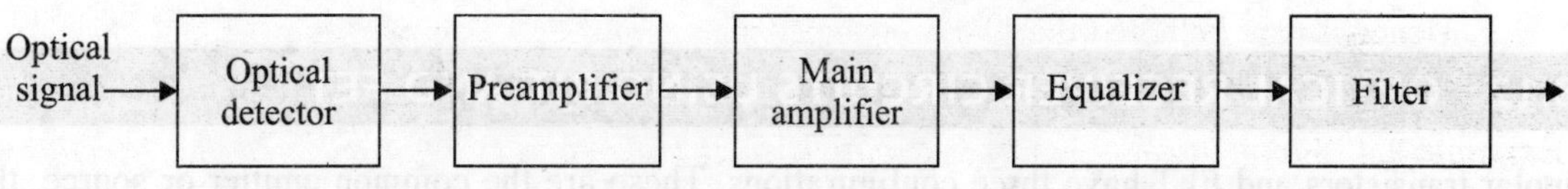

Fig. 8.7 Block schematic showing major elements of an optical fiber receiver

Detector

The optical detector is either a *p-i-n* photo diode or an avalanche photo diode and it performs the function of linear conversion of the received optical signal into an electrical current signal. The electrical current signal has the same waveform as that of the envelope of the optical signal.

Preamplifier

After the detector, there is a preamplifier. The output electrical signal from the detector is quite small and is amplified by the preamplifier for getting a suitable signal level. Thus, by the initial amplification additional noise is kept to a minimum so that the corruption of the received signal is avoided. As the noise sources within the preamplifier may be dominant, its design and configuration are important factors in deciding the receiver sensitivity.

Main Amplifier

The main amplifier is capable of providing additional amplification with a low noise in the signal to provide increased signal level for the following stage. Although the optical devices are very linear devices and do not themselves introduce considerable distortion on the signal, other components within the optical fiber communication system may exhibit nonlinear behaviour. For example, the received optical signal may be distorted due to the dispersive mechanism of the optical fiber itself. The combined transfer function of the preamplifier and main amplifier combination may be such that the input signal is considerably distorted in the case of the high impedance front end preamplifier.

Equalizer

In order to compensate for the distortion and to provide a suitable signal shape for the filter an equalizer is frequently included in the receiver. The equalizer may precede or follow the main amplifier or may be incorporated in the functions of the amplifier and filter. Of course in Fig. 8.7 the equalizer is shown as a separate element following the main amplifier and preceding the filter.

Filter

The filter follows the equalizer and is the final element of the receiver. The function of this final element is to maximize the value of the signal to noise ratio (SNR), while preserving the essential feature of the signal. In the digital system the function of the filter is essentially to reduce the intersymbol interference (discussed in Fig. 3.10 in chapter-III), whereas in an analog system it is generally required to hold amplitude and phase response of the received signals within certain limits. The filters are also designed to reduce noise bandwidth and the noise levels.

The general receiver consisting of the above elements as shown in Fig. 8.7 is generally referred to as a linear channel because all operation on the received optical signal may be considered to be linear.

8.12 OPTICAL RECEIVER CIRCUITS USING BJT AND FET

Bipolar transistors and FET have three configurations. These are the common emitter or source, the common base or gate and the emitter or source follower for the bipolar transistor and FET respectively. Each connection has characteristics which will be suitable to a particular preamplifier configuration. It is therefore necessary to discuss three basic preamplifier structures which are low impedance, high impedance and transimpedance front ends.

Low Impedance Voltage Preamplifier

Figure 8.8(*a*) shows a *p-i-n* photodiode with a grounded emitter low impedance voltage preamplifier using a bipolar transistor. The circuit is quite popular because it may be designed with a low impedance and can operate over a moderate bandwidth without the need of an equalizer. Of course, this is achieved at the expense of an increased thermal noise because it offers a low effective resistance to the detector circuit. However, it is possible to reduce the thermal noise contribution of the preamplifier by choosing a transistor with a high current gain at low emitter current in order to maintain bandwidth of the stage.

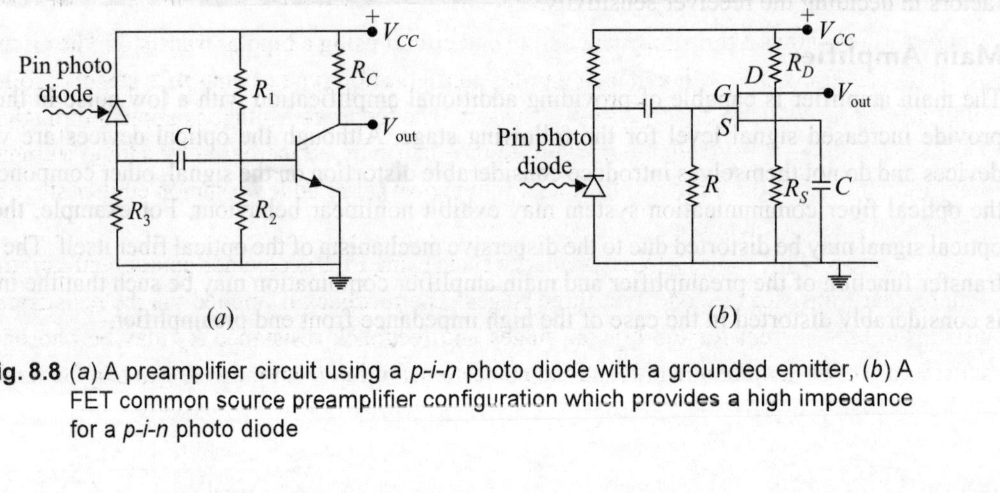

Fig. 8.8 (*a*) A preamplifier circuit using a *p-i-n* photo diode with a grounded emitter, (*b*) A FET common source preamplifier configuration which provides a high impedance for a *p-i-n* photo diode

Common Source Preamplifier using FET

The preferred preamplifier configuration for a low noise operation is either a high impedance integrating front end or a transimpedance front end amplifier. Figure 8.8(*b*) shows a FET common source preamplifier configuration. This circuit is capable of providing high impedance for the *p-i-n* photodiode. When the FET circuit operates in common source configuration, then both power gain and output impedance are considerably high. The high value of the output impedance tends to reduce the noise contribution from the following stages. This is especially true, when the voltage gain of the common source amplifier is minimized so that the gate-to-drain Miller's capacitance of the FET is decreased. This may be achieved by following the common source stage with a stage having a low value of input impedance. Figure 8.9(*a*) shows a grounded source FET followed by common emitter amplifier where shunt feedback is applied through resistance R_1 Fig. 8.9(*b*) shows a cascode amplifier where the second stage is a common base amplifier and with the initial grounded FET forms the cascode amplifier. Both circuits have high input impedance.

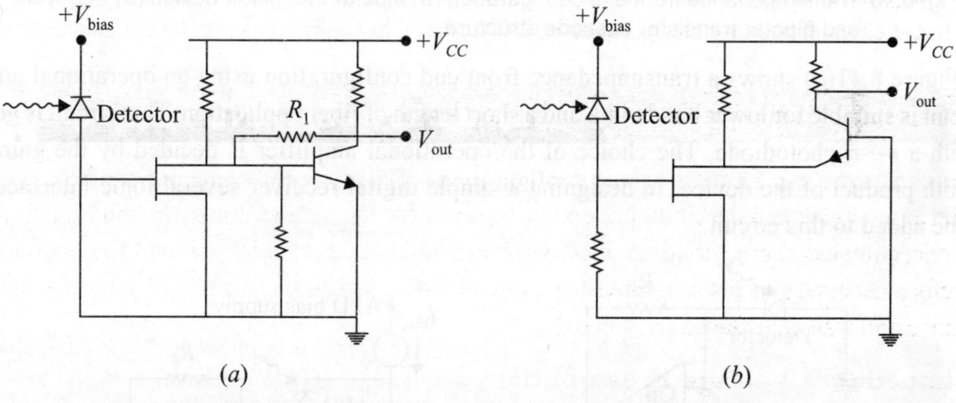

Fig. 8.9 High impedance preamplifier configuration: (*a*) a grounded source FET followed by a common emitter connection with shunt feedback, (*b*) A cascode connection. The separate bias voltage indicates the use of either a *p-i-n* or an avalanche cascode photodiode

8.13 TRANSIMPEDANCE FRONT END CIRCUITS

Figure 8.10(*a*) shows a transimpedance front end circuit using a bipolar transistor. The circuit shows an *n-p-n* bipolar transistor with common emitter configuration type of amplifier stage T_2 followed by an emitter follower T_1 using second *n-p-n* transistor. The feedback voltage is applied through resistor R_F. The output voltage level from the transimpedance pair may be increased by adding a common emitter amplifier stage after the emitter follower T_1 (not shown). The added common emitter stage must not be included in the feedback loop.

Figure 8.10(*b*) shows a transimpedance stage using FET in common source configuration followed by a bipolar transistor cascode over three stages. The feedback is applied via the resistance R_F. In this circuit the bias current for the bipolar stages and feedback resistance R_F may be chosen to provide satisfactory open-loop bandwidth and the noise contribution from these stages may be negligible.

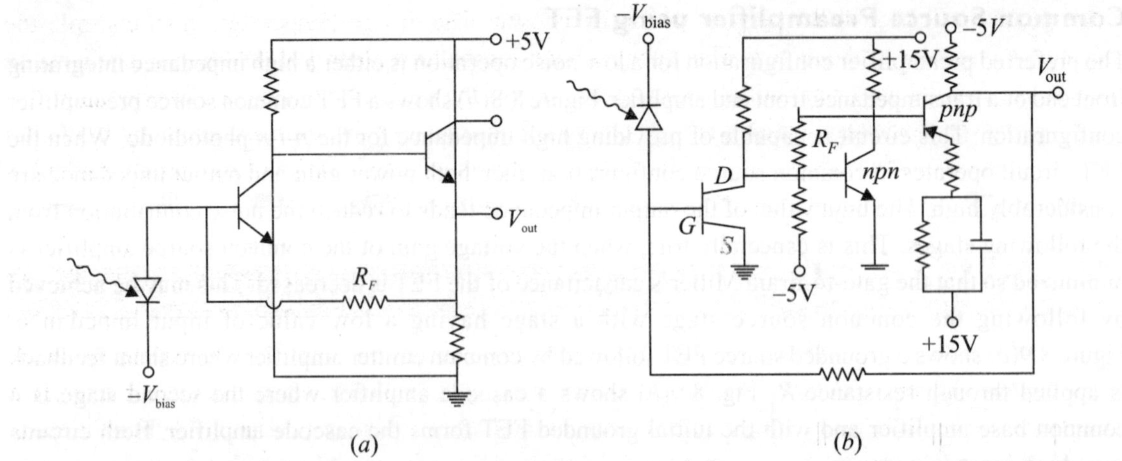

Fig. 8.10 Transimpedance front end configuration (*a*) bipolar trtansistor design (*b*) FET front end bipolar transistor cascode structure

Figure 8.11(*a*) shows a transimpedance front end configuration using an operational amplifier. The circuit is suitable for lower bandwidth and a short length of fiber application. This circuit is generally used with a *p-i-n* photodiode. The choice of the operational amplifier is decided by the gain versus bandwidth product of the device. In designing a simple digital receiver several logic interface stages should be added to this circuit.

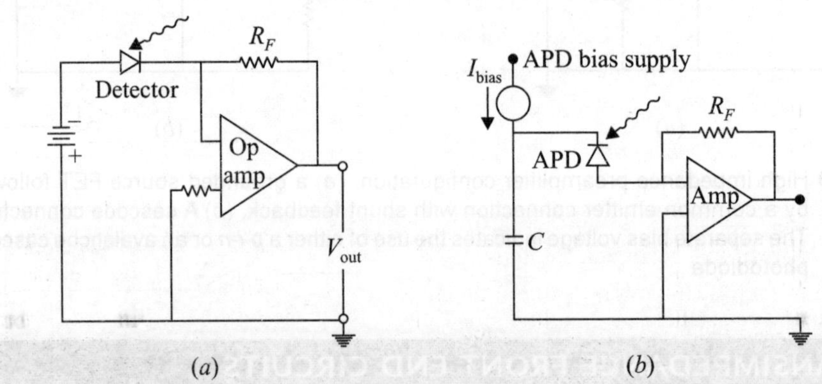

Fig. 8.11 (*a*) A typical circuit for an operational amplifier transimpedance front end. (*b*) The bias of an APD with a constant current source

8.14 AUTOMATIC GAIN CONTROL (AGC) CIRCUIT

It is essential that a receiver circuit must provide a steady reverse voltage to the photodetector for proper biasing. For a *p-i-n* photodetector this condition is not critical and a voltage ranging from 5 to 80 volts with an extremely low current can serve the purpose. The avalanche photodiode detector requires much more bias voltage lying between 100 and 400 volts. The multiplication fact or *M* of the avalanche photodiode is determined by the bias voltage. In order to minimize the signal-to-noise ratio the optimum value of multiplication factor M_{opt} is generally chosen. The device temperature also controls the multiplication factor of the device. Thus the provision for the time control of the bias voltage must be included so that the optimum multiplication factor M_{opt} of the avalanche diode is maintained.

However, the multiplication factor M can be kept constant by an automatic gain control (AGC) circuit. An additional advantage of AGC is that it reduces the dynamic range of the signals applied to the preamplifier providing increased optical dynamic range of the receiver.

8.15 CONSTANT CURRENT SOURCE TO PROVIDE SIMPLE AGC

One simple method of providing automatic gain circuit is simply to bias the avalanche photodiode (APD) with a constant d.c. current source I_{bias} as shown in Fig. 8.11(*b*). The constant source is decoupled by a capacitor C at all signal frequencies to prevent the gain modulation (Variation). When the mean optical power is known, the mean current to the APD is defined by the bias which provides a constant multiplication factor (gain) at all temperatures. Any variation in the multiplication factor will produce a variation in the electric charge on C thus adjusting the biasing of the APD back to the required multiplication factor. Therefore, the output current from the photodetector is only determined by the input current source giving full automatic gain control. However this simple AGC technique depends on a constant mean optical input power level and takes no account of the dark current generated within the photodetector.

Peak Detection and Feedback to Provide AGC

A more widely used method which allows for the effect of variations in the detector dark current while providing critical AGC is to peak detect the ac coupled signal after suitable low noise amplification as shown in Fig. 8.12. The signal from the final stage of the main amplifier is compared with a present reference level and feedback to adjust the high voltage bias supply in order to maintain a constant signal level. Thus effectively created a constant current source with dark current subtracted.

A further advantage of this technique is that it may be used to provide AGC for the main amplifier giving full control of the receiver gain.

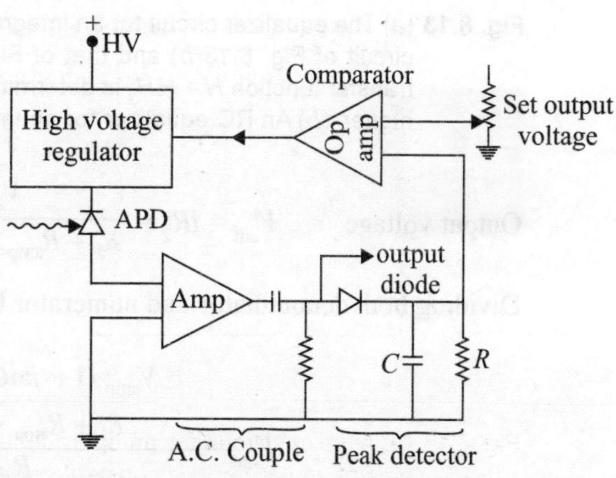

Fig. 8.12 The bias of an APD by peak detection and feedback to provide automatic gain control (AGC)

8.16 EQUALIZER

Figure 8.13(*a*) shows the configuration of an equalizer which is used in the high impedance front end amplifier. The main function of an equalizer is to compensate the loss in gain of a high impedance front end amplifier at high frequency range. Let the impedance of the parallel combination of R_1 and C of Fig. 8.13(*b*) be denoted by Z so that

$$Z = \frac{R_1}{1 + j\omega CR_1} \tag{8.16.1}$$

The current I through the circuit is

$$I = \frac{V_{\text{amp}}}{R_{\text{amp}} + Z + R_2} \tag{6.16.2}$$

Substituting the value of Z from Eq. (8.16.1), we get

$$I = \frac{V_{\text{amp}}}{R_{\text{amp}} + R_2 + \dfrac{R_1}{1 + j\omega CR_1}} = \frac{V_{\text{amp}}(1 + j\omega CR_1)}{R_1 + R_{\text{amp}} + R_2 + j\omega CR_1(R_{\text{amp}} + R_2)} \tag{8.16.3}$$

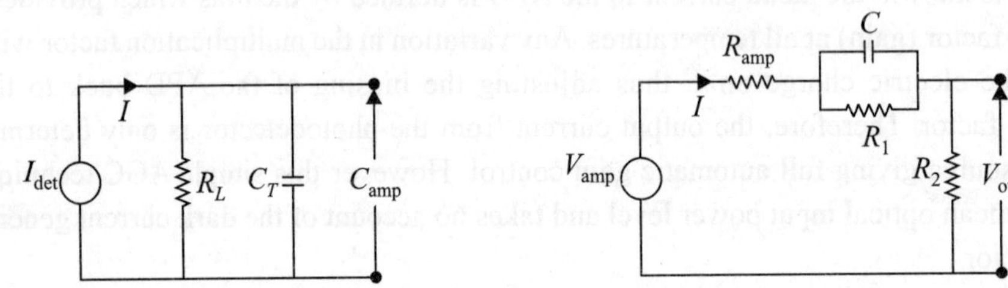

(a) (b)

Fig. 8.13 (*a*) The equalizer circuit for an integrating front end. The transfer function H_1 of the circuit of Fig. 8.13(*b*) and that of Fig. 8.13(*a*) H_2 are found out. Their combined transfer function $H = H_1H_2$ is determined. When $R_1 >> (R_{\text{amp}} + R_2)$, the bandwidth is higher, (*b*) An RC equalizer for an integrating front end circuit

Output voltage
$$V_{\text{out}} = IR_2 = \frac{V_{\text{amp}}(1 + j\omega CR_1)R_2}{R_1 + R_{\text{amp}} + R_2 + j\omega CR_1(R_{\text{amp}} + R_2)} \tag{8.16.4}$$

Dividing both denominator and numerator by $(R_1 + R_{\text{amp}} + R_2)$, we get

$$V_{\text{out}} = \frac{\dfrac{V_{\text{amp}}(1 + j\omega CR_1)R_2}{R_1 + R_{\text{amp}} + R_2}}{1 + j\omega CR_1 \dfrac{R_{\text{amp}} + R_2}{R_1 + R_{\text{amp}} + R_2}} \tag{8.16.5}$$

The transfer function

$$H_1 = \frac{V_{\text{out}}}{V_{\text{amp}}} = \frac{(1 + j\omega CR_1)R_2/(R_1 + R_{\text{amp}} + R_2)}{1 + j\omega CR_1 \dfrac{(R_{\text{amp}} + R_2)}{R_1 + R_{\text{amp}} + R_2}} \tag{8.16.6}$$

Figure 8.13(*b*) shows the equivalent circuit of the high impedance amplifier where R_L and C_T are the load resistance and total shunting capacitance so that their equivalent value Z_L is given by

$$Z_L = \frac{1}{1 + j\omega C_T R_L} \tag{8.16.7}$$

Transfer function

$$H_2 = \frac{V_{amp}}{I_{det}} = \frac{1}{1 + j\omega C_T R_L} \tag{8.16.8}$$

Where H_2 is the transfer function of the front end amplifier combined transfer function

$$H_1 \cdot H_2 = \frac{1}{1 + j\omega C_T R_L} \times \frac{(1 + j\omega C R_1)\dfrac{R_2}{R_1 + R_{amp} + R_2}}{H + j\omega C R_1 \dfrac{R_{amp} + R_2}{R_1 + R_{amp} + R_2}} \tag{8.16.9}$$

If we choose R_1 and C such that

$$\frac{1}{R_1 C} = \frac{1}{R_L C_T} \tag{8.16.10}$$

The combined transfer function

$$H = \frac{\dfrac{R_2}{R_1 + R_{amp} + R_2}}{1 + j\omega C R_1 \dfrac{R_{amp} + R_2}{R_1 + R_{amp} + R_2}} \tag{8.16.11}$$

At a frequency $\omega = \omega_2$ the right part of denominator is 1, where ω_2 is the 3dB cutoff frequency

so that
$$\omega_2 C R_1 \frac{(R_{amp} + R_2)}{R_1 + R_{amp} + R_2} = 1 \tag{8.16.12}$$

and
$$\omega_2 = \frac{R_1 + R_{amp} + R_2}{C R_1 (R_{amp} + R_2)} \tag{8.16.13}$$

Picking $R_1 >> (R_{amp} + R_2)$ will ensure a higher bandwidth than the integrating front end alone.

8.17 DIGITAL RECEIVER PERFORMANCE

In a digital receiver the amplified and filtered signal coming out from the equilizer is compared to a threshold level once per unit time slot to determine whether or not a pulse is present at the photodetector at that time slot. In an ideal case the output signal $V_{out}(t)$ would always exceed the threshold voltage when a "1" is present and would be the less than the threshold voltage when no pulse (*a* 0) was sent. But in actual systems, there are deviations from the average value of $V_{out}(t)$. These deviations are caused by various noises, interference from adjacent pulses and conditions wherein the light source is not completely extinguished during zero pulse.

There exists several standard methods of measuring the rate of error occurrences in a digital data stream. One common approach is to divide the number of errors N_e occurring over a certain interval t by the total number of pulses N_t (ones and zeros) transmitted during this interval. This is called either the error rate or bit error rate which is commonly abbreviated as BER. Thus we have

$$\text{BER} = \frac{N_e}{N_t} = \frac{N_e}{B_t} \tag{8.17.1}$$

Where $N_t = B_t$ and $B_t = 1/T_b$ is the bit rate pulse transmission rate. The bit error rate is expressed by a number such as 10^{-6} for example, which states that on the average one error occurs for every million pulses sent. The typical error rates for optical fiber communication system range from 10^{-6} and 10^{-10}. This error rate depends on the signal-to-noise ratio of the receiver.

8.18 RECEIVER SENSITIVITY AND BIT-ERROR RATE

The important function of a digital receiver is to take a weak optical signal and convert it into an electrical signal. Then it is to decide whether the electrical signal output represents a logical 1 or 0 with the help of a comparator (Fig. 8.15) and to generate an electronically suitable voltage respresentative of the logic state. Since the output of the detector-preamplifier combination is contaminated by noise, the next step in our analysis is consideration of the threshold voltage of the comparator (*i.e.*, the detection level). The bit error rate (BER) will be partially determined by the value of threshold voltage. By removing the amplification effect of the preamplifier, the threshold voltage can be converted into an equivalent threshold current at the output of a diode. We are interested to find whether the instantaneous current contaminated by noise is above or below the threshold current.

Figure 8.14 illustrates the concept of recovery of a signal. The top curve (*a*) represents a noise-free output from an ideal detector and the threshold current is shown to be I_D. The noise-free signal Fig 8.14(*a*) gives the current logical values which represent the perfect recovery of the transmitted signal. The middle curve (*b*) shows the noise contaminated signal. For the threshold shown in the bottom curve is produced at the comparator output with one of the output bit in error. The number of bit error in the recovered signal is a function of threshold current I_D.

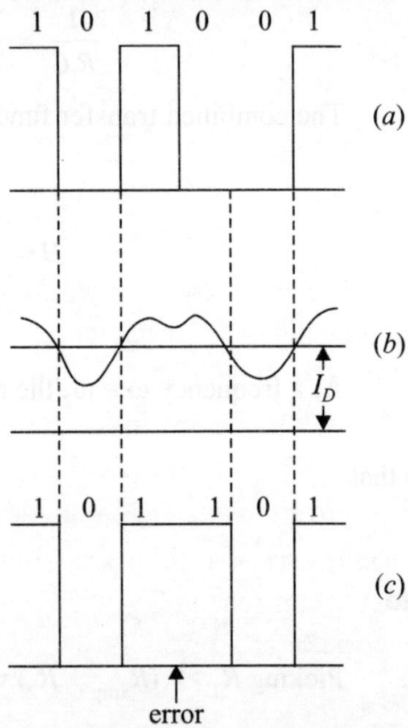

Fig. 8.14 The concept of recovery of a signal. (*a*) A noise-free output from an ideal detector. (*b*) Noise contaminated signal. (*c*) Recovered signal with the help of a comparator with a bit error

We will use the simplest assumption to illustrate the design procedure. First assumption is that the output currents in the detector are the Gaussian random variables. The Gaussian probability distributions for the current is expressed by

$$P(i_N)\, di_N = 1/\sqrt{2\pi}\,\sigma \exp(-i_N^2/2\sigma^2)\, di_N \tag{8.18.1}$$

where $P(i_N)\, di_N$ is the probability of the output noise current being between the values i_N and $(i_N + di_N)$ and the standard deviation σ is a measure of the width of the probability distribution. The mean square noise current predicted from consideration of noise generation mechanism, is equal to the square of standard deviations *i.e.*,

$$\sigma^2 = \langle i_N^2 \rangle \tag{8.18.2}$$

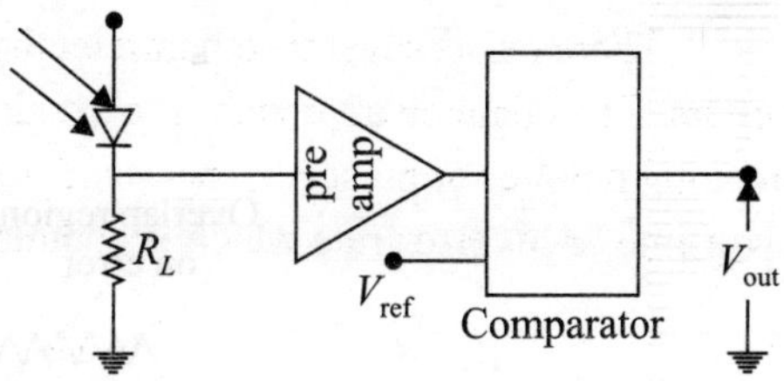

Fig. 8.15 Block diagram of a comparator for recovering correct signal from the contaminated signal. The output voltage from the preamplifier is applied as the input voltage of the comparator

we shall represent now the mean value of the receiver current with an incident logical "1" with $\overline{i(1)}$ with a variance of σ_1; Similarly, we shall denote the mean value of the current when a "0" is received by $\overline{i(0)}$ with a variance σ_o^{-2}. (For simplicity, we shall assume that $i(0) = 0$, since this is correct in an optical fiber link.)

If a logical "0" is sent, we find that an error will be made if the noise current is positive and of amplitude

$$i_N > I_D \tag{8.18.3}$$

Similarly, a given input is logical "1", we would have an incorrect output, if the noise current contribution i_N were negative and met the condition $i_N < I_D$, where i_N is the noise current

$$i_N < I_D \tag{8.18.4}$$

The probability of making one error or the other depends on the statistics of the data stream. The total probability of making an error is the sum of probability of calling a "1" as a "0" when a "1" was sent plus the probability of calling a "0" as a "1" when a logic a "0" was sent.

Mathematically this (total probability of making an error) is written as

$$P_e = P(0 \mid 1)\, P(1) + P(1 \mid 0)\, P(0) \tag{8.18.5}$$

where $P(0 \mid 1)$ is the probabity of deciding that the output is a "0" when a "1" was sent. $P(1)$ is the probability when a "1" is sent. Similarly $P(1 \mid 0)$ is the probability of deciding that the output is a "1" when a "0" is sent and $P(0)$ is the probability that a "0" is sent. If $P_1(1)$ is "a" and $P(0)$ is "b", then the combined error probability is P_e. P_e will be given by

$$P_e = aP[i_N < -(1-K)\ \overline{i(1)}\,] + bP(i_N > K\overline{i(1)}\,] \tag{8.18.6}$$

where $P(.)$ is the probability of an event occurring and K is a fraction less than 1.

Now a simplifying assumption is made that most messages can assuming logical "1"s or "0"s are equally probable so that $a = b = 1/2$ and

$$P_e = \frac{1}{2} P(i_N < -(1-K)\,[\overline{i(1)} + \frac{1}{2} P(i_N > Ki(1)] \tag{8.18.7}$$

From Fig. 8.16 we find that the error of Eq. (8.18.7) is the area underneath the tail of the upper probability distribution below the threshold I_D is shown by shaded area.

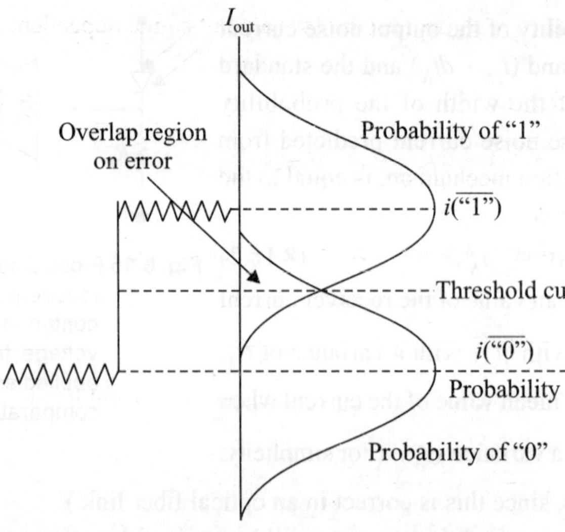

Fig. 8.16 Probability density curve for a logical "1" and a logical "0"

From Fig. 8.16 we noticed that Eq. (8.17.7) is the area beneath the tail the of the upper probability distribution below the threshold current I_D that

$$P(0\mid 1)=\frac{1}{\sigma_1\sqrt{2\pi}}\int_{-\infty}^{I_D}\exp\left(\frac{I-\overline{i(1)}^2}{2\sigma_1^2}\right)dI=\frac{1}{2}\,\text{erfc}\,\frac{\overline{i(1)}-I_D}{\sigma_1\sqrt{2}} \tag{8.18.8}$$

where the complimentary error function erfc(x) is defined as

$$\text{erfc}\,(x)=\frac{2}{\sqrt{\pi}}\int_x^{-\infty}e^{-y^2}\,dy \tag{8.18.9}$$

The error $P(1\mid 0)$ is the area under the tail of lower distribution curve above the threshold current I_D. It is given as

$$P(1\mid 0)=\frac{1}{\sigma_1\sqrt{2\pi}}\int_{i_D}^{\infty}\exp\left(-\frac{I_D-\overline{i(0)}^2}{2\sigma_o^2}\right)dI=\frac{1}{2}\,\text{erfc}\left(\frac{I_D-\overline{i(0)}}{\sigma_o\sqrt{2}}\right) \tag{8.18.10}$$

Hence the probability of error which is referred to as bit error rate BER is given as

$$P_e=\text{BER}=\frac{1}{4}\left[erfc\left(\frac{\overline{i(1)}-I_D}{\sigma_1\sqrt{2}}\right)+\text{erfc}\left(\frac{I_D-\overline{i(0)}}{\sigma_o\sqrt{2}}\right)\right] \tag{8.18.11}$$

our final aim is to select the value of I_D (threshold current) to minimize the error. It can be shown that the optimum value is one that satisfied the following relationship.

$$\frac{\overline{i(1)}-I_D}{\sigma_1}=\frac{I_D-\overline{i(0)}}{\sigma_0} \tag{8.18.12}$$

or

$$I_D=\frac{\sigma_0\overline{i(1)}+\sigma_1\overline{i(0)}}{\sigma_0+\sigma_1} \tag{8.18.13}$$

In general it is found that $\sigma_1 > \sigma_0$ because larger value of signal dependent shot noise for a "1" is greater than that for a "0".

We define the parameter Q as

$$Q = \frac{\overline{i(1)} - I_D}{\sigma_0} = \frac{I_D - \overline{i(0)}}{\sigma_0} \tag{8.18.14}$$

With the selection of optimal threshold setting Eq. (8.18.11) is

$$\text{Bit error rate BER} \quad = \frac{1}{2}\,\text{erfc}\left(\frac{Q}{\sqrt{2}}\right) \tag{8.18.15}$$

When it is assumed that $Q > 3$, a useful approximation to the right hand side of (Eq. 8.18.15) is

$$\text{BER} = \frac{1}{2}\,\text{erfc}\left(\frac{Q}{\sqrt{2}}\right) \gg \frac{e^{-Q^2/2}}{Q\sqrt{2\pi}} \text{ for } Q > 3 \tag{8.18.16}$$

This approximation is correct for $Q > 3$. A graph of BER versus the value of Q is shown in Fig. 8.17.

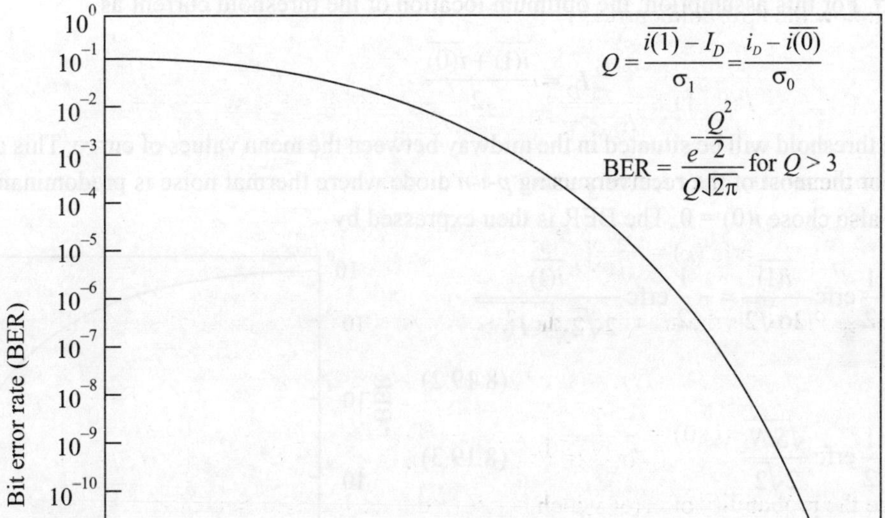

Fig. 8.17 A curve for the approximation for BER versus Q

A plot of this approximation is shown in Fig. 8.17. The values of Q from the approximation for several values of BER used fiber communication systems are given in Table 8.1.

Table 8.1

BER	*Q exact*	*Q approximate*
10^{-6}	4.753	4.76
10^{-7}	5.199	5.21
10^{-8}	5.614	5.62
10^{-9}	6.000	5.998
10^{-10}	6.36	6.361

Table 8.1 values of several values of BER (using both exact formula and the approximate formula.

8.19 SIMPLIFIED MODEL

We now make another simplifying assumption that the probability distribution have equal width so that $\sigma_0 = \sigma_1 = \sigma$. For this assumption, the optimum location of the threshold current as

$$I_D = \frac{\overline{i(1)} + \overline{i(0)}}{2} \tag{8.19.1}$$

The threshold will be situated in the midway between the mean values of curve. This assumption is correct for the most of the receivers using *p-i-n* diode where thermal noise is predominant.

We also chose $i(0) = 0$. The BER is then expressed by

$$\text{BER} = \frac{1}{2}\,\text{erfc}\,\frac{\overline{i(1)}}{2\sigma\sqrt{2}} = -\frac{1}{2}\,\text{erfc}\,\frac{\overline{i(1)}}{2\sqrt{2}\sqrt{<I_N^2>}} \tag{8.19.2}$$

$$= \frac{1}{2}\,\text{erfc}\,\frac{\sqrt{S/N}}{2\sqrt{2}} \tag{8.19.3}$$

We have noted that $\dfrac{\overline{i(1)}}{\sqrt{<1_N^2 i>}}$ is $\sqrt{S/N}$ for our definition of signal to noise ratio S/N. A graph of Eq. (8.19.3) is shown in Fig. 8.18 where the error probability is plotted vertically and the corresponding $\sqrt{S/N}$ is plotted horizontally.

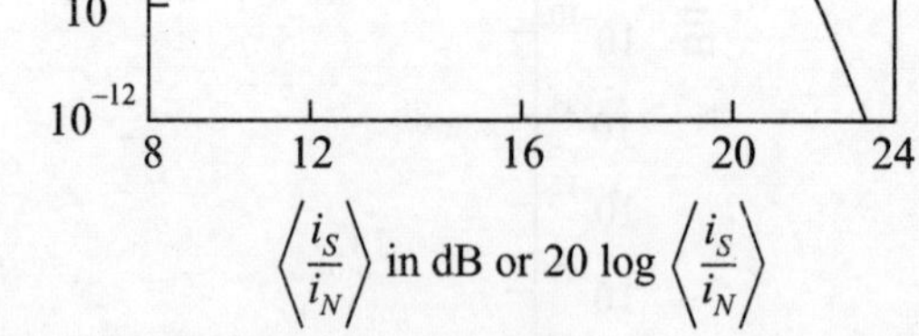

Fig. 8.18 Dependance of bit error rate or signal-to-noise ratio when $\sigma_0 = \sigma_1 = \sigma$ and $\Gamma_{(0)} = 0$ (Here $i_S = \overline{i(1)}$)

To use this curve one begins we assume a desired error probability (BER) and determine signal to noise ratio. For example, a BER of 10^{-9} is a probability-erf-error of 10^{-9} and is frequently used in modern digital communication. From the graph we estimate the required ratio of $i_s\sqrt{< I_N^2 >}$ as 21.5 dB or a value of 11.89. From the knowledge of the limiting noise source of the detector (*i.e.*, the thermal noise in a *p-i-n* diode or shot noise for an APD) we can calculate $\sqrt{<I_N^2>}$. Knowing the required signal-to-noise ratio $\sqrt{S/N}$ we find the minimum RMS signal current from

$$<I_s>_{mm} = \sqrt{(S/N)_{min}}\sqrt{<I_N^2>} \quad (8.19.4)$$

The signal current is product of the optical power and the responsivity R of the detector (*i.e.*, is $= RP$) giving

$$P_{min} = \frac{\sqrt{(S/N)_{min}}\sqrt{<I_N^2>}}{R} \quad (8.19.5)$$

This optical power must be available at the detector to achieve the desired error rate. We can then use this value to investigate possible trade-offs between the fiber loss, connector loss coupling efficiency and source power levels to complete the design.

Example 8.4: *Calculate the minimum power required to maintain a bit error of 10^{-6} achieved with a photodiode detector with a responsivity 0.4 A/W. Assume that the signal-to-noise ratio is limited by-thermal noise with 50 Ω. load, 400 K noise temperature and a 10 MHz noise bandwidth.*

Solution: For BER = 10^{-6} we find from Fig. 8.13

$$20 \log \frac{<i_S>}{<i_N>} = 19.6 \text{ dB}$$

$$\frac{<i_S>}{<i_N>} = 10^{dB/20} = 10^{19.6/20} = 1^{0.98} = 9.55$$

Since $<I_N> = \sqrt{<I_N^2>}, B = 10 \text{ MHz}$

$$<i_S> = 9.55 <I_N^2> = 9.56\sqrt{\frac{4KT\,B}{R}} = 9.55\sqrt{\frac{4\times1.38\times10^{-2}\times400\times10^{7}}{50}}$$

$$= 6.35 \times 10^{-7} \text{ A} = 635 \text{ nA}$$

We then find P_{min} as

$$P_{min} = \frac{<i_S>}{R} = \frac{635\times10^{-9}}{0.4} = 1.587\times10^{-6} \text{ W}$$

$$= 1.587 \ \mu\text{W}$$

Various simplifying assumptions have been made in our analysis. The assumption of Gaussian noise reduced by the complexity of problem, but produces an $1/\sqrt{<i_N^2>}$ value that is too low.

We have also assumed that $i(0) = 0$ that a logical 0 are equally probable. This assumption caused us to assume that the threshold was midway between $i(0)$ and $i(1)$.

Furthermore we have implicately assumed a noise-free amplifier and that is a negligible pulse spreading in the received pulses.

This last effect can lead to a form of noise termed Inter Symbol Interferences (ISI) Chapter 3. As may be observed in Fig. 8.10 that any energy that splits out of the expected pulse position is lost to the receiver for detecting the pulse of interest. If large enough, this energy might trigger a false detection in the adjacent slot.

While this inter symbol interference (ISI) may be partially eliminated by designing a proper equalization amplifier at the expense of increased thermal noise. It has been found that the intersymbol interference is negligible if the bit period T_b is kept larger than 4 times the r.m.s. pulse spreading of the fiber Γ_{rms}.

$$T_b \geq 4\Gamma_{rms}.$$

8.20 OEIC (OPTOELECTRONIC INTEGRATED CIRCUITS)

Introduction

It is a well recognized fact that the marvellous developments in the electronics field are due to the key technology of the monolithic integration of silicon devices. The progress in opto electronic devices will also follow that of silicon device integration processes. It is reasonable to expect, as the next technological step, an effort to be made in the development of monolithic integration of opto-electronic devices as well as electronic devices.

Generally, most optical elements should be connected with electronic circuits. However, stray capacitances and inductances appear at the connected regions. It is expected that OEICs, which are monolithically integrated optical elements and electronic circuits, will resolve this crucial problem. It is clear that OEICs offer the advantage of small packaging, high reliability and low cost for practical use. Furthermore, it is anticipated that the development of the OEIC will produce superior performance for each optical element and electronic circuit.

The optoelectronic integrated circuit (OEIC) receiver is an emergent technology for monolithic integration of optoelectronic and electronic devices on the same semiconductor substrate. Initial research work in the area of OEIC receivers was confined to the development of hybrid integrated circuits (IC's) in which discrete devices on separate functional blocks or chips are connected using electronic (leads) or optical (fiber) interconnects. Later on, attention was focused on the development of a monolithically integrated photoreceiver which offers significant advantages over hybrid forms.

History

In 1978 Lee reported the first OEIC with the integration of the Laser and the Gunn diode. Since then, monolithic integration of Laser or photo diode with electronic circuits on a single substrate has been considered as an effective approach in the realization of high-speed, highly reliable and compact systems. Later monolithic integration of an LD and field-effect transistors (FETs) or a PD and FETs was possible. However, most efforts have been focused on AlGaAs/GaAs systems which operate at a wavelength of about 0.8 μm, probably because the materials of AlGaAs/GaAs are widely used and because metal-semiconductor (MES) FETs can readily be fabricated on those materials. However, in view of the smaller transmission loss of optical fibers, OEICs operating in the 1.3 to 1.6 μm range would be more useful in fibre-optic communication systems which uses InGaAsP/InP systems.

Material for OEICs

The integration of several optical components on a single substrate is still a technological challenge. Several problems are being encountered in the development of high quality and reliable integration on a single substrate. The main requirements of the material for an OEIC are; transparency and good optical properties in the appropriate part of electromagnetic spectrum. (such as generation and detection of light in that part of electromagnetic spectrum), feasibility of fabrication of low loss thin film waveguides for optical interconnects, good electrical, electrooptical and acousto-optical properties. Obviously, some of these requirements are incompatible with each other. Hence, all of the above mentioned requirements cannot be satisfied in a single material. However, III-V semiconductor materials and their alloys (*i.e.*, GaAs, InP, InGaAsP, InGaAs) satisfy best the above mentioned requirements. Although, earlier optical fiber communication systems were designed at GaAs emission wavelength, however now InP based materials are being considered as more important material according to recent researches.

Types of OEIC

Technologies for integrating optoelectronic devices and electronic circuitry can be classified as either hybrid or monolithic.

Hybrid: This combines the best passive and active components from different materials as conceptually shown in Fig. 8.19. Hybrid integration involves combining optoelectronic devices and integrated circuits in the same package or substrate. This can be done using traditional hybrid techniques for simply combining packaged devices on a ceramic substrate, but the device density of the resulting optoelectronic integrated circuit (OEIC) is very low and many of the advantages of using optics are lost.

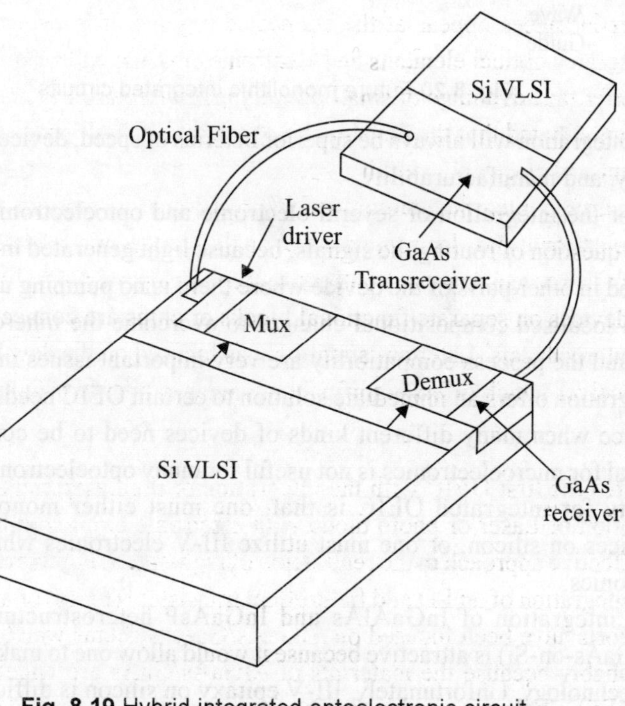

Fig. 8.19 Hybrid integrated optoelectronic circuit

Consequently, alternative hybrid techniques have been proposed to increase device density while maintaining manufacturability. Recently, one of the most useful techniques involve flip-chip/solder-ball or -bump integration of discrete optoelectronic devices on multi-chip modules or directly on silicon IC chips. The disadvantages of these techniques are they consume a large amount of area.

Monolithic: The second approach is to use monolithic integration. This requires that all the components be fabricated on a single substrate, essentially from the same material as schematically shown in Fig. 8.20. This results in the low value of parasitic capacitance.

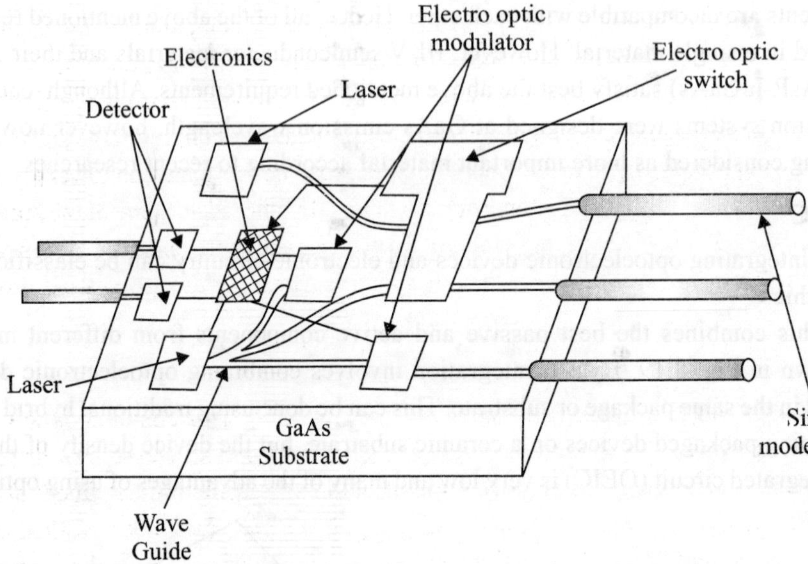

Fig. 8.20 Future monolithic integrated circuits

Monolithic integration will always be superior in terms of speed, device density, system reliability, ultimate complexity, and manufacturability.

However, for the integration of several electronic and optoelectronic components on a single substrate raises the question of routing the signals, because light generated in one part of the device will be strongly absorbed in other parts of the device where there is no pumping unless sophisticated wafers are fabricated with localised compositional changes to overcome the inherent absorption of photons. Also the structure and the process compatibility are very important issues in this approach.

Hybrid integration offers an immediate solution to certain OEIC needs, and it may always be the technology of choice when many different kinds of devices need to be combined. Silicon, the most widely used material for microelectronics is not useful for many optoelectronic applications. Therefore, the major difficulty for integrated OEIC is that, one must either monolithically integrate III-V optoelectronic devices on silicon, or one must utilize III-V electronics which is less developed than silicon microelectronics.

Monolithic integration of InGaAlAs and InGaAsP heterostructures on silicon (a process generically called GaAs-on-Si) is attractive because it would allow one to make use of the well developed silicon BiCMOS technology. Unfortunately, III-V epitaxy on silicon is difficult as the lattice constants of GaAs and Si differ by 4%, and that the thermal expansion coefficients differ by almost 50%. The lattice mismatch can be overcome by special growth initiation techniques and interfacial buffer layers, but the thermal mismatch is more troublesome. Significant numbers of defects (and even cracks) are introduced as GaAs-on-Si structures are cooled from the growth temperature to room temperature, and the III-V layers are under severe tension, which reduces device lifetimes dramatically, reduces reliability, and in general adversely affects performance. The most promising solution to this incompatibility involves introducing an AlAs layer between the growth-initiation layers and the active device heterostructures, and subsequently etching this AlAs layer away to physically separate the III-V device structure from the Si, but otherwise leave it monolithically integrated on the chip.

The difficulty of fabrication on silicon and advances in III-IV compounds related studies has led to a shift of emphasis to integration using GaAs or InP. There are two approaches. The first one is known as "epitaxy-first" or "stacked-heterostructure" methods. It involves first growing all of the epitaxial layers needed for both the optoelectronic devices and electronic devices on a suitable substrate. The top layers are then removed selectively in areas where devices are to be fabricated on lower layers and each of the various device types is formed in the appropriate layers. In other idea, some of the epitaxial layers are grown, they are etched away where they will not be needed, and more epitaxial layers are grown in newly cleared areas; this approach results in a more planar surface but the growth and processing are much more complex.

Applications of OEICs

(i) *Applications of OEICs in communication systems:* Most optical communication systems in practical use are based on optical-electrical and electrical-optical conversion or regeneration for transmission, detection and amplification at both ends, while optical fibers are used as a transmitting medium for optical signal transmission. OEICs have been studied extensively due to their potential use in future optical communication systems. Using these optoelectronic circuits, conventional time division multiplexing (TDM) and frequency division multiplexing (FDM) techniques can be implemented. The more advance approach would be to take advantage of wavelength division multiplexing (WDM) technique. In such a system specific wavelengths are assigned to separate channels for the transport of large number of signals via fiber by using wavelength selectivity to sort and to route signals to their final destinations in a subscriber loop.

(ii) *Applications of OEICs in optical fiber local area networks:* As computing environment evolves toward the distributed network scheme, with data sharing among remote locations, the overall operating efficiency increasingly relies on the efficiency of data links. Due to this reason, activities in the field of local area network (LAN) in general, and more recently in the field of optical fiber local area network (OFLAN) has intestified during the last few years as a result of increasing demand for information technology. In recent years there has been tremendous interest in OEICs that found its application in Optical Fiber LAN technology.

(iii) *Application of OEICs in optoelectronic processor:* A two-dimensional OEIC processor structure would consist of holographic memory along with all necessary signal processing and deciding circuitry. In this type of OEICs, by making use of WDM technique, multiple holographic layers can be used for large data storage which will find its application in computing system.

REVIEW QUESTIONS

8.1 Describe with a relevant diagram with the signal path through optical data link via transmitter, fiber and receiver giving the nature of the signal wave forms.
What is a bit period? The bit frequency of the link is 10^7 Hz. On the average one error is encountered in a second, find the value of the BER (bit error rate)

8.2 Give the block diagram of a receiver showing different types of noise generated giving the expression for each type of noise.

8.3 A high input impedance type of preamplifier is used in an optical receiver and the effective input resistance is 3 MW. The amplifier is matched with a detector having the bias resistance of 3 MΩ estimate

(*a*) Maximum bandwidth that can be achieved without equalization if the total capacitance is 5 pf.

(*b*) The mean square thermal noise current generated by the impedance R_{TL} (parallel combination of two 3 MΩ when the operating temperature is 27°C for unity bandwidth.

(*c*) Calculate the values calculated in (*a*) and (*b*) with those obtained from high impedance is replaced by a transimpedance front and amplifier with feedback resistance R_F of 100 KΩ. and the open loop gain of the amplifier is 500. It may be assumed that $R_F \approx R_{TL}$ and the total effective capacitane is 5 pf.

8.4 Give the theory of equalization circuit. Show that after equalization the high frequency amplification is improved and the high frequency response curve is flat.

8.5. Calculate the minimum optical power required to maintain the bit error rate (BER) at 10^{-7} of a *p-n* photodiode detector whose responsivity is 0.5 A W^{-1}.

CHAPTER 9

Optical Fiber Measurements

CHAPTER FOCUS

- *Introduction: Attenuation measurement*
- *Optical time domain reflectometer (OTDR)*
- *Time domain measurement of dispersion*
- *Frequency domain dispersion measurements*
- *Sweep frequency techniques for frequency domain dispersion measurements*
- *Numerical aperture measurement using scanning photo detector and rotating stage*
- *Fiber numerical aperture measurements*
- *Interferometric Method: Refractive index profile measurements*
- *Outer diameter by shadow method*
- *Core diameter*
- *Measurement of losses in splice and connectors*
- *Review Questions*

9.1 INTRODUCTION: ATTENUATION MEASUREMENT

The main emphasis in this chapter will be on the description of different techniques for the measurement of several important parameters of an optical fiber. These are:

1. Measurement of attenuation of signal in an optical fiber by cutback technique.
2. Measurement of fiber attenuation and splice or connector loss by optical time domain reflectometer (OTDR) technique.
3. Measurements of time domain fiber dispersion.

4. Measurements of frequency domain fiber dispersion.
5. Refractive index profile of fiber.
6. Fiber numerical aperture measurement.
7. Fiber diameter measurement.
8. Measurement of signal attenuation in a splice or connector.

Signal Attenuation Measurements

There are two basic methods for the measurement of attenuation of a signal. These are: (1) Cutback Technique and (2) Optical Time Domain Reflectometer (OTDR) Technique.

Cut-back Technique

Figure 9.1 illustrates the experimental arrangement for the measurement of attenuation of a signal by "Cut-back Technique." The method is a destructive one requiring access to the input and output ends of the fiber.

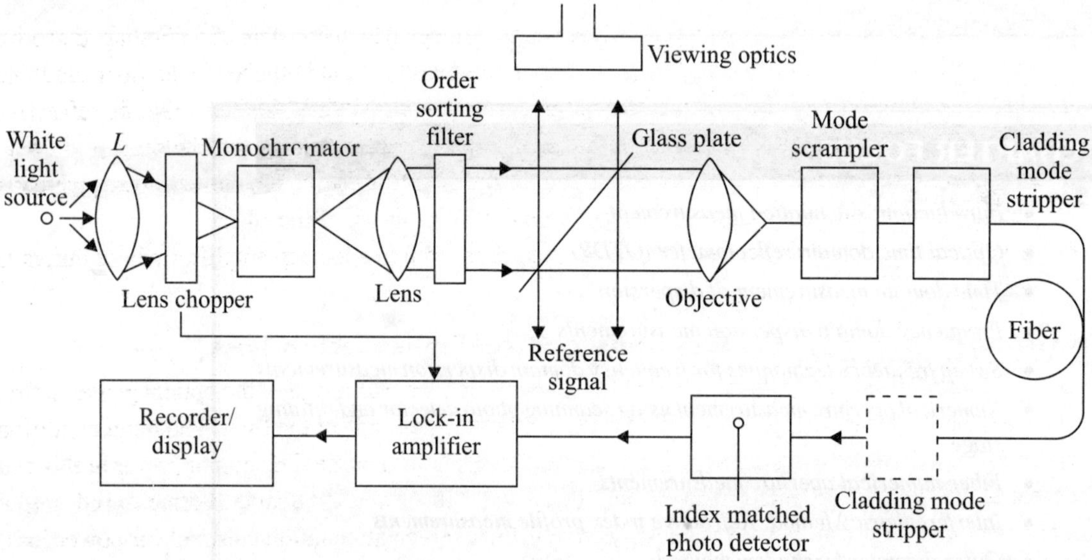

Fig. 9.1 A typical experimental arrangement for the measurement of attenuation of a signal in an optical fiber using the cutback technique

Description of Experimental Set-up

1. The apparatus consists of a white light source usually a tungsten halogen or Xenon arc lamp. The light is focussed by lens *L* and mechanically chopped at a low frequency of few hundred cycles per second so that the light path becomes open or closed. This enables the lock-in-amplifier in the receiver to make the phase sensitive detection.
2. The chopped light is applied to a monochromator which is a prism or diffraction grating arrangement to select the desired wavelength at which the attenuation of the signal is to be measured.
3. Thus, the light is filtered out and focussed on the fiber by means of a microscope objective lens.
4. A beam splitter which consists of a glass plate making an angle of 45° with the direction of the light beam may be used before the fiber to provide light for viewing optics and a reference signal is used to compensate the output power fluctuations.

5. When the measurement is carried on a multimode fiber, it is very much dependent on launch condition of the signal and this will be illustrated in Fig. 9.2.

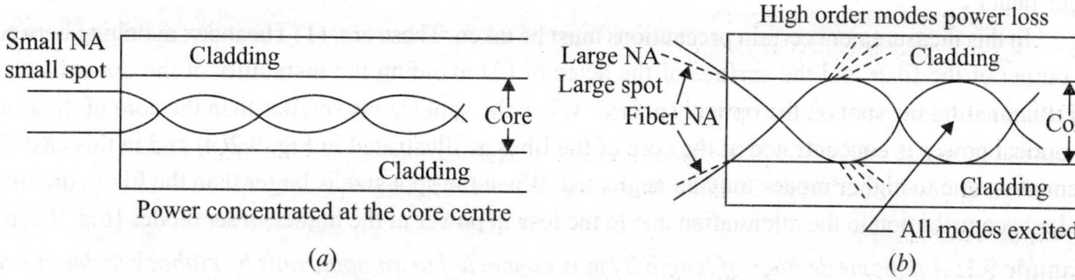

Fig. 9.2 Effect of launch on numerical aperture and spot size on the modal distribution (*a*) under filling fiber excites only lower-order modes (*b*) an over filled fiber has excess attenuation for higher order mode loss

6. The fiber is usually put through a cladding-mode stripper which consists of an *S*-shaped groove cut in teflon and filled with glycerine. The device removes the light launched in the fiber-cladding through radiation into the index matched glycerine having slightly higher value of refractive index than the material of the fiber, so that light cannot escape through the cladding.
7. A mode stripper may be put at the output end of the fiber to remove the optical power which is scattered into the core from the cladding throughout the length of the fiber.
8. Finally, the detected output from the photodetector is fed into a lock-in amplifier whose output is recorded.

Determination of Attenuation Constant (or Transmission Loss)

In order to determine the attenuation constant (or transmission loss) of a signal, the optical power is first measured at the output end of the fiber which is known as the "Far end" of the fiber. The input condition of the fiber is not disturbed and the fiber is cut few metres from the source. The output power at this end is measured and this end is referred to as the "Near end". If P_F and P_N denote the measured output optical power at the far end and near end respectively, the average attenuation constant (or power loss) in dB per Kilometer (denoted by αdB) is given by

$$\alpha dB = \frac{10}{L}\log_{10}\frac{P_N}{P_F} = \frac{10}{L_1 - L_2}\log_{10}\frac{P_N}{P_F} \tag{9.1.1}$$

where L is the difference in length in Km between the far end and near end L_1 and L_2 respectively, where $L = L_1 - L_2$. L_1 and L_2 are measured in Kilometers. αdB has the units in dB per Km.

Eq. (9.1.1) can also be written in the form

$$\alpha dB = \frac{10}{L_1 - L_2}\log_{10}\frac{V_N}{V_F} \tag{9.1.2}$$

where V_N and V_F are the measured electrical voltages at the near end and far end respectively. It may be noted that the electrical voltage ratio V_N/V_F is directly substituted in Eq. (9.1.2) for the optical power ratio P_N/P_F of Eq. (9.1.1). The reason is that optical power is directly proportional to electrical voltage.

The reason for using the destructive cutback technique is that it is extremely difficult to determine the amount of the optical power launched into the fiber.

By adopting the cutback technique it is assumed that the optical power emerging from a fiber of short length L_2 is essentially the input power of the fiber with length L (= $L_1 - L_2$) where L_1 is much larger than L_2.

In this measurement certain precautions must be taken. These are: (1) The index matching between the output of the fiber and the surface of the detector (2) avoiding the instability of the optical source and illuminating the spot on the optical surface. When the spot size is smaller than the core of the fiber, the optical power is concentrated at the core of the fiber as illustrated in Fig. 9.2(*a*) and in this case the attenuation due to higher modes may be neglected. When the spot size is larger than the fiber core there is a large contribution to the attenuation due to the loss in power at the higher-order modes [Fig. 9.2(*b*)].

Example 9.1: *A multimode fiber of length 2 km is connected to an apparatus by cutback technique for the attenuation measurement. The measured output voltage from the photodetector using 2 km length is found to be 2.1 volts at the wavelength of 0.85 μm the fiber is cutback to a length of 2 metres and the measured output voltage increased to 10.5 volts. Calculate the attenuation per Kilometer in dB/km at the wavelength of 0.85 μm.*

Solution: Using Eq. (9.1.2) for αdB, we get

$$\alpha dB = \frac{10}{L_1 - L_2} \log_{10} \frac{V_N}{V_F}$$

Data given: $L_1 = 2$ km, $L_2 = 2$ metre, $V_N = 10.7$ volts and $V_F = 2.1$ volts. Substituting the values we get

$$\alpha dB = \frac{10}{2 - 0.002} \log_{10} \frac{10.5}{2.1} = \frac{10}{1.998} \log_{10} 5 = 3.5 \text{ dB/km}.$$

Example 9.2. *A spot measurement of fiber attenuation is performed on a 1.5 km length of the optical fiber at a wavelength of 1.1 μm. The measured optical output power from 1.5 km length of the fiber is 50.1 μW. When the fiber is cut back to 2 metre in length the measured output optical power is 385.4 μW. Determine the attenuation per km for the fiber in dB/km at the wavelength of 1.1 μm.*

Solution: From Eq. (9.1.1), we get

$$\alpha dB = \frac{10}{15 - 0.002} \log_{10} \frac{385.4}{50.1} = 5.92 \text{ dB/km}.$$

9.2 OPTICAL TIME DOMAIN REFLECTOMETER (OTDR)

The optical time domain reflectometer (OTDR) method is not a destructive one and requires the access to one end of the fiber. It operates on the fundamental principle of radar system. Fig. 9.3 shows the block diagram of the experimental set up for the OTDR technique.

Description of the Experimental Set-up

1. The sharp optical pulses from an injection laser in the transmitter is launched into one end of the test fibers to propagate in the forward direction down the length of the fibers up to the other end of the fiber #2 via fiber #1 and the connectors.

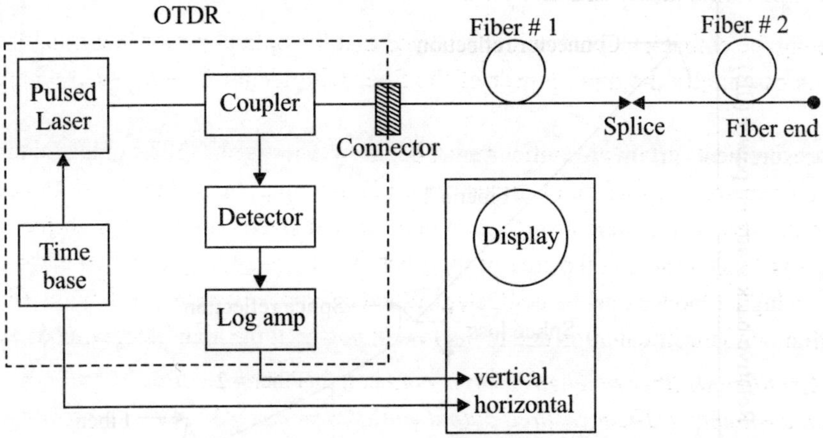

Fig. 9.3 Schematic diagram of an optical fiber time domain reflectometer (OTDR)

2. A directional coupler is used to change the direction of the light pulse. A portion of the light pulses may be directed to a photodetector which is blanked to avoid saturation when the pulses propagate in the forward direction. (Alternatively, sometime a system of lenses with beam splitter is used to change the direction of the beam of light).
3. The light pulses are reflected from the other end of the fiber #2 Via fiber #1 and connector. During the return period the reflected light pulses are directed to the detector which transform the light pulses into the electrical pulses. In the case of an ideal fiber, Rayleigh scattering will occur uniformly along the whole length of both the fibers. For fibers with uniform losses the returned pulses decline (decrease) with constant slope when the measured power is expressed in decibels. In the case of perfectly lossless fibers the returned pulses will be constant in time as the excitation light pulses traverse the length of both fibers.
4. The coupler which is connected to the photodetector allows the returned pulses to be detected by the photodetector. The detected electrical pulses pass through a logarithmic amplifier. The output of the logarithmic amplifier is applied to the vertical plates of the oscilloscope and the time base controlling the light pulses in the laser is applied to the horizontal plates of the oscilloscope so that a fixed pattern is displayed on the screen. The loss per unit length can be determined by the measurement of the slope of the pattern on the screen of the scope.

Fig. 9.4 illustrates the representive result of the experiment of Fig. 9.3. The trace starts from the detection of light reflected from the connector where the fiber under test connects the optical time domain reflectometer (OTDR) (in some system the detector is saturated by the partial reflection of the strong outgoing pulse from the coupler). During this saturation time the detector cannot detect back scatter light or any reflected light pulses. The length of the fiber corresponding to the detector saturation time is known as the OTDR "dead time zone". A length of fiber is generally included within the OTDR between the coupler and the OTDR connector to allow sufficient time for the detector to come out of the saturation. However, the loss suffered by this piece of the fiber decreases the power available to the test fiber). The back scattered light is measured from the fiber no #1 and diminishes with time and distance. The slope of the segment #1 is the loss factor or attenuation αdB for the fiber expressed in dB/km. At the splice, the discontinuity will give rise to a pulse in the reflected light. The difference between the optical power level expressed in dBm or dBμ just before the splice and the power level just after the splice is a measure of the loss introduced by the splice.

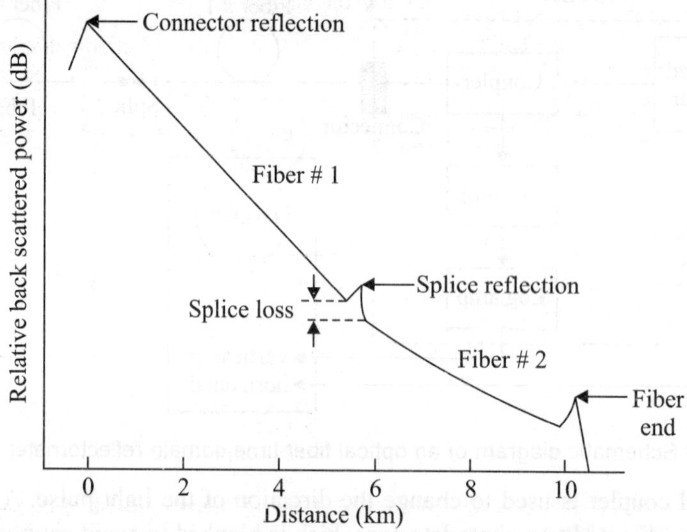

Fig. 9.4 A typical result from an OTDR experiment of Fig. 9.3

The light continues to be back scattered from the segment #2 of the fiber in same way as in the segment #1. The different fiber losses or attenuation in two fibers will be reflected as different slopes. The discontinuity at the end produces a large reflected pulse. Other fiber segments can also be added. The limit on the total losses that can be measured is determined by the dynamic range of the detector. The dynamic range may be defined as the difference between the maximum power (in dBm or dBμ) that can be detected reliably without saturation and the minimum power that can be detected as different from the receiver noise. One problem with current techniques is the limited dynamic range available. This range is approximately 20 dB in commercially available OTDR and the range 20 dB limits the length of the fiber whose attenuation can be measured.

It may be noted that the accurate information on position is also available from this trace. The technique can be used to localize breaks, abrupt dislocations and localized sources of loss. Of course, the position resolution is obtained by the pulse width of the optical source. The pulsed laser with the pulse width of the order of 10 ns are generally used in OTDR technique.

9.3 TIME DOMAIN DISPERSION MEASUREMENTS

Fig. 9.5 shows an experimental set up for the most common method of time domain measurement of optical pulse dispersion in multimode fibers.

Different parts of the experimental arrangements are the following:

1. A suitable optical source such as AlGaAs DH injection laser with fast drive circuit is used to launch a narrow optical pulse of duration ranging from 100 to 400 ps (1ps = 10^{-12} sec.) into one end of the fiber.
2. The optical pulses travel down the one kilometer length of the fiber under test and are broadened by different dispersion mechanisms. When the laser pulses with narrow spectral width are used in the case of a multimode fiber, it is possible to separate different types of dispersion mechanisms.

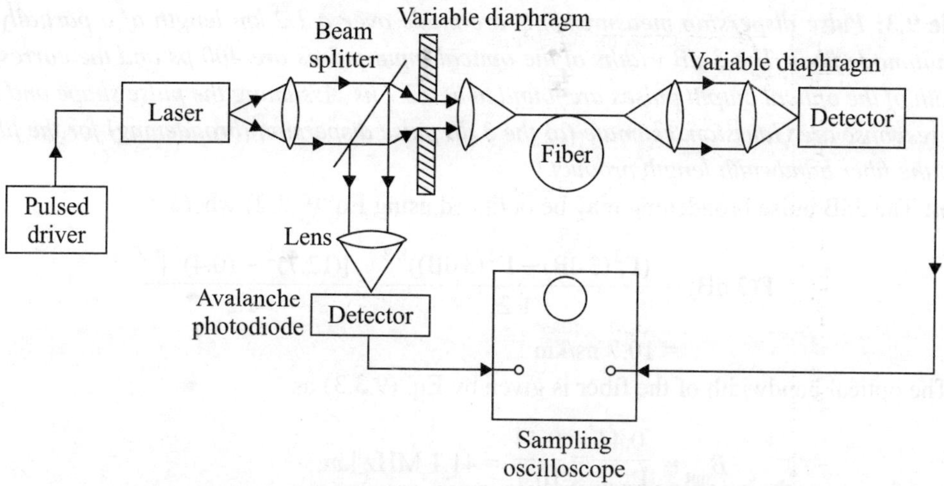

Fig. 9.5 A typical experimental arrangement for making measurement of dispersion in a multimode fiber in time domain

3. The emerging pulses from the other end of the fiber are received by a high speed photodetector such as an avalanche photo diode or a *p-i-n* photo diode and the detected electrical pulses are displayed on the screen of a fast sampling oscilloscope.
4. A beam splitter is used for triggering the oscilloscope and for the input pulse measurement.
5. After the beam splitter a variable diaphragm is used to control the width of the light beam.

The fiber dispersion can be found out from the measurement of the pulse widths. The width is measured at any convenient fraction of their amplitude. If $P_i(t)$ and $P_o(t)$ are the input and output power respectively and assuming the pulse shape to be Gaussian in shape, then we can write

$$\Gamma_0^2\,(3\text{ dB}) = \Gamma_o^2\,(3\text{ dB}) + \Gamma_i^2\,(3\text{ dB}) \tag{9.3.1}$$

where Γ_i(3 dB) and Γ_0(3 dB) are the width of the fiber input and output pulses respectively, Γ(3 dB) is the width of the fiber impulse response measured in the half power maximum amplitude. The pulse Dispersion in the fiber which is commonly referred to as the pulse broadening when we consider 3 dB pulse width expressed in ns.km^{-1} is given by

$$\Gamma(3\text{ dB}) = \frac{[\Gamma_o^2\,(3\text{ dB}) - \Gamma_i^2\,(3\text{ dB})]^{1/2}}{L}\ \text{ns.km}^{-1} \tag{9.3.2}$$

where Γ(3 dB), Γ_i (3 dB) and Γ_0 (3 dB) are measured in ns and L is the fiber length in km. It must be noted that if the length of the fiber is cutback to a short length in order to take the input pulse width measurement, then L corresponds to the difference between the two fiber lengths in km. When the launched optical pulses are Gaussian, then the 3 dB optical bandwidth for the fiber B_{opt} may be calculated using

$$\begin{aligned} B_{apt} \times \Gamma(3\text{dB}) &= 0.44\ \text{GHz ns} \\ &= 0.44\ \text{MHz ps.} \end{aligned} \tag{9.3.3}$$

Hence the value of the optical bandwidth for the fiber may be obtained from the measurement of the pulse broadening.

Example 9.3: *Pulse dispersion measurements are made over a 1.2 km length of a partially graded index multimode fiber. The 3 dB widths of the optical input pulses are 400 ps and the corresponding 3 dB width of the optical output pulses are found to be 12.7 ns. Assuming the pulse shape and the fiber impulse response are Gaussian, estimate (a) the 3 dB pulse dispersion (broadening) for the fiber in ns km^{-1} (b) the fiber bandwidth length product.*

Solution: The 3dB pulse broadening may be obtained using Eq. (9.3.2) where

$$\Gamma(3\text{ dB}) = \frac{(\Gamma_0^2(3\text{ dB}) - \Gamma_i^2(3\text{ dB}))^{1/2}}{1.2} = \frac{[(12.7)^2 - (0.4)^2]^{1/2}}{1.2}$$

$$= 10.7\text{ ns/km}$$

The optical bandwidth of the fiber is given by Eq. (9.3.3) as

$$B_{\text{opt}} = \frac{0.44}{\Gamma_{3\text{ dB}}} = \frac{0.44}{10.7} = 41.1\text{ MHz km.}$$

9.4 FREQUENCY DOMAIN DISPERSION MEASUREMENTS

Frequency domain dispersion measurements yield information on the output amplitude versus frequency of the signal applied to the fiber. This is because the *base band* frequency response $H(w)$ for a fiber may be obtained directly from this measurement using the relationship.

$$H(w) = \frac{P_0(w)}{P_i(w)} \tag{9.4.1}$$

without making an assumption for the Gaussian waveform. Hence $P_0(w)$ and $P_i(w)$ are the output and input powers respectively at a frequency w. The optical bandwidth of a multimode fiber is determined from the frequency domain measurements. There are two methods for the frequency domain dispersion measurements.

Pulsed Source Method

This method uses a similar type of pulsed source as that is used for the time domain measurement as shown in Fig. 9.5 but in this case the sampling oscilloscope is replaced by a spectrum analyzer. The spectrum analyzer makes the Fourier transform of the pulses in the time domain. The experimental arrangement is illustrated in Fig. 9.6.

1. A suitable optical source such as AlGaAs DH injection laser with a fast drive circuit is used to launch short optical pulses of duration ranging from 100 to 400 picoseconds (10^{-12} sec) into one end of the fiber.
2. The launched optical pulses travel down the one kilometer length of the fiber under test and are broadened due to different dispersion mechanisms.
3. The emerging pulses from the other end of the fiber are received and detected by a high speed photodetector such as a *p-i-n* photo diode or an avalanche photo diode. The optical pulses are transformed into electrical pulses.
4. A beam splitter is used for triggering the oscilloscope and for the measurement of the input pulses.

5. After the beam splitter a variable diaphragm is used to control the width of the light beam.
6. A spectrum analyzer takes the Fourier transforms of the pulses in the time domain and displays the constituent frequencies of the pulses.

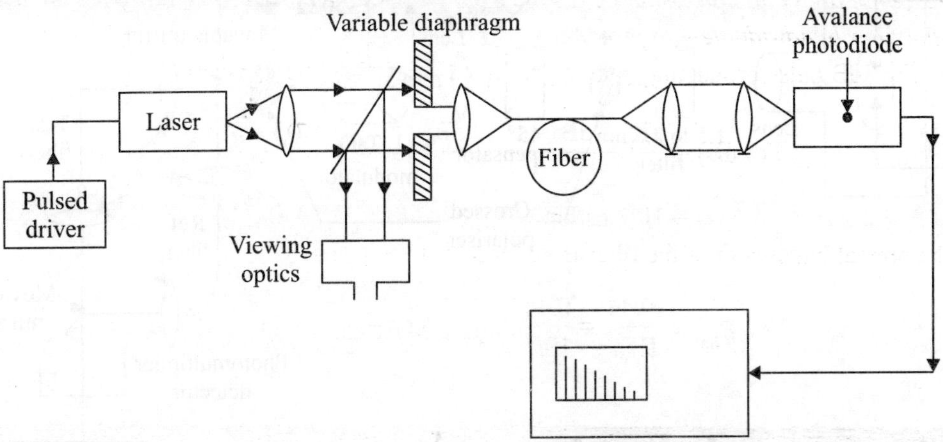

Fig. 9.6 An experimental arrangement for determining base band frequency response. Frequency domain measurement of dispersion in multimode fibers

Frequency Domain Dispersion Measurement (Frequency and Phase Response)

In this frequency domain response measurements we shall get information on amplitude versus frequency and amplitude versus phase responses. The data are more useful for the system designers than the time domain pulse dispersion measurements, especially if equalization techniques are to be performed on the detected signal at the receiver. The dispersion measurements can be made by sinusoidally modulating a continuous-wave (C.W.) light signal about a fixed level. The base band frequency response is then found from the ratio of the sinewave amplitudes at the beginning and end of the fiber. An experimental set up for determining for fiber frequency response is shown in Fig. 9.7.

After passing through a set of narrow band interference filters, light from a xenon arc lamp is focussed into a $LiTaO_3$ electro optic crystal. The intensity of the incoherent Xenon light carrier is modulated by a frequcncy tunable sinusoidal voltage which acts as the envelope of the Xenon light signal. The modulated signal is then focussed into either the fiber under test or into a short length of (approximately two metres) reference fiber. A photo multiplier tube detects the fiber output and the base band modulation components are shown on a spectrum analyzer.

There are several advantages in the frequency domain measurement technique

1. The fiber transfer function is found directly without taking the Fourier transform of time domain data. To find the impulse response in the time domain requires a cumbersome deconvolution process of the output pulse with input pulse.
2. Photodetector linearity over a wide range is not required as in the time domain measurement because of the small signal modulation about a constant light level.
3. It is easier to modulate sinusoidally an optical source at high frequencies than it is to generate a series of narrow output pulses.

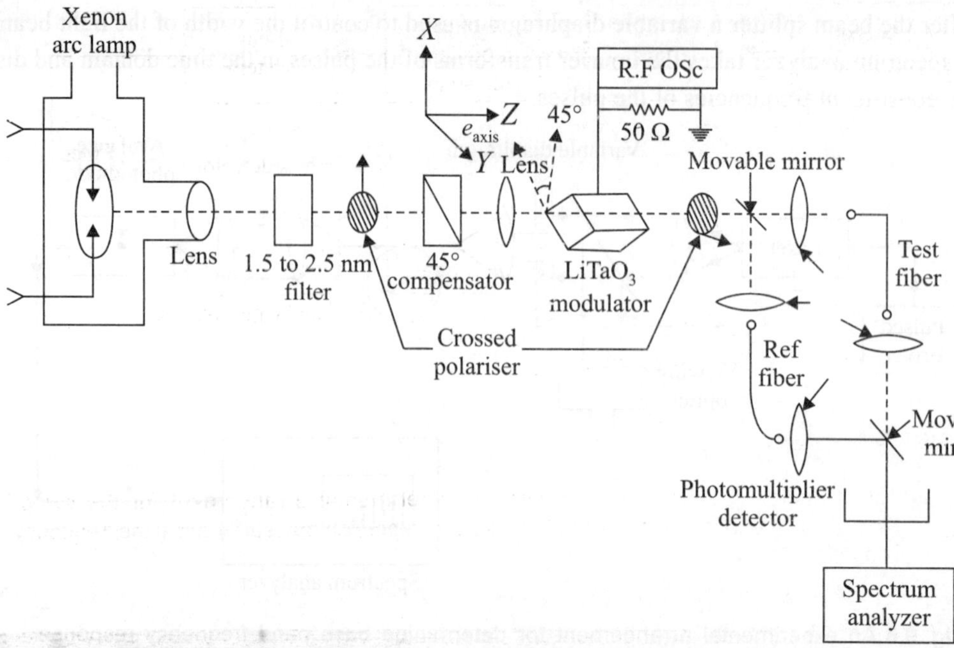

Fig. 9.7 Experimental arrangement for the measurement of dispersion in the frequency domain using pulsed laser source

The components at the modulating frequency from the reference fiber is taken as $P_{in}(f)$ and the same component in the test fiber is $P_{out}(f)$. The reduction of the sinewave amplitude. P_{out}/P_{in} resulting from the transmission of the fiber yields

$$\frac{P_{out}(f)}{P_{in}(f)} = |H(f)| \tag{9.4.2}$$

where $H(f)$ is the magnitude of the power transfer function.

9.5 SWEEP FREQUENCY TECHNIQUES FOR FREQUENCY DOMAIN DISPERSION MEASUREMENTS

In this technique a sinusoidally modulated optical signal at different selected modulating frequencies is launched at different selected frequencies using sweep oscillator. Therefore, the signal energy is concentrated in a very narrow frequency band in the base band region. Fig. 9.8 shows an experimental arrangement for sweep frequency measurement technique. The optical source consists of D H injection laser which may be directly modulated by a sweep oscillator. In order to display the signal a spectrum analyzer may be used. Again Eq. (9.5.2) is used to obtain base band frequency response using either a cutback technique or substitution procedure in a similar manner to that used in the time domain measurement. However, the spectrum analyzer gives no information on the phase of the received signal. Therefore a vector voltmeter or ideally a network analyzer may be used to provide both frequency and phase information.

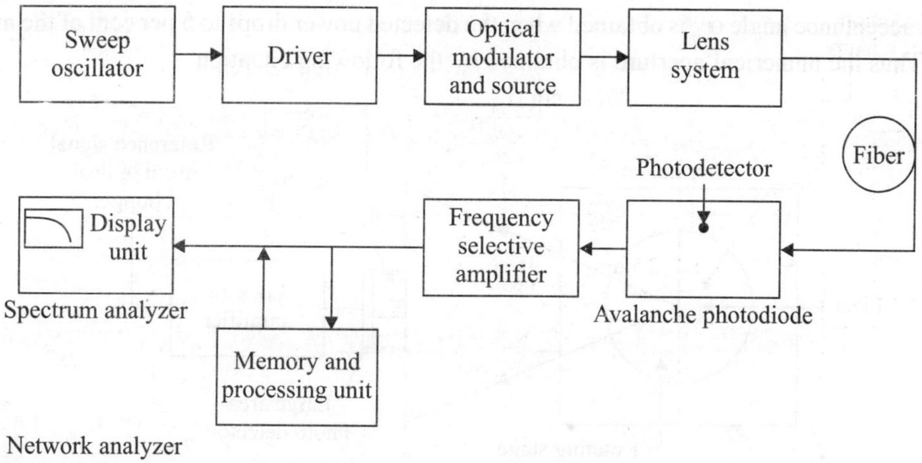

Fig. 9.8 Block schematic diagram showing an experimental arrangement for the swept frequency measurement method to give fiber dispersion measurement in the frequency domain

9.6 NUMERICAL APERTURE MEASUREMENT USING SCANNING PHOTODETECTOR AND ROTATING STAGE

One of the most important parameters of an optical fiber is its numerical aperture. Several characteristics of a fiber such as light gathering capacity and normalized frequency V of the fiber depend on it $V = \left[\frac{2\pi a}{\lambda}\right] NA$. Again V determines the number of modes propagating within the fiber. The fiber attenuation α also depends on numerical aperture (NA). The numerical aperture for a step index fiber is already defined as

$$NA = \text{numerical aperture} = \sin^{-1} \alpha_m = (n_1^2 - n_2^2)^{1/2}$$

where α_m is the maximum acceptance angle, n_1 and n_2 are the refractive indices of the core and cladding respectively. It is assumed that light is incident on the fiber from air whose refractive index is unity.

A Scanning Photo Detector with a Rotating Stage

Fig. 9.9 shows the experimental set up for the measurement of numerical aperture using a scanning photodetector and a rotating stage. A graded index fiber having a length of 2 metres has its both faces prepared in order to ensure square smooth terminations. The output of the fiber is positioned on the rotating stage with its end parallel to the photodetector input and as such its output is perpendicular to the axis of rotation of the rotating stage. Light at an wavelength of 0.85 μm is launched into the fiber at all possible angles (over filling the fiber).

The photo detector may be either a small area device or an aperture of large area device and is placed 10 to 20 cms. from the output end of the fiber. At the start of measurement the photodetector is positioned to get the maximum value of signal with no rotation (0° degree rotation). When the rotating stage is turned at different angles, the limits of the far field pattern may be recorded from the detector. The output powers at different angles of rotations are measured and plotted as the function of angle. The maximum acceptance angle α_m is obtained when the detected power drops to 5 per cent of the maximum intensity. Thus the numerical aperture is obtained by the following equation

$$NA = \sin \alpha_m \tag{9.6.1}$$

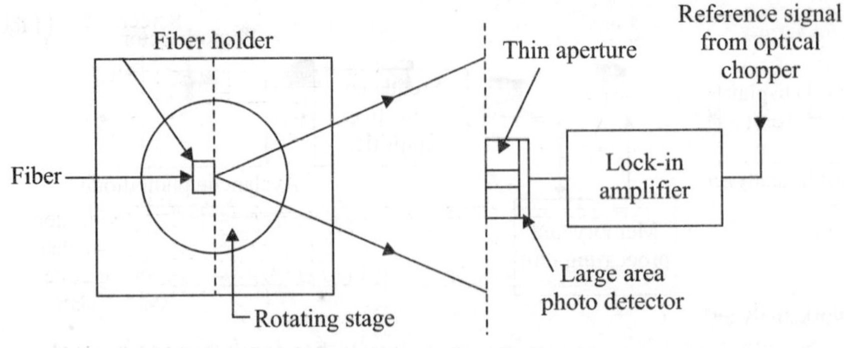

Fig. 9.9 An arrangement for measurement of numerical aperture (NA) using a scanning photodetector and rotating stage

Example 9.4: *Numerical aperture measurements are performed by using a scanning photodetector and a rotating stage on an optical fiber. The angular limit of the far field pattern is found to be 26.1° and then displayed on a screen where the size is measured as 16.7 cm. Determine the numerical aperture of the fiber and the distance of the fiber output end face from the screen.*

Solution. Numerical aperture $NA = \sin \alpha_m = \sin 26.1 = 0.44$ since the length of the picture is 16.7. Half of the picture subtends an angle of 26.1°.

$$\tan 26.1 = \frac{\text{half the picture}}{\text{the distance from the screen}}$$

$$AB = \text{Distance from the screen} = \frac{8.35}{\tan 26.1} = \frac{8.35}{0.49} = 17 \text{ cm}.$$

9.7 FIBER NUMERICAL APERTURE MEASUREMENTS

A rough measurement of the numerical aperture can be made from the far field pattern by trigonometric means. Fig. 9.10 shows the experimental arrangement. In this set up the end prepared fiber is located in an optical base plate or slab. The light is launched into the fiber under test over the full range of numerical aperture and the far field pattern from the fiber is displayed on a screen which is situated at a known distance D from the output end face of the fiber. The test fiber is then aligned so that the optical intensity on the screen is maximized. Finally, the pattern size on the screen A is measured using a calibrated vernier caliper. The numerical aperture can be obtained from the simple trigonometrical relationship where

$$NA = \sin \alpha_m = \frac{A/2}{\left[(A/2)^2 + D^2\right]^{1/2}} = \frac{A}{\left[A^2 + 4D^2\right]^{1/2}} \tag{9.7.1}$$

The accuracy of the measurement depends on the visual assessment of the far field pattern from the fiber. The above measurement methods are generally used with multimode fibers only, because the far field patterns from the singlemode fibers are affected by diffraction phenomena.

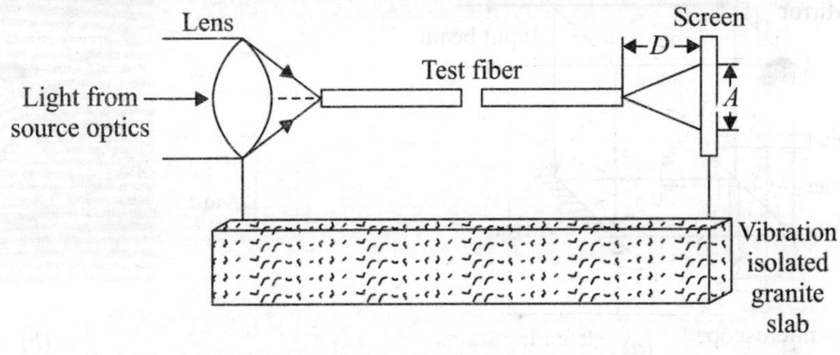

Fig. 9.10 An apparatus for trigonometric fiber aperture measurement

Example 9.5: *In order to determine the numerical aperture of a step index fiber the trigonometric measurement technique is adopted. The screen is placed 10.0 cm from the fiber end face. When the fiber is illuminated from a wide angled visible source, the measured output pattern size is 6.0 cm. Estimate the approximate numerical aperture of the fiber.*

Solution: Eq. (9.7.1) gives the value of *NA* of the fiber as

$$NA = \frac{A}{\left[A^2 + 4D^2\right]^{1/2}}. \text{ Data given } A = 6.0; \text{ and } D = 10.0 \text{ cm.}$$

$$NA = \frac{6}{[36 + 4 \times 100]^{1/2}} = \frac{6.0}{21} \approx 0.29.$$

Refractive Index Measurements

9.8 INTERFEROMETRIC METHOD

Interferometry is an accurate method for determining the refractive index profile of both fibers and preforms. There are two methods. These are (1) Slab method and (2) Transverse method, we shall discuss the slab method.

Slab Method

In slab method of determining the refractive index profile of a fiber, a thin circular slice is cut out of the fiber or preform to be tested. The faces of the slab must be polished to make it flat and accurately parallel. The light passing through the thin slab undergoes a phase shift which depends on the optical path length. The optical path length is equal to the product of slab thickness and the refractive index $n(r)$ of the slab. Fig. 9.11(*a*) shows the apparatus for the slab measurement and it is known as Mach Zahnder interferometer. Light from the microscope travels normal to the faces of the prepared fiber slice and the differences in the refractive index result in the different optical path lengths. The situation is illustrated in the Mach Zahnder interferometer. When the phase of the incident light is compared with the phase of the emerging light, a field of parallel interference fringes are observed. Fig. 9.11(*b*) shows the typical field pattern.

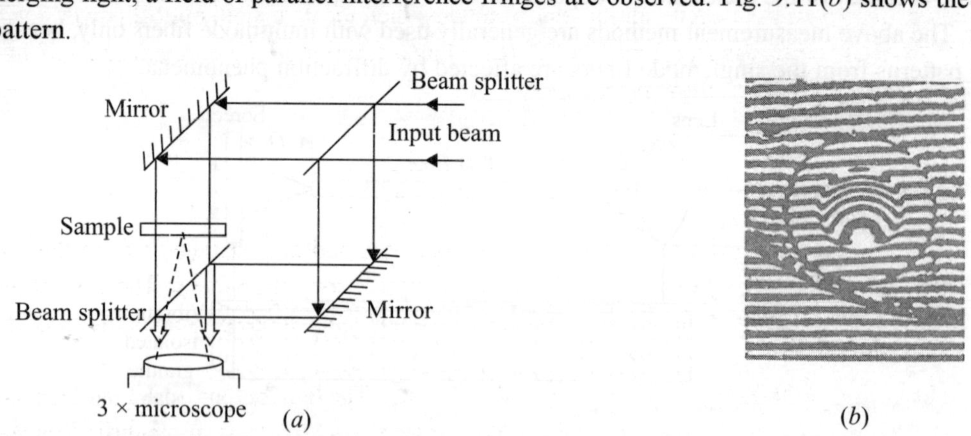

Fig. 9.11 (*a*) The principle of Mach Zahnder interferometer (*b*) Fringe pattern observed by the microscope

The field displacements for the points within the fiber core are then measured using the parallel fringes outside the fiber core in the cladding. The refractive index difference between the core and cladding can be calculated from the fringe shift $S(r)$ and the parallel fringe spacing D (in the cladding) according to the relationship (Fig. 9.12)

$$n(r) - n_2 = \frac{\lambda S(r)}{Dd} \tag{9.8.1}$$

where λ is the wavelength of the measuring light, d is the slab thickness. The parameter $S(r)$ is the central field deviation at a distance r measured from the base line connecting the same cladding fringe at both sides of the core as shown in Fig. 9.12.

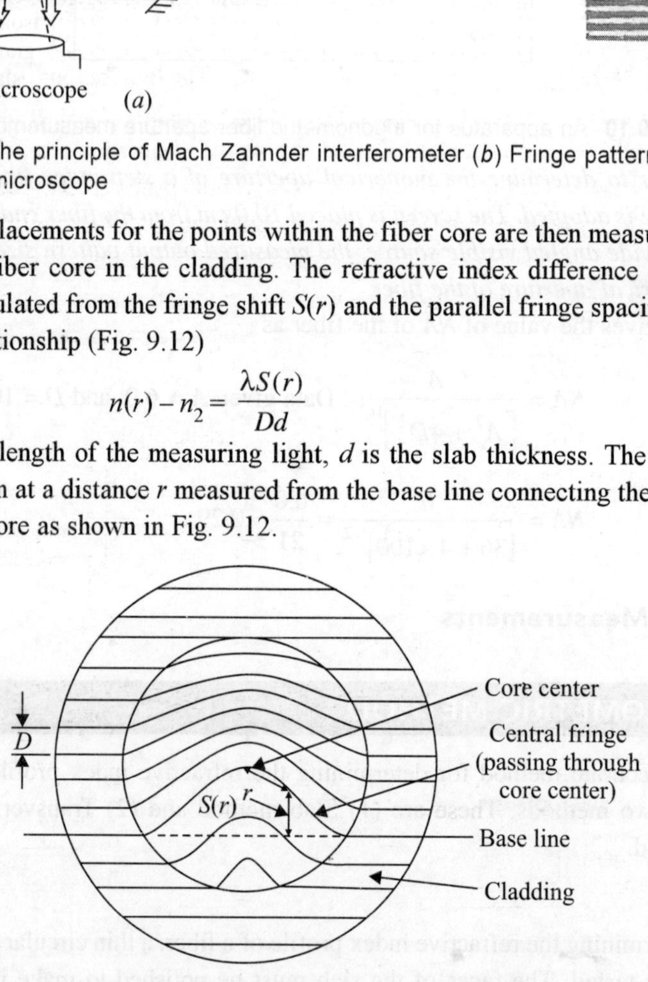

Fig. 9.12 Fringe pattern to determine the refractive index

Near Field Scanning Method

There exists a close resemblance between field intensity distribution and the refractive index profile for a fiber when all the guided modes are equally illuminated. This property is utilized in the Near field scanning method. This is also referred to as "Transmitted near field method." It provides a straight forward and rapid method for acquiring the knowledge of the refractive index profile. When a tungsten filament or a LED is used to excite all the guided modes, then the near field optical power density at a distance r from the core axis $P_D(r)$ may be expressed as a fraction of the core axis near field optical power density $P_D(O)$ by the following equation.

$$\frac{P_D(r)}{P_D(O)} = C(r, z)\left[\frac{n_1^2(r) - n_2^2}{n_1^2 - n_2^2}\right] \tag{9.8.2}$$

where $n_1(O)$ and $n_1(r)$ are refractive indices of the core axis and at a distance r from the core axis respectively, n_2 is the cladding refractive index and $C(r, z)$ is a correction factor. The correction factor $C(r, z)$ which is included in the equation to compensate for any leaky mode present in the short test fiber may be obtained analytically.

An experimental arrangement is shown in Fig. 9.13(*a*). The light output from a lambartian source is focussed on to the one end of the fiber using a microscope objective lens. A magnified image of the fiber output end is displayed in the plane of a small active area photodetector which may be a silicon *p-i-n* photo diode. The photo detector which scans the optical field transversely receives the amplification from the phase sensitive combination of an optical chopper and lock-in amplifier. Hence the profile may be plotted directly by an *X-Y* recorder. However, the profile may be corrected with regard to the correction factor $C(r, z)$. The test fiber is generally 2 meters in length in order to laminate any differential mode attenuation and mode coupling *A* practical index profile for a practical step index fiber measured by spin scanning method is Fig. 9.13(*b*).

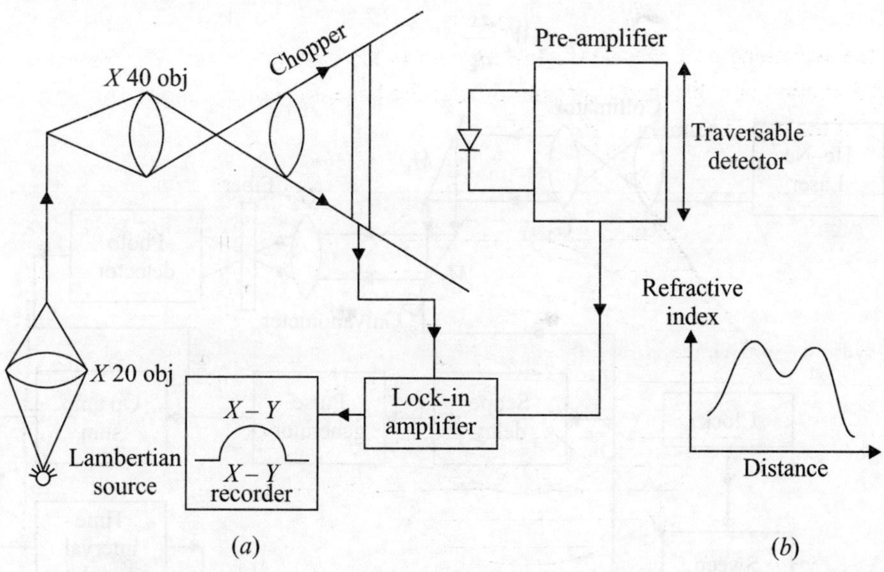

Fig. 9.13 (a) Experimental setup for near field Scanning measurement of refractive index profile (b) Refractive index profile for a practical step index fiber measured by near field scanning method

Fiber Diameter Measurements

9.9 OUTER DIAMETER MEASUREMENT BY SHADOW METHOD

The most common on-line measurement technique utilizes the fiber image projection (shadow method). The experimental arrangement is shown in Fig. 9.14. In this method a laser beam is swept at a constant velocity transversely across the fiber and a measurement is made of the time interval during which the fiber intercepts the beam and casts a shadow on a photodetector. In the apparatus illustrated in Fig. 9.14. The beam from a laser operating at a wavelength of 0.6328 μm is used and collimated using two lenses G_1 and G_2. The beam is then reflected off by two mirrors (M_1 and M_2) and the second mirror M_2 is driven by a galvanometer. The galvanometers make the mirror M_2 to rotate through a small angle with a constant angular velocity before returning back to its original starting position. Therefore, the laser beam which is focussed in the plane of the fiber by a lens G_3 is swept across the fiber by the oscillating mirror and is incident on the photodetector, unless it is blocked by the fiber. The velocity ds/dt of the fiber shadow, thus, creates at the photodetector is directly proportional to the mirror velocity $\frac{d\phi}{dt}$ following

$$\frac{ds}{dt} = L\frac{d\phi}{dt} \tag{9.9.1}$$

where L is the distance between the mirror and photodetector. Moreover, the shadow is registered by the photodetector as an electrical pulse of width W_e which is related to the outer diameter of the fiber do as:

$$d_o = W_e \frac{ds}{dt} \tag{9.9.2}$$

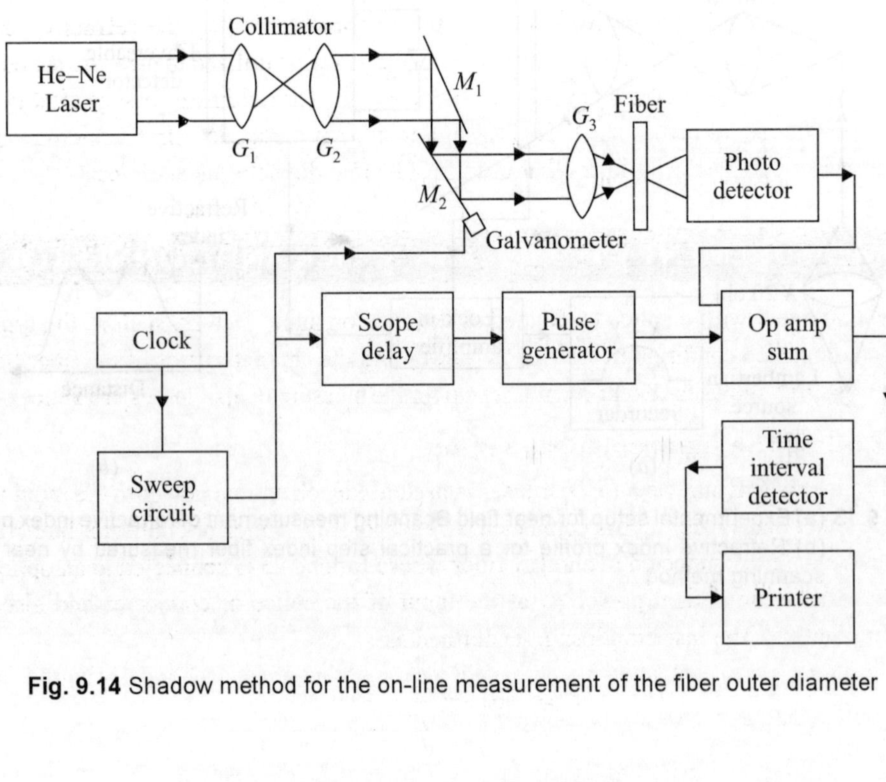

Fig. 9.14 Shadow method for the on-line measurement of the fiber outer diameter

Thus the fiber diameter may be quickly determined and recorded on the printer. The measurement speed is largely determined by the inertia of the mirror rotation and its accuracy by the rise time of the shadow pulse.

Example 9.4: *The shadow method is used for the on-line measurement of the outer diameter of an optical fiber. The apparatus uses a rotating mirror with an angular velocity of 4 rad sec^{-1} which is located 10 cm from the photo detector. At a particular instant of time, a shadow pulse of width 250 μm is recorded by the photo detector. Determine the outer diameter of the fiber in μm at this instant of time.*

Solution: The shadow velocity $\left(\frac{ds}{dt}\right)$ is given by Eq. (9.9.1) as

$$\frac{ds}{dt} = L\frac{d\phi}{dt} = 0.1 \times 4 = 0.4 \text{ met/sec}$$

$$= 0.4\ \mu\text{m}/\mu\text{s}^{-1}$$

Hence the outer diameter of the fiber d_o in μm is given by (Eq. 9.9.2)

as
$$d_o = W_e \frac{ds}{dt} = 250\ \mu\text{s} \times 0.4\ \mu\text{m}\ \mu\text{s}^{-1}$$

$$= 100\ \mu\text{m}.$$

9.10 CORE DIAMETER

The core diameter for step index fiber is defined by the step change in the refractive index profile in the core cladding interface. Thus the technique employed for determining the refractive index profile (interferometric, near field scanning and refracted ray etc.) may be utilized to measure the core diameter.

Core diameter measurement is also possible from the near field pattern of a suitably illuminated fiber where all the modes are excited. The measurement may be made using a microscope equipped with a micrometer eye piece similar to that used for off time diameter measurement.

9.11 MEASUREMENT OF LOSSES IN SPLICE AND CONNECTORS

The losses associated with a splice or connector depends on many factors, such as the type of optical sources used, launched condition of the optical power and the characteristics of the fiber on both sides of the joint. Fig. 9.15 shows the experimental set-up for the measurement of losses in a splice or connector.

Description of the Experimental Set-up

An optical source which may be a *LED* or laser is used to launch optical power to the input end *A* of the fiber and the other end of the fiber is connected to one end of the splice or connector under test at point #1. The other end #2 is connected to another fiber whose far end #3 is connected to an optical detector.

We measure the optical power P_1 at the input of the splice or connector and also the output power P_2 at point #2. The insertion loss L_s is defined as

$$L_s = -10 \log \frac{P_2}{P_1} \tag{9.11.1}$$

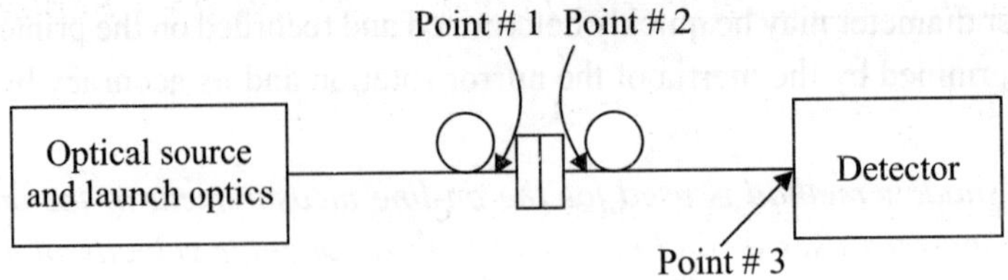

Fig. 9.15 Experimental set-up for the measurement of losses in a splice or connector

When the length of the fiber is short, we find that the loss depends on the launching condition at the source. However, with a long length of the fiber, we observed that the connector has been isolated from the launching conditions due to the mode conversion effect. Hence, we require that an equilibrium mode distribution is established in the illuminated fiber. Two techniques have evolved for this.

In one technique, a long length of a fiber as a "Pigtail" is included to allow equilibrium pattern to be established before the light reaches the connector. In the other technique an equilibrium mode stimulator is included and it is usually a short length of the fiber undergoing a serpentine path that allows an ample opportunity for mode mixing. A frequently used mode stimulator is one metre or so in length of fiber wrapped around a 1.6 diameter spool which is often called Mandrel. On the other hand, splices can have effects on the measured length of a multimode fiber. The splice will redistribute the power in the mode of the fiber.

Observation of this effect involves the measurement of power (P_1 and P_3) in the points #1 and #3 respectively in the set-up of Fig. 9.15. The measured loss – 10 log P_3/P_1 is usually greater than the sum of the insertion loss and the fiber loss. Since this extra loss is due to the splice rather than the fiber loss, the loss is properly added to the splice loss rather than the fiber loss. Thus the total loss of the splice or connector is

$$L_{s\ \text{total}} = -10 \log \frac{P_3}{P_1} - \alpha L \tag{9.11.2}$$

where αL is the product of loss coefficient (or attenuation constant) and the length of the fiber. The excess loss αL_s excess of the splice is the difference between $L_{s\ \text{total}}$ and insertion loss L_s of the splice.

$$\delta L_{s\ \text{excess}} = L_{s\ \text{total}} - L_s = -10 \log \frac{P_3}{P_1} - \alpha L - 10 \log \frac{P_2}{P_1} \tag{9.11.3}$$

In the case of a splice having reasonable extrinsic losses (good alignment) between identical fiber $L_{s\ \text{total}}$ is greater than L_s by a factor 2 or so. The conclusion is that while the insertion loss is the well defined local quantity, the true total splice loss $L_{s\ \text{total}}$ is highly variable and depends on not only the splice but also the length, mode making property and the modal attenuation of the receiving fiber. Usually L_s is the parameter quoted for a splice is measured with a long transmitting fiber and a short receiving fiber. The value may be regarded as the partial representative of the splice effect in the system.

Example 9.5. *In measuring the connectors loss we find the power of the fiber at the input of the connector is 100 μW and the output power immediately after the connector is 83.2 μW (a) Calculate the insertion loss of the connector and (b) When 1.8 km. of 1.5 dB/km is added after the connector the measured output power at the end of the added fiber is 35.5 μW. Calculate the excess loss of the connector.*

Solution. Data supplied $P_1 = 100\ \mu W$, $P_2 = 83.2\ \mu W$, $P_3 = 35.5\ \mu W$, $\alpha = 1.5$ dB/km and $L = 1.8$ km.

$$(a)\ L_s = -10 \log \frac{P_2}{P_1} = -10 \log \frac{83.2}{100} = -10 \times 0.08 = 0.8 \text{ dB}$$

(*b*) When 1.8 km of 1.5 dB/km is added to the connector

$$\delta L_{s\ \text{excess}} = -10 \log \frac{P_3}{P_1} - L_s - \alpha L$$

$$\delta L_{s\ \text{excess}} = -10 \log \frac{35.5}{100} - 0.8 - 1.9 \times 1.8$$

$$= 4.5 - 0.8 - 2.7 = 1 \text{ dB}.$$

REVIEW QUESTIONS

9.1 Describe with necessary experimental arrangement for the measurement of the signal attenuation in an optical fiber by cutback technique.
A multimode fiber of length 1.8 km is connected to an apparatus for the measurement of attenuation coefficient by cutback technique. The output voltage from the photodetector using the 1.8 km fiber length is found to be 2.5 volts at an wavelength of 0.9μm. Without disturbing the system the fiber is then cut back to a length of 2 metres, and the measured output voltage from the photodetector is found to be 11.5 volts. Estimate the value attenuation per km. in dB.

9.2 Describe and explain the experimental set up used in Optical Time Domain Reflectometer (OTDR) technique. How is the attenuation measured by this technique?
Give the representative results obtained from this experiment. Explain a typical curve obtained on the scope giving representative results of the experiment. What is the limitation of this technique?

9.3 Describe the arrangement for experimental set up for the time domain fiber dispersion measurement. Find the relationship between the pulse broadening and the widths of the input and output pulses with the assumption that the shapes of pulses are Gaussian. Pulse broadening measurement experiment is done on a 2 km length of a multi mode step index fiber. The 3dB width of the input and output pulses are found to be 300 ps and 12.6 ns respectively. Estimate.
(*a*) the 3dB pulse broadening for the fiber in ns/km, and
(*b*) the bandwidth length product of the fiber. Assume that the input and output pulses are Gaussian in shape.

9.4 Describe with the experimental arrangement for the frequency domain dispersion measurement.

9.5 Discuss a method for the refractive index profile measurement. Give the diagram for MachZahnder microscope arrangement for refractive index profile measurement by (*a*) the measurement of refractive index from the interference pattern photograph and (*b*) by counting the number of fringes in the photograph.

9.6 How can you measure the numerical aperture of a fiber by (*a*) scanning photodetector and rotating stage method and (*b*) trigonometric method.

9.7 Describe the methods of the measurement of diameter of the fiber core and cladding.

9.8 Describe with necessary experimental set up for measurement of attenuation of signal in a splice or connector.

CHAPTER 10

Optical Fiber Sensors

CHAPTER FOCUS

- *Introduction*
- *Phase and polarization of fiber sensors*
- *Ring interferometer with multiturn fiber coil*
- *Optical fluid level detector*
- *Optical fiber flow sensors*
- *Optical displacement sensors*
- *Optical displacement moire fringe modulation sensors*
- *Microbend optical fiber sensors: Introduction*
- *Intrinsic fiber sensors*
- *Current measurement by single mode optical fiber sensors*
- *Fluorooptic temperature sensors*
- *Photoelastic pressure sensors*
- *Laser Doppler Velocimeter using optical fiber*

10.1 INTRODUCTION

We know that a microphone is a transducer which is capable of converting sound energy to electrical energy in a measurable form. Similarly, a sensor which may be regarded as a transducer can convert one physical quantity into another one in a measurable form. Sensors are generally electrical in construction and provide convenient form of electrical or optical signal.

Optical Fiber Sensors

Optical fiber sensors find many applications such as in the measurement of temperature or pressure, flow rate, liquid level etc. Temperature and pressure changes can give rise to the change in effective length of the fiber and this change in length produces a small phase change. When the fiber is long, even this small phase change gets considerably magnified over the whole length of the fiber and provides a measurable amount of phase change. As this phase change is due to the change in temperature or pressure, so that it automatically measures temperature or pressure respectively.

Generalized Optical Fiber Sensor System

The modulation (or variation in parameter) of light beam transmitted through a feed fiber takes place in the modulation zone of the generalized optical fiber sensor system as shown in Fig. 10.1. The light beam can be modulated in five of its basic properties: optical intensity, phase, polarization, wavelength and spectral distribution. Most fiber sensors utilize the first three, although there is a growing interest in the last two. Moreover, it is possible to categorise the device into two distinct groups based solely on their use of intensity and phase modulations. These are (*a*) Intrinsic and (*b*) Extrinsic devices.

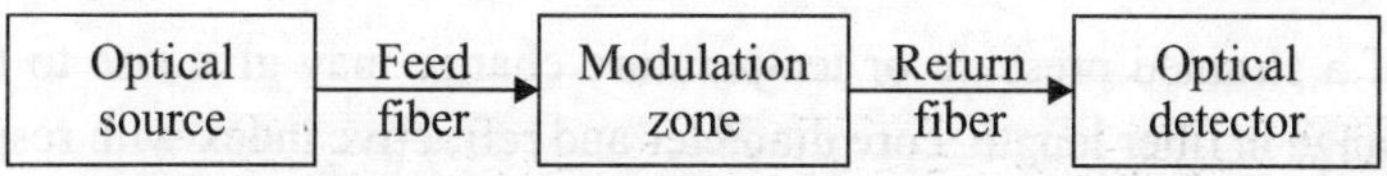

Fig. 10.1 Block diagram for the arrangement for basic optical fiber sensor system

Intrinsic Devices

The intrinsic type is arranged such that the physical parameter to be sensed acts on the fiber itself to produce a change in the transmission characteristics.

Extrinsic type uses the fiber as the light guide to and from the sensors which is configured to allow the measurand to change to coupling characteristics of feed and the return fiber.

10.2 PHASE AND POLARIZATION FIBER SENSORS

Interferometric technique is generally used in a very sensitive single mode fiber as a phase sensors. Fig. 10.2(*a*) illustrates the basic principle for interferometric arrangement for the optical fiber sensor.

1. A laser source emits a monochromatic beam which passes through a beam splitter which is a glass plate put making an angle of 45° to the direction of the laser beam. The beam splitter is used to divide the beam in two directions at right angle. The main beam passes through lens L_1 and then passes through a reference fiber which is maintained isolated from any external influence. From the other end of the reference fiber the beam impinges to lens L_2.
2. The splitted beam at right angle to the main beam passes to lens L_3 and from lens L_3 then beam is launched into the test fiber which is kept in the environment which is to be measured and finally this beam passes to lens L_2 from the other end of the test fiber.
3. The device causes the interference of the monochromatic light beam propagating in a strained or temperature varying fiber with the light beam from the laser source guided by the reference fiber isolated from any external influence at the focal plane of L_2.

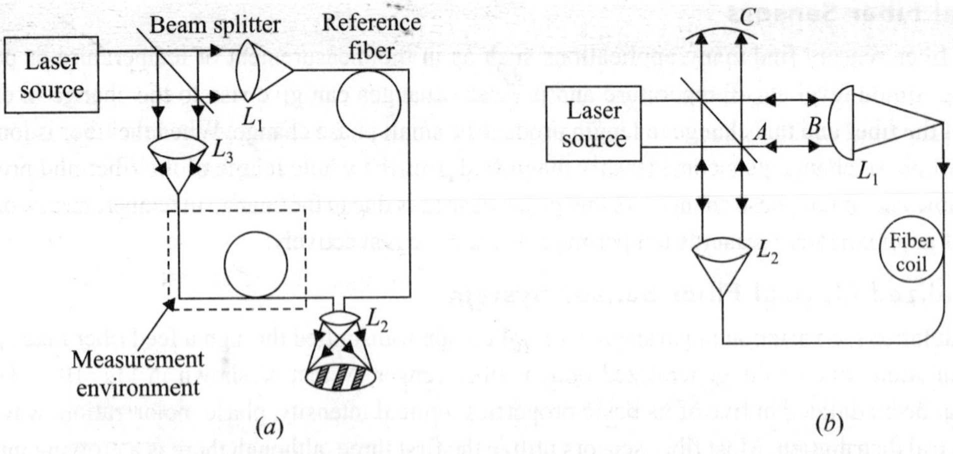

Fig. 10.2 (*a*) Singlemode fiber sensor system using interferometer technique. Two arm interferometer (Mac Zahnder type) (*b*) Ring interferometer system using multiturn fiber coil (single mode fiber interferometer Sagnac type)

The effect of a strained pressure or temperature change may give rise to the differential path length due to the change in fiber length, core diameter and refractive index with respect to the reference fiber. This provides a phase difference of the lights emitted from the two fibers giving interference pattern as shown in Fig. 10.2(*a*). A very accurate measurement of temperature or pressure may be obtained from this interference pattern. It may be noted that the interference sensor arrangement of Fig. 10.2(*a*) is that of Mach Zehnder configuration. For example, on using fused silica glass in such a two arm interferometer, it is found that the temperature sensitivity is about 107 radians per °C^{-1} m^{-1}.

10.3 RING INTERFEROMETER WITH MULTITURN FIBER COIL

Another common singlemode fiber interferometric sensor is the fiber "Gyroscope" which gets extensive applications. The device is based on the classical Sagnac ring interferometer and the fiber version of this is illustrated in Fig. 10.2(*b*).

Description of the Apparatus

1. Light beam from a laser source is divided into two paths at right angle with the help of a beam splitter and the main beam is launched on one end of a multiturn fiber via lens L_1 and the emerging end of the beam moves through lens L_2. Thus, the beam into the multiturn fiber gets divided into two waves propagating in the opposite directions, one from A to B and the other one from B to A. The two waves return in phase at after moving through the same path in the opposite directions. When the fiber coil rotates about an axis perpendicular to the plane of the coil, the lengths of the paths between two counter propagating waves differ. The difference produces a phase shift which can be measured by the interferometric technique. So that the value of rotation of the fiber coil can be accurately determined.

10.4 OPTICAL FLUID LEVEL DETECTOR

Fig. 10.3 shows an arrangement for an optical fluid level detector. It consists of a laser source and a detector, and in between these two an optical 'Dip stick' is put. The fluid whose level is to be sensored has a refractive index greater than that of the glass forming the optical dipstick. When the lower end of the dipstick does not touch the fluid level as shown in Fig. 10.3(*a*) then the light beam from the laser source undergoes number of total internal reflections on the surface of the dipstick and manages to reach the detector to get detected. The detection of the signal indicates that the fluid level is not touched.

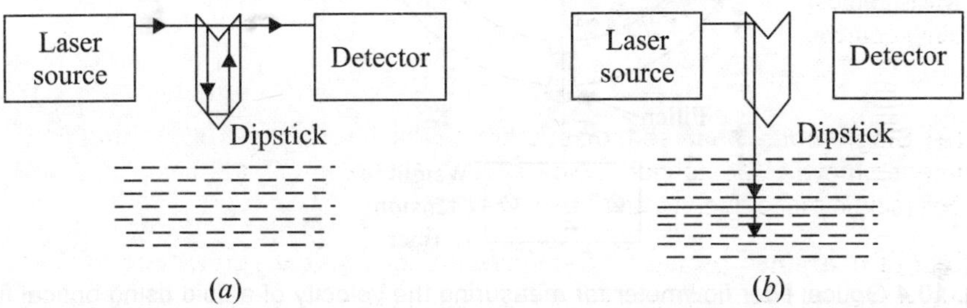

Fig. 10.3 Optical fiber fluid level detector (*a*) when the dipstick is not touching the fluid and (*b*) when the dipstick touches the fluid

When the level of the fluid whose refractive index is greater than that of the glass forming the optical dipstick, touches the lowest end of the dipstick, the total internal reflection ceases and the light beam is transmitted into the fluid. The detector does not receive any laser beam and the indication of the fluid level is obtained at the optical detector as shown in Fig. 10.3(*b*). Although the system is somewhat crude and will not provide continuous measurement of a fluid level. It is simple and safe for use with inflammable liquids.

10.5 OPTICAL FIBER FLOW SENSORS (EXTRINSIC)

Fig. 10.4 shows an intrinsic optical fiber flow sensor system. A circular pipe is bent in the form of a circle and a liquid flows through this pipe. A multimode stretched optical fiber is inserted across the tube pipe. When the liquid flows through this pipe, the flow of the liquid occurs transversely across the stretched fiber.

The turbulence resulting from the presence of the fiber causes the fiber to oscillate at a frequency roughly proportional to the flow rate of the liquid in the pipe. This results in a corresponding oscillation in the mode power distribution within the fiber giving a similarly modulating intensity profile at the optical receiver. The technique has been used to measure the flow rate from 0.3 to 3 ms^{-1}. However, it cannot measure flow rates below those in which turbulence occurs.

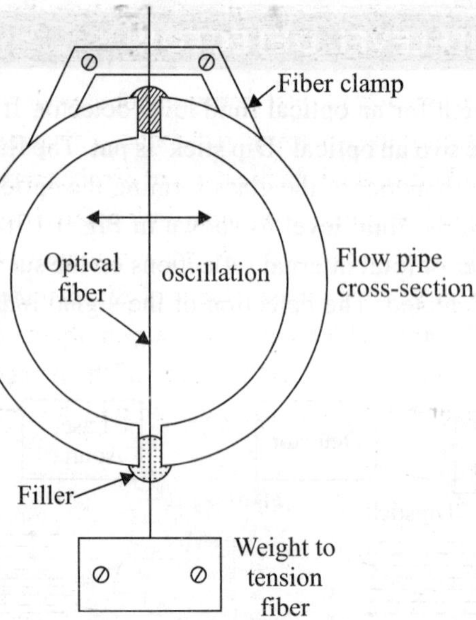

Fig. 10.4 Optical fiber flowmeter for measuring the velocity of a fluid using optical fiber

10.6 OPTICAL DISPLACEMENT SENSORS (EXTRINSIC)

Intensity modulation of the transmitted light beam is used in the extrinsic reflective or photonic optical sensor to make a measurement of displacement. The arrangement of the sensor system is shown in Fig. 10.5. In this device the light reflected from the target is collected by a return fiber and finally detected by the detector. The detected power is a function of the distance between the fiber end and the target. Hence the position or displacement of the target may be registered at the optical detector. Moreover, the sensitivity of the sensor may be improved considerably by placing the axes of the feed fiber and the returned fiber at an angle to one another and also to the target (Fig. 10.5). Unfortunately, the technique has the drawback with regard to the stability of the optical components which is a common feature in the intensity modulated fiber sensors.

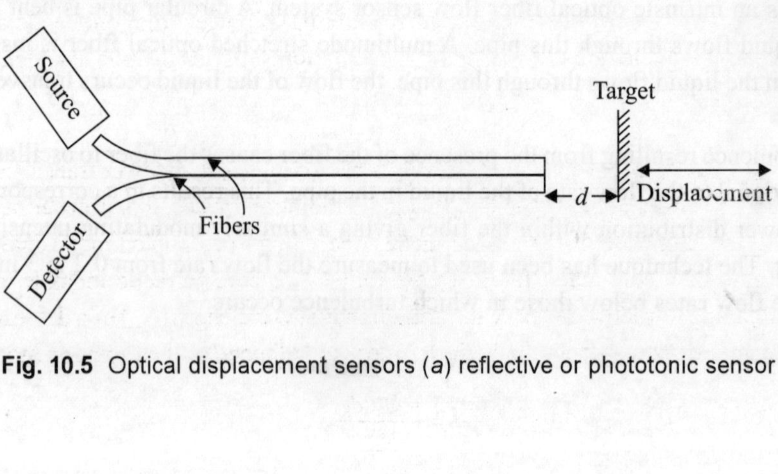

Fig. 10.5 Optical displacement sensors (*a*) reflective or phototonic sensor

10.7 OPTICAL DISPLACEMENT MOIRE FRINGE MODULATION SENSORS

A method of eliminating the drawback of instability encountered with the optical fiber intensity modulation sensors is to use a digital measurement technique. The arrangement in such a technique is shown in Fig. 10.6 and in this technique the measurement of displacement is done using a Moire Fringe modulator. In this case the opaque line gratings produce dark Moire fringes. The transverse movement of one grating with respect to the other one causes the fringes to move up and down. Thus, a count of fringes as the gratings are moved gives a measurement of displacement. The counting of fringes is independent of instabilities with the system components which affects the optical intensity. However, the mechanical vibration may severely affect the accuracy of the measurement and proves to be very difficult to eliminate fully. There are also problems involved with loss of counts if the optical power to the sensor is interrupted.

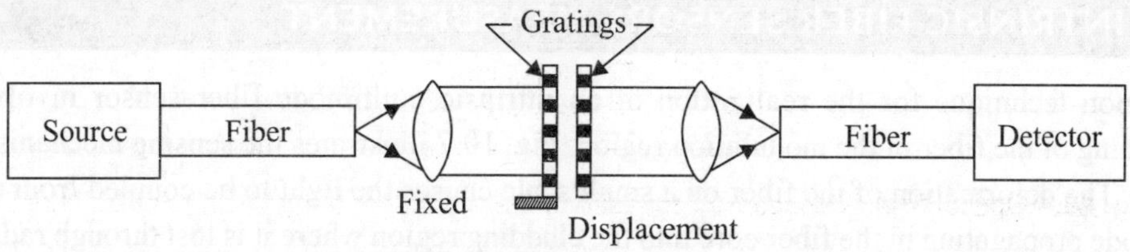

Fig. 10.6 Moire-fringe modulation sensors

10.8 MICROBEND OPTICAL FIBER SENSORS: INTRODUCTION

When microbends are induced in an optical fiber deliberately, an excessive transmission loss is introduced to the fiber. The property of loss is used to detect or measure displacement, acoustic pressure, strain, and temperature. It is known that if a portion of a fiber is forced by a deformer to go through a continuous succession of small bends on a microscopic scale, the fiber would suffer excessive loss. The fiber may be sandwitched between a pair of loothed or serrocated plates to induce microbending. When a fiber is simply kept between two pieces of sand paper under pressure microbending may be induced in the fiber. Such regular or random meandering of fiber axis gives rise to the redistribution of guided optical power between different modes of the fiber and also coupling of power from one mode or mode group to another. If the spatial wavelength (Λ) of bendic deformation satisfies the phase matching condition:

$$\beta_p - \beta_q = \frac{2\pi}{\Lambda} \tag{10.8.1}$$

where β_p and β_q represent modal propagation constants of two modes and power transfer will take place from pth to qth mode. In case the qth mode happens to be radiation modes, this transfer of optical power will cause a net transmission loss of the guided mode. According to a coupled mode analysis, it can be shown that in the case of a step-index fiber of core radius 'a', core refractic index n_1 and difference between core and cladding refractive index Δ, the critical spatial wavelength (Δ_{cr}) of a deformation required to induce heavy transfer of power from the highest order guided mode to the radiation mode will be

$$\Lambda_{cr} = \frac{\pi a}{\sqrt{\Delta}} = \frac{\sqrt{2}\pi a n_1}{NA} \tag{10.8.2}$$

For example, a step index fiber with core radius $a = 62.5$ μm has the numerical aperture $NA = 0.30$, the spatial deformation periodicity required to induce heavy microbending loss would be ~0.69 mm, whereas for a 200 μm core similar fiber of $NA = 0.18$ (plastic clad silica fiber) the required periodicity to induce large transfer of power from the highest order of guided modes to the radiation mode is ~3.6 mm. From the point of view of easier fabrication of microbend inducer, the use of large core and large NA multimode fibers would diviously be more suitable for constructing microbend type fiber sensors. It has been shown that microbend induced loss in bare multimode fibers is given by

$$\text{loss} = \frac{Ka^4}{\Delta^3 b^6} \qquad (10.8.3)$$

where K is a proportionality constant and b = radius of fiber cladding.*

10.9 INTRINSIC FIBER SENSORS MEASUREMENT

The common technique for the realization of an intrinsic multimode fiber sensor involves in the microbending of the fiber in the modulation region. Fig. 10.7 illustrates the sensing mechanism for this technique. The deformation of the fiber on a small scale causes the light to be coupled from the guided optical mode propagating in the fiber core into the cladding region where it is lost through radiation into the surrounding region around it. When the spatial wavelength L of deformation is correctly chosen

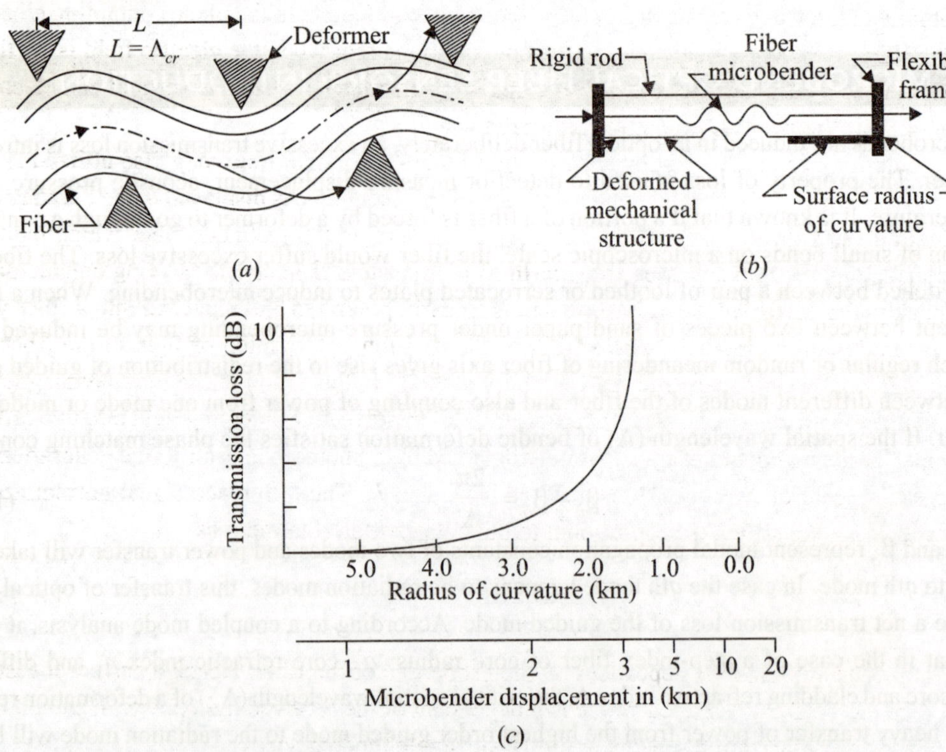

Fig. 10.7 Fiber microbending sensor (*a*) sensing mechanism (*b*) structure and (*c*) test data from (*a*)

* Bishaup Pal, Fundamental of Fibre Optics, NewAge International (P) Limited, Publisher, New Delhi.

(Fig. 10.7*a*) the optical power coupled from the core of the fiber to the radiation mode in the cladding is high. The result is that high sensitivity to the pressure applied to the deformer is provided in the direction perpendicular to the core axis.

Moreover, if this deformation is caused by the measurands (*e.g.*, pressure, vibration, sound etc.), the fluctuation of the light intensity in either core or cladding is directly proportional to the value of the measurand for small deformation. Thus, monitoring the fiber core or cladding allows the detection of the measurand.

Fig. 10.7(*b*) shows the configuration of microbending fiber sensor for monitoring the structural deformation. The test data corresponding to this configuration is given in Fig. 10.7(*c*) for 1.4 mm spatial of 3 wavelengths. However, it should be noted the requirement of this type of sensor is the removal of the cladding mode by a cladding mode stripper both immediately before to and after the modulation zone. Moreover, the fiber damage resulting from deformation does not appear to be a severe problem and a setting period may be observed before the attainment of the constant characteristics of the fiber.

10.10 CURRENT MEASUREMENT BY SINGLE-MODE OPTICAL FIBER SENSORS

When a plane electromagnetic wave propagates, its electric and magnetic field vectors are at right angle to each other and the wave propagates in the direction at right angle to the plane containing electric and magnetic field vectors. In communication the convention is that when the electric field is vertical, the wave is termed as a vertically polarized wave and when the magnetic field is vertical and electric field is horizontal, it is known as horizontally polarized wave.

Modulation of the polarization state of light within an optical fiber may also utilized to make a physical measurement. A successful implementation of this technique is displayed in the Faraday rotation current monitor as shown in Fig. 10.8. The device consists of a singlemode fiber which maintains a single polarization. The fiber passes from earth to loop around the current carrying conductor before passing back to the earth. A He-Ne Laser beam is linearly polarized and launched into a fiber which is stripped of any cladding mode. The direction of the rotation polarization of light in the fiber core is rotated by the longitudinal magnetic field around the loop via the action of the Faraday magneto-optic effect. A Wallaston prism is used to sense the rotation and resolved the emerging light into two orthogonal components. These components are separately detected by a photodiode prior to the generation of sum and difference signal of two intensities denoted by I_1 and I_2. The difference signal normalized to the sum gives a parameter which is proportional to the polarization rotation ρ as

$$K\rho = \frac{I_1 - I_2}{I_1 + I_2} \tag{10.10.1}$$

where K is a constant which is dependent on the properties of the fiber. Hence a current measurement (either d.c. or a.c.) may be obtained which is independent of the received light power. It should be noted that the phase and polarization sensors are intrinsic single mode devices.

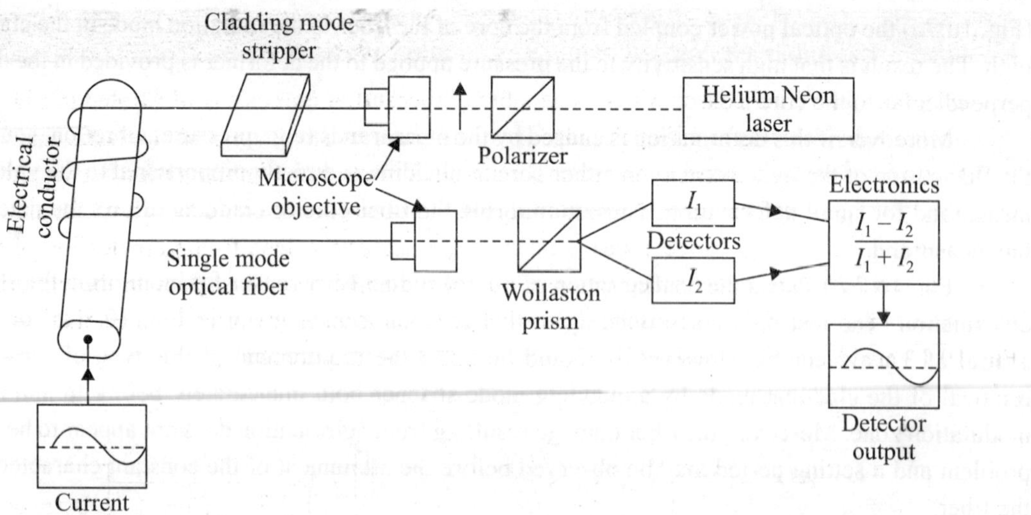

Fig. 10.8 Single mode optical fiber sensor for the measurement of electric current

10.11 FLUOROPTIC TEMPERATURE SENSORS

Another technique for the measurement of temperature utilized in the fluoroptic temperature sensors. The device has been illustrated in Fig. 10.9. The device uses a form of wavelength modulation which occurs when incoherent light is transmitted along a multimode fiber with a large core to a phosphorescent material. The light excites the rare-earth phosphor which emits a number of lines at different wavelengths. In this case two of the spectral lines at wavelength of 540 nm and 630 nm are selected with the use of filter in the receiver. The ratio of intensities of these two chosen lines (540 and 630 nm) is a single valued function of the temperature of the phosphor. The measurement of this ratio provides an accurate measurement of the temperature. The resolution of this instrument is 0.1 °C over the range of –50 to 250°C and is independent of light intensity. However, the instrument has rather complex processing electronics and requires calibration during thermal stabilization. Since the sensitivities of the optical detectors even in matched condition do not track together at this two different emitted wavelengths as the temperature changes.

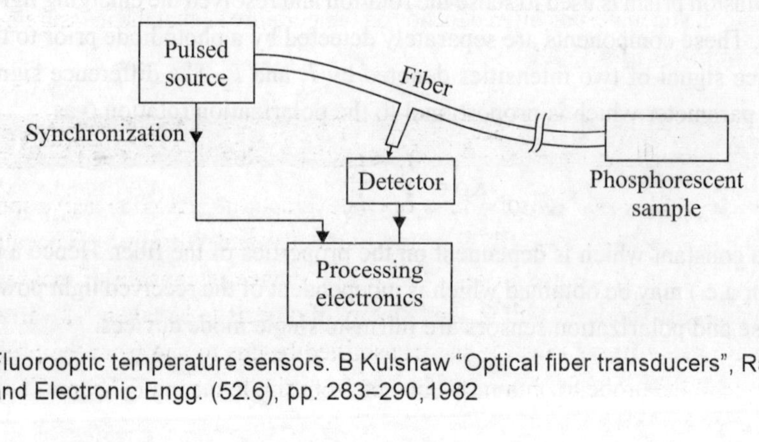

Fig. 10.9 Fluorooptic temperature sensors. B Kulshaw "Optical fiber transducers", Radio and Electronic Engg. (52.6), pp. 283–290,1982

10.12 PHOTOELASTIC PRESSURE SENSORS

A multimode optic fiber sensor which is capable of measuring pressure is illustrated in Fig. 10.10. In this device the photoelastic effect is induced by mechanical stress on a photoelastic materials (*e.g.* piezooptic glass, polyurethane, epoxy resin) and used to rotate optical polarization between the pairs of crossed polarizer. This phenomenon is referred to as 'birefringence'. This occurs with the application of mechanical stress (force) to the transparent isotropic materials. Due to the application of the stress isotropic materials are transformed into optically anisotropic one. This gives to the variation of the transmitted light through the sensor. The response of the sensor is typically sinusoidal as shown in Fig. 10.10 and a resolution of 10^{-4} of the full range has been observed.

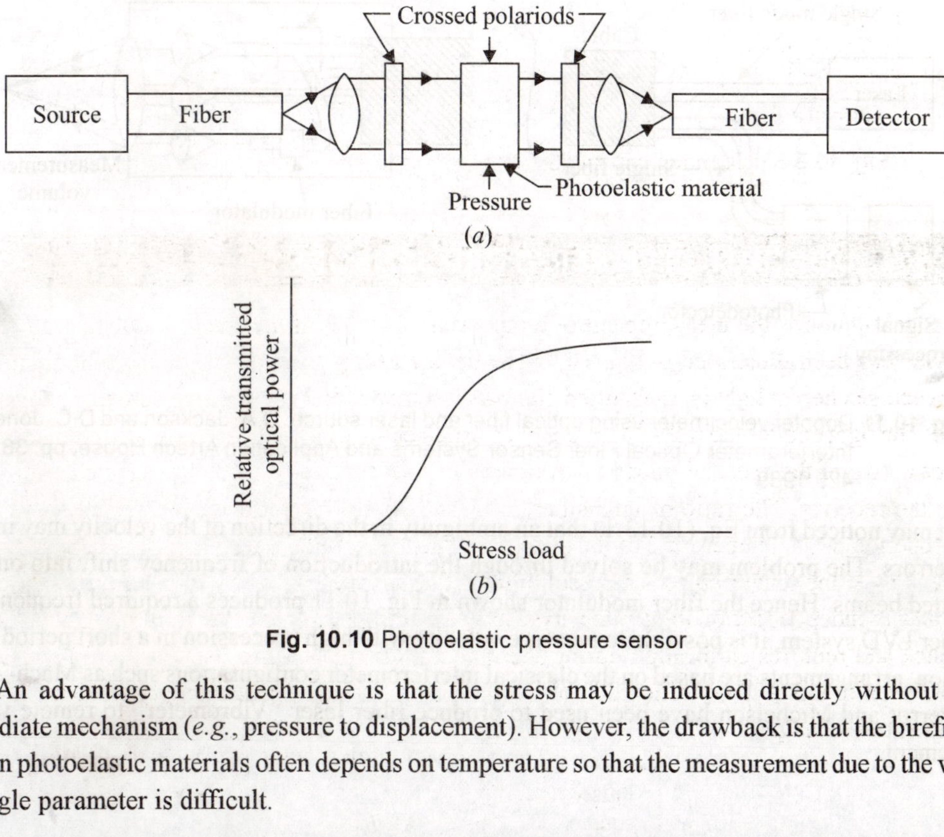

Fig. 10.10 Photoelastic pressure sensor

An advantage of this technique is that the stress may be induced directly without need of intermediate mechanism (*e.g.*, pressure to displacement). However, the drawback is that the birefringence shown in photoelastic materials often depends on temperature so that the measurement due to the variation of a single parameter is difficult.

10.13 LASER DOPPLER VELOCIMETER USING OPTICAL FIBER

Extrinsic single-mode fiber sensors have been developed to get nonevasive measurement of several physical measurands (velocity, fluid surface velocity and vibration). The all-fiber laser Doppler velocimeter (LVD) is illustrated in Fig. 10.11. The device is capable of making the measurement of velocity in gases and fluids as well as the velocity of objects to be taken. The arrangement in the figure uses two single mode fibers to guide the transmitted beams to and from the probe. A fiber directional coupler is used at the probe to obtain two beams from single incoming beam. The measurement volume is formed by the region of intersection of two coherent optical beams which are independently scattered and Doppler shifted. A Doppler difference technique is then used because the frequency shift in each beam is different as they are travelling in different directions. The two frequency shifts beat together to produce a frequency (δf_{DO}) which is proportional to the component of velocity of the scattering particle v perpendicular to the mean direction of the incident beam and in their plane so that

$$\delta f_{DO} = [(2v/\lambda)n \sin \theta/2)] \tag{10.13.1}$$

where λ is the wavelength of the laser source, n is the refractive index of the measurement volume and θ is the angle of convergence between the two input beams and δf_{DO} = frequency.

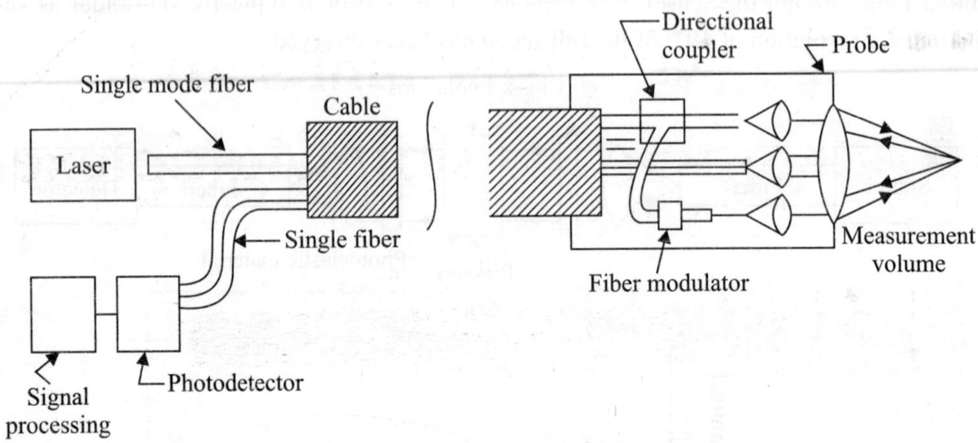

Fig. 10.11 Doppler velocimeter using optical fiber and laser source. D.A. Jackson and D.C. Jones, Interferometer Optical Fiber Sensor Systems and Application Artech House, pp. 381–429,1989

It may noticed from Eq. (10.13.1) that an ambiguity in the direction of the velocity may introduce serious errors. The problem may be solved through the introduction of frequency shift into one of the transmitted beams. Hence the fiber modulator shown in Fig. 10.11 produces a required frequency shift. With fiber LVD system, it is possible to measure velocity with high precession in a short period of time. In addition, arrangements are based on the classical interferometer configurations such as Mach-Zehnder, Fabry-Perrot and Michelson have been used to produce fiber laser "Vibrometer" to remote vibration measurement.

CHAPTER 11

Optical Fiber Links and Wavelength–Division Multiplexers

CHAPTER FOCUS

- *Introduction: Point-to-point links*
- *Choice of components in optical link system*
- *Link power budget*
- *Specific design example*
- *Line coding*
- *System rise time*
- *Dispersion and attenuation limited transmission distance*
- *Material dispersion limited transmission distance*
- *Modal dispersion limited transmission distance*
- *Attenuation limited transmission distance*
- *Plotting curves between maximum transmission distance vs data rate*
- *Wavelength division multiplexing system*
- *Unidirectional wavelength division multiplexer*
- *Bidirectional WDM link*
- *Insertion loss, channel width and crosstalk*
- *Classification of wavelength division multiplexer*
- *Angularly dispersive WDM system*
- *WDM system using multilayer thin film reflector*
- *Reflecting grating WDM system*
- *Filtering devices*
- *Sources in WDM system*

11.1 INTRODUCTION: POINT-TO-POINT OPTICAL FIBER LINKS

In chapter 1 we have given the schematic diagram of an optical fiber link. In this chapter we shall discuss in details the simplest type of a point-to-point link and turn our attention to more complex link architectures such as, wavelength-division multiplexer systems, multi-point and multi-wavelength transmission network systems.

Fig. 11.1 shows the diagram for a point-to-point optical fiber link. It has a transmitter at one end and a receiver at other end of the optical fiber; but this simplest system does not have much demand in the optical fiber technology. Of course, it helps in understanding a more complex system.

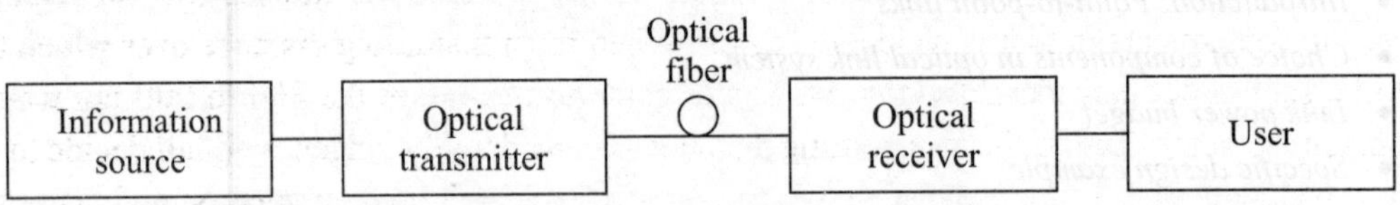

Fig. 11.1 Schematic diagram for a point-to-point optical fiber communication link

In order to design an optical fiber link we must consider many inter-related quantities. These are usually the operating characteristics of the optical sources (such as LED or laser) and those of photo detector (such as *p-i-n* or avalanche diode). In making successful analysis several interactions with the device characteristics are needed. The main requirements needs in making analysis in the design of the link are:

1. the possible transmission distance
2. the data rate which controls the channel bandwidth and
3. the bit error rate (abbreviated as BER)

Two analyses are usually considered to ensure that the desired performance can be achieved. These are:

(*a*) Link power budget

(*b*) Rise time of the system.

To meet the desired requirements, the designer has the choice of following components and their characteristics.

1. Multimode or single mode fiber

   (*a*) Core size

   (*b*) Core refractive index profile (step index or graded index)

(*c*) Band width
(*d*) Attenuation
(*e*) Numerical aperture (NA)

2. Optical Sources (LED or laser diode)
(*a*) Wavelength emitted
(*b*) Spectral line width
(*c*) Output optical power
(*d*) Effective radiating area
(*e*) Emission pattern

3. Detector
(*a*) Responsivity
(*b*) Range of operating wavelength
(*c*) Sensitivity
(*d*) Speed of response

11.2 CHOICE OF COMPONENTS IN OPTICAL LINK SYSTEM

Transmitting Wavelength

In order to decide the link power budget we shall first determine the transmitting wavelength and choose the components in that wavelength region. When the transmitting distance over which the data are to be sent is not considerably long, we shall decide to operate in the 800 to 900 nm wavelength region. On the other hand, if the transmitting distance is considerably longer, we shall decide to operate at the wavelength of 1.3 μm, because in this region of wavelength both attenuation and dispersion are low.

Source and Photodetector

After choosing the operating wavelength we shall now interrelate the system performance of the receiver, transmitter and optical fiber. Normally, we shall choose the characteristics of the transmitter and those of the receiver and compute the characteristics of the optical fiber to find that the required performance of the system is achieved. The procedure followed here is to select a photodetector and an optical source. Then we examine that how far the data can be transmitted through a particular fiber before a repeater is used in the line to boost up the power level of the optical signal. In choosing a particular photodetector we mainly need to determine the maximum optical power that must fall on the photodetector so that the bit error rate (BER) requirement is satisfied at the specified data rate. We know that the *p-i-n* photodiode receiver is simpler, more stable with the change of temperature and less costly than the avalanche photodiode receiver. Besides these advantageous features, the bias voltages of a *p-i-n* photo diode are usually less than 50 Volts. But these advantages of a *p-i-n* photodiode may be overruled by the higher sensitivity of an avalanche photodiode. In case a low optical power level is to be detected an avalanche photo diode is more suitable to detect the signal. The parameters, such as signal dispersion, data rate, transmission distance and cost would decide whether a LED or laser diode will be used. The spectral output of the laser has much narrower width than that of the LED. This is an important factor in the wavelength region of 800 to 900 nm, and as such, a laser diode must be used at these wavelength range. At wavelengths around 1.3 μm the signal dispersion is low and bit rate distance product of at least 1500 Mb/s.km are achievable with a LED. In the case of InGaAsP-type lasers the bit rate-distance product is greater than 25 G bit/sec.km.

In typical cases laser diodes couple 10 dB to 15 dB more optical power into the fiber than LED's and a greater transmission distance without using a repeater can be attained by using a laser diode. This advantage and a lower value of dispersion of signals of a laser diode may be offset by its high cost. In addition to this, the electronic circuits of a laser diode are much more complicated, because the lasing threshold voltage has to be dynamically controlled as a function of temperature and device aging.

Optical Fibers

In making choice of the optical fiber we have to select it between single mode or multimode type, either of which has a stepindex or graded index core. This choice is to be decided by the type of light source used and the amount of signal dispersion that can be tolerated. Multimode fibers must be used with a LED source, whereas a very small amount of optical power can be coupled into a singlemode fiber from a LED source. We have seen in Chapter-I that the optical power that can be coupled into a fiber from a LED depends on the core-cladding refractive index difference $\Delta = n_1 - n_2$ where n_1 and n_2 are the refractive index of the core and cladding of the fiber respectively and the numerical aperture (NA) of the fiber which depends on Δ. As the value of Δ increases, the optical power coupled to the fiber increases correspondingly. Again, the dispersion of the signal increases with the increase of Δ. Therefore, a compromise should be made between the optical power that can be launched into the fiber and maximum tolerable dispersion. Either a singlemode or a multimode fiber can be used with a laser diode. A singlemode fiber can provide the ultimate bit rate-distance-product with a value of 30 Gb/s.km. The disadvantage of a singlemode fiber is that a small core size (10 to 15 μm in diameter) makes the fiber splicing more difficult and critical than that for a multimode fiber having a core diameter of 50 μm or more.

When we choose the attenuation characteristics of a cabled fiber, we should consider the extra loss that results from cabling process, in addition to the attenuation of the fiber itself. This must also include the connector and splice losses as well as environmental induced losses that will arise from the temperature variations, radiation effects, and dust and moisture on the cable conductor.

11.3 LINK POWER BUDGET

Fig. 11.2 shows an optical power loss model for a point-to-point link. The amount of the optical power falling on the photodetector depends on the amount of light coupled to the optical fiber, loss occurring in the fiber itself and at the connector and splices. The link power budget is derived from the total power loss suffered by all elements in the link. Each of these loss components is expressed in decibels (dB).

The total loss is defined as the ratio between the optical power received by the photodetector and the optical power received by the fiber. Thus, the total loss expressed in dB is

$$\text{Total loss} = 10 \log \frac{P_{out}}{P_{in}} \tag{11.3.1}$$

where P_{in} and P_{out} are the optical power falling on the fiber and that falling on the photo detector respectively.

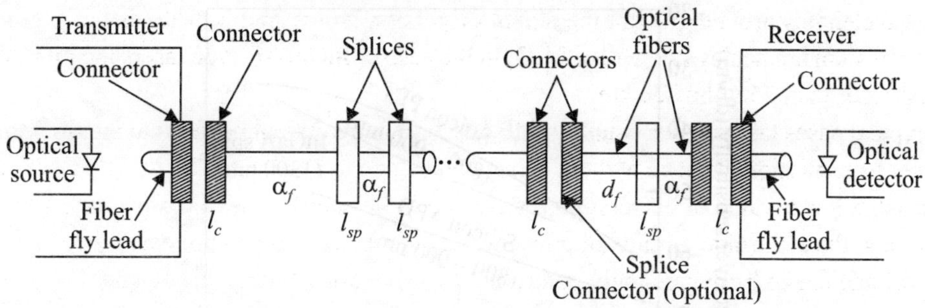

Fig. 11.2 Optical power loss for a point-to-point link: losses occur at the connector l_C, splice l_{sp} and in the fiber with an attenuation (or loss) of α_f per km

In addition to the power loss in the link as illustrated in Fig. 11.2 a link power margin is provided in the analysis for a link power budget. The power margin is allocated for the power loss due to component aging (using for a considerable period), temperature fluctuation and losses that will be added from the components that might be connected to the link in future. A link margin of 6 to 8 dB is generally used for the systems.

In computing the link power loss budget we simply consider the total optical power loss p_T that is allowed between the light source and photodetector. This loss may be considered due to the cable attenuation, connector loss, splice loss and the power margin allocated to the system. If, thus, P_S represents the optical power coming out after the end of the fly lead attached to the light source and if P_R is the receiver sensitivity, then the relationship between P_S and P_R is given by

$$P_T = P_S - P_R$$
$$= 2\,l_c + \alpha_f L + \text{system margin} \qquad (11.3.2)$$

where l_c is the connector loss due to two connectors, α_f is the attenuation constant for the fiber and L is the length of the transmission distance so that the loss in the fiber is $\alpha_f L$ where $\alpha_f L$ is expressed in dB/km. The system margin is usually taken as 6 to 8 dB. If we assume that the cable length of L has two connectors at the two ends and none in between, the splice loss is included in the cable loss. $2l_c$ is the loss due to two connectors.

11.4 SPECIFIC DESIGN EXAMPLE

In order to understand how a link power budget is calculated we shall consider a specific design example. Suppose we assume that a data rate of 20 Mb/s and a bit error rate (BER) of 10^{-9} which means that one error can occur for every 10^9 bits transmitted. We shall choose a *p-i-n* photodiode receiver operating at the wavelength of 850 nm. Fig. 11.3 shows that the receiver sensitivity as a function of bit rate for a bit error rate (BER) of 10^{-9}. From the figure it appears that for the *p-i-n* photodiode the receiver requires an input signal of 42 dBm (which means that 42 dB below one milliwatt.) We next choose a GaAlAs LED which can couple a 50 μW (–13 dBm) average optical power level into a fiber flylead. The core diameter of the fiber is 50 μm. Therefore, we find that the allowable power loss is –(42 – 13) = –29 dB (Fig. 11.3). We assume that a 1 dB power loss occurs when the fiber flylead is connected to the cable and another 1dB connector loss occurs at the cable to photodetector interface. Considering a 7 dB system margin the possible transmission distance L with an attenuation of $\alpha_f L$ dB/km can be found from following equation.

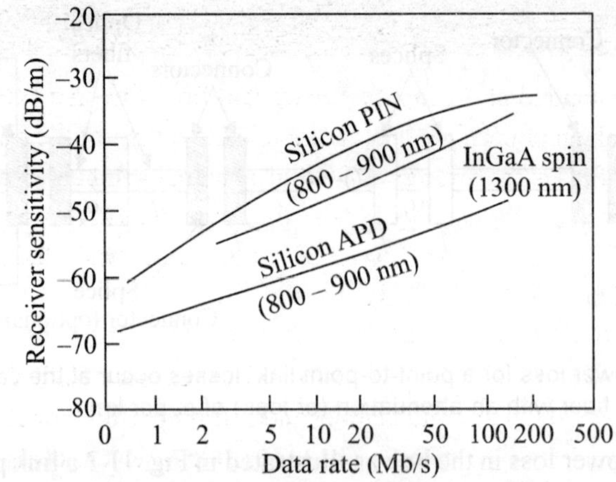

Fig. 11.3 The sensitivity of a receiver as a function of bit rate for a BER of 10^{-9}

$$P_T = P_S - P_R = 29 \text{ dB} = 2(1 \text{ dB}) + 7 \text{ dB} + \alpha_f L \; 29 - 2 - 7 = \alpha_f L$$

when $\alpha_f = 3.5$ dB/km, the transmission distance

or $$L = \frac{20}{\alpha_f} = \frac{20}{3.5} = 5.7 \text{ km}$$

Thus a 5.7 km transmission path is possible.

Fig. 11.4 illustrates the link power budget graphically. The vertical axis represents the optical power loss allowed between transmitter and receiver, while the horizontal axis represents the transmission distance.

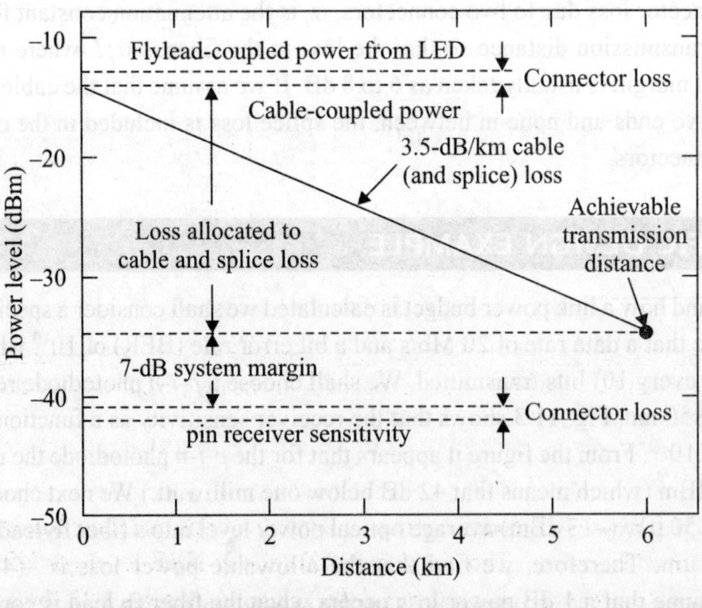

Fig. 11.4 Illustration of a link-power budget graphically: The vertical axis represents the optical power loss allowed between the transmitter and receiver, while the horizontal axis represents the transmission distance

11.5 LINE CODING

The digital signal should be applied at the input of an optical fiber link in a certain format (pattern). In designing an optical fiber link an important consideration is the format of the transmitted optical signal. Signal encoding uses a set of rules for arranging the digital signal symbols in a particular pattern. The process of arranging is referred to as "channeling" or "line coding". There are three basic types of two-level binary line codes. These codes are used in optical fiber transmission links. These codes are:

1. **The non-return-to zero (NRZ) formats**
2. **The return-to-zero (RZ) format, and**
3. **The phase encoded (PE) format or The Optical Manchester Code format**

In NRZ codes, a transmitted data bit occupies the full bit period. For RZ code format the pulse width is less than a full bit period. In PE (Phase Encoded) format which, is also known as Manchester code format both full width and half width data pulses are present. Various code formats are shown in Fig. 11.5.

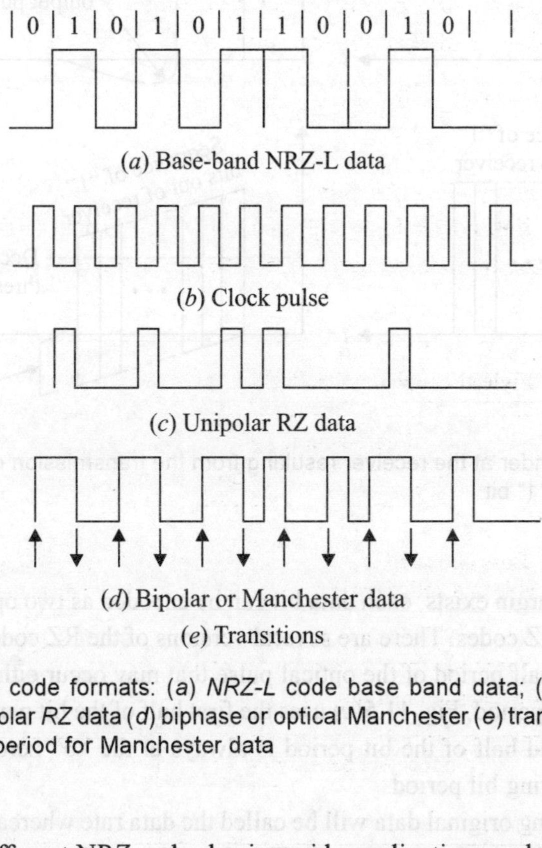

Fig. 11.5 Various code formats: (*a*) *NRZ-L* code base band data; (*b*) clock signal pulse (*c*) unipolar *RZ* data (*d*) biphase or optical Manchester (*e*) transition occurring within the bit period for Manchester data

The NRZ Codes

There are number of different NRZ codes having wide applications and their bandwidths serve as the reference for all other code groups. The simplest NRZ code is NRZ-level (or NRZ-L) as shown in Fig. 11.6. For a serial data stream an on-off (or unipolar) signal represents a "1" by a pulse of current or light filling an entire bit period, whereas for a "0" no pulse is transmitted. It is very simple to generate and decode these codes. The shortcomings of these codes are that they possess no error monitoring and correcting capabilities and they have no self clocking (or timing) feature.

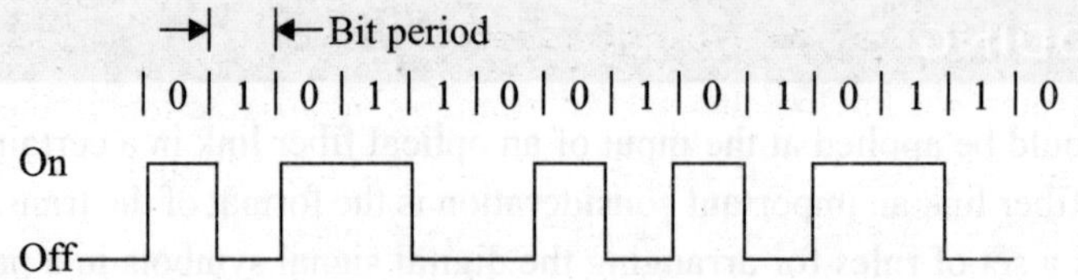

Fig. 11.6 Example of an NRZ data pattern

The code requires minimum bandwidth. The average input power to the receiver depends on the data pattern. For example, the high level of received power occurring in a long string of consecutive "1" bits can cause a base line wander as shown in Fig. 11.7.

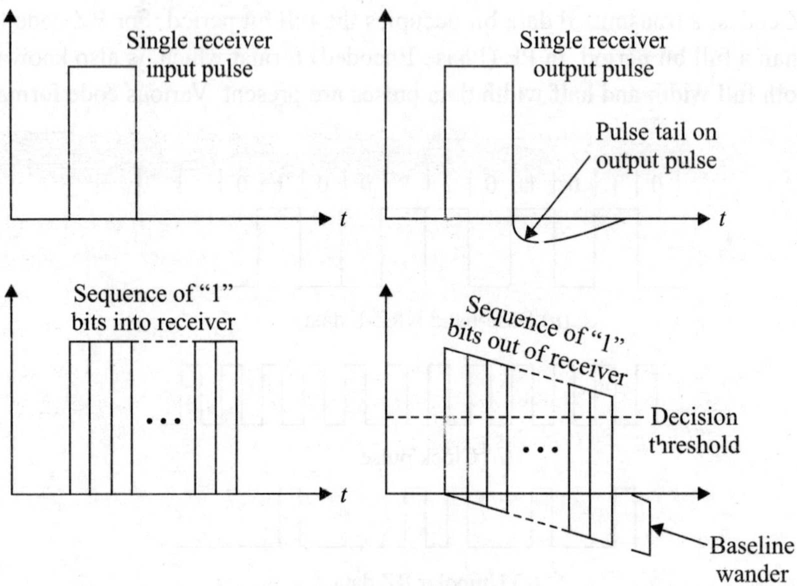

Fig. 11.7 Baseline wander at the receiver resulting from the transmission of long strings of NRZ one "1" bit

The RZ Code

If an adequate bandwidth margin exists, each data bit can be encoded as two optical line code bits. This is the basic principle of an RZ codes. There are several versions of the RZ codes. In unipolar RZ data a "1" bit is represented by a half period of the optical pulse that may occur either in the first-half or the second half period of the bit period. Fig. 11.5(*c*) uses the first half of the bit period to represent the level "1" of the bit and the second half of the bit period is always at the "0" zero level. A zero "0" bit is represented by no signal during bit period.

The rate of transmitting original data will be called the data rate whereas the rate of transmitting encoded data will be referred to as the "baud rate". The channel capacity required will be equal to baud rate.

The disadvantage of a unipolar RZ code format is that a long string of "0" bit can cause the loss of synchronization.

The Phase encoded format or Manchester Code

The common data format not having the loss of timing synchronization is the biphase or optical Manchester code as shown in Fig. 11.5(*d*). It may be noted that it is a unipolar code which is in contrast with the conventional bipolar Manchester code used in wire lines.

The optical Manchester code signal format can be obtained by direct module-2 addition of the base band NRZ-L signal of Fig. 11.5(*a*) and the clock signal of Fig. 11.5(*b*). In this code there is a transition at the center of each bit period. A negative transition at the center of each bit period indicates a "1" bit whereas the positive going transition represents that a "0" bit is sent (Fig. 11.5*c*). It is very simple to generate and decode the Manchester code. Since this is an RZ type code, it requires twice the bandwidth of an NRZ code. Moreover, it has no inherent error-detecting and correcting capability. (Integrated circuits are available that automatically encode or decode the data stream using the Manchester code)

11.6 SYSTEM RISE TIME

The system rise time can be determined from individual rise time of each component in the system. These components are the transmitter (Optical source such as LED or laser diode), the fiber cable, and the receiver (detecting components such as *p-i-n* or avalanche photo diode).

Fig. 11.8(*a*) shows that a step input voltage of amplitude V is applied to a low pass filter circuit Fig. 11.8(*b*) which is basically the high-frequency equivalent circuit of an RC coupled amplifier. The voltage rise in the output of the filter is exponential with time in nature as shown in Fig. 11.9. The output voltage V_o across the capacitor C is given by

$$V_o(t) = V\left[1 - \exp\left(\frac{-t}{RC}\right)\right] \tag{11.6.1}$$

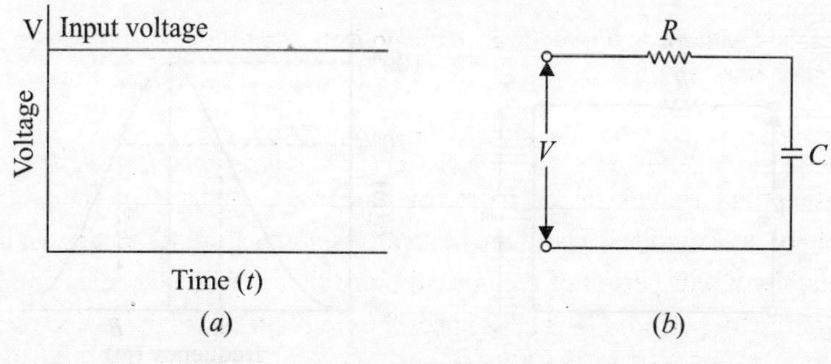

Fig. 11.8 Rise time of a signal in a filter: A step index voltage of amplitude V is applied to a filter circuit consisting of a resistance R and a capacitance C

where V is the amplitude of the step index voltage, RC is the time constant of the filter circuit. We notice that the output will attain the maximum value V theoretically at infinite time, ($t = \infty$). The rise time t_r is defined in terms of Gaussian response as 10 to 90% rise (or fall) of the voltage. From Fig. 11.9 it appears that the rise time is equal $t_r = t_2 - t_1$ where t_1 and t_2 are the instants at which the output voltage is 10% and 90% respectively of the final voltage V. From Eq. (11.6.1)

$$V_1(t) = V\left(1 - \exp\frac{-t_1}{RC}\right) \qquad (11.6.2)$$

or $\quad t_1 = 0.1\ RC$ when $V_1(t) = 0.1$ V

Similarly $\quad V_2(t) = V\left(1 - \exp\frac{-t_2}{RC}\right) \qquad (11.6.3)$

or $\quad t_2 = 2.3\ RC$ when $V_2(t) = 0.9$ V

Here $V_1 = 0.1$ V and $V_2 = 0.9$ V represent 10% and 90% of V respectively.

The rise time, $\quad t_r = t_2 - t_1 = 2.3RC - 0.1RC = 2.2RC \qquad (11.6.4)$

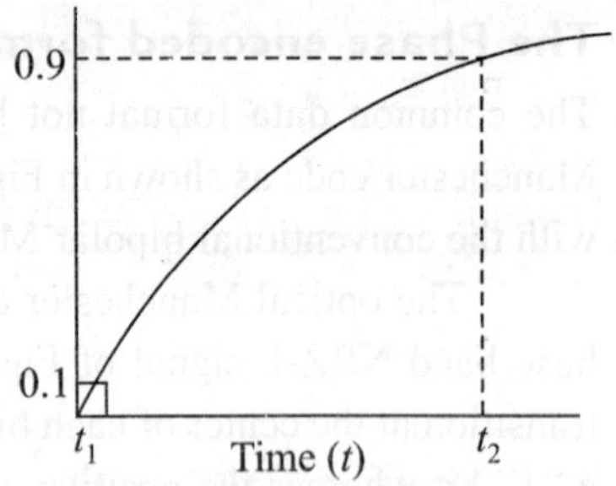

Fig. 11.9 The output voltage in the filter rises exponentially. The output voltage V_o is equal to 0.1 V at the time $t_1 = t_1$, and output voltage is 0.9 V, when $t = t_2$ and rise time $t_r = t_2 - t_1$

Bandwidth and Transfer Function of the Filter Circuit

The transfer function of the filter circuit of Fig. 11.10(*a*) can be obtained in the following way:

$$V_o = \frac{V\dfrac{1}{j\omega C}}{R + \dfrac{1}{j\omega C}} = \frac{V}{1 + j\omega RC} \qquad (11.6.5)$$

or $$\frac{V_o}{V} = \frac{1}{1 + j\omega RC}$$

or $$\left|\frac{V_o}{V}\right| = \frac{1}{(1 + \omega^2 R^2 C^2)^{1/2}} = \frac{1}{(1 + (2\pi)^2 f^2 R^2 C^2)^{1/2}} \qquad (11.6.6)$$

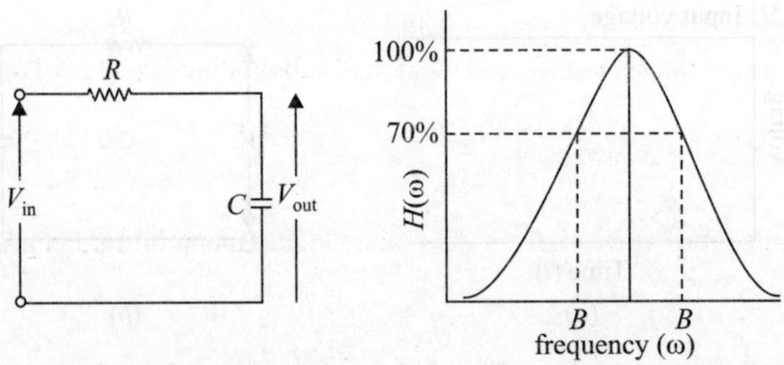

(*a*) (*b*)

Fig. 11.10 The calculation of the transfer function $H(\omega) = (V_{out}/V_{in})$ of the filter circuit of Fig. 11.8(*b*) $H = (1/1+\omega^2 C^2 R^2)$ so that bandwidth $B = f = \dfrac{1}{2\pi RC}$

Therefore, 3 dB band width of the circuit can be obtained putting second term in denominator equal to 1 and $f = B$.

Bandwidth $$f = B = \frac{1}{2\pi RC} \quad (11.6.7)$$

on combining Eq. (11.6.4) and Eq. (11.6.7) yields

Rise time $$t_r = 2.2RC = 2.2\,\frac{1}{2\pi f} = \frac{0.35}{f} = \frac{0.35}{B} \quad (11.6.8)$$

Total System Rise Time

The total system rise time can be determined from the individual rise times of the different components of the system. The fiber cable 10 to 90% rise time may be separated from the rise time arising from inter modal t_{mod} and intramodal (or material dispersion Δt_{mat})

The total system rise time t_{syst} is given by

$$\Delta t_{syst} = 1.1\,(\Delta t_s^2 + \Delta t_{mod}^2 + \Delta t_{mat}^2 + \Delta t_D^2)^{1/2} \quad (11.6.9)$$

where Δt_{mod} and Δt_{mat} are the modal and material dispersions respectively, Δt_s and Δt_D are the source and detector 10 to 90% rise times respectively. All the rise times are measured in nanoseconds. Comparison of rise time edge with the overall pulse dispersion results in the weighting factor 1.1.

Calculation of Maximum Bit Rate B_T (max)

The result for the 10 to 90% rise time as given in Eq. (11.6.4) is of general validity, but different constant term may be obtained with different filter circuits. However, for the calculation of the rise time of the optical fiber system, the constant 0.35 is usually used and hence, in Eq. (11.6.8), $t_r = t_{syst}$. Alternatively, if an ideal (unrealizable) filter with arbitrary sharp cut off is considered, the constant term in Eq. (11.6.8) becomes 0.44. The conservative estimate obtained using a constant term of 0.35 is accepted for the use of rise time calculation. To conserve the shape of the pulse with reasonable fidelity through the *RC* circuit the 3 dB bandwidth must be at least large enough to satisfy the condition $B_T = \Gamma$ where Γ is the width of the pulse. Combining this relation with Eq. (11.6.8) gives

$$\Delta t_{syst} = t_r = 0.35\,\Gamma \quad (11.6.10)$$

For *RZ* pulse format the bit rate $B_T = B = 1/\Gamma$ and substituting Eq. (11.6.10), we get

$$B_T(\max) = \frac{0.35}{\Delta t_{syst}} \quad (11.6.11)$$

For the *NRZ* pulse format $B_T = B/2 = 1/2\,\Gamma$ and the maximum bit rate is given by

$$B_T(\max) = 0.7/\Delta t_{syst} = \frac{0.7}{\Delta t_{syst}} \quad (11.6.12)$$

11.7 DISPERSION AND ATTENUATION LIMITED TRANSMISSION DISTANCE

We now consider an optical fiber link with a small rise time of the source and detector so that the rise time does not limit the transmission distance of the link. When the transmission distances of the link are considerably long or the data rates are high the transmission distance can be limited by the three factors such as either the material dispersion or the modal dispersion or the attenuation (or loss) of the signal in the fiber

1. When the transmission distance in a link is limited by the material dispersion only, not by the modal dispersion or attenuation, the distance is referred to as "Material dispersion limited transmission distance".
2. When the transmission distance is limited by modal dispersion only, not by the material dispersion or attenuation, the distance is known as "Modal dispersion limited transmission distance".
3. When the transmission distance in the link is limited by attenuation (or loss) only, not by material or modal dispersion, the distance is known as "Attenuation limited transmission distance".

11.8 MATERIAL DISPERSION LIMITED TRANSMISSION DISTANCE

We now consider the material dispersion limited transmission distance in a link. When a signal in *RZ* coding is applied we know that

$$\Delta t_{mat} \leq \Gamma \leq 0.35\, T_B \tag{11.8.1}$$

where Δt_{mat} is material dispersion delay time and T_B is the bit-period which is reciprocal of bit rate B_R so that $T_B = 1/B_R$, Γ is the width of the applied pulse.

Again we know that putting $Z = L$ in Eq. (3.15.11) and Δt_{mat} becomes

$$\Delta t_{mat} = \left(\frac{L}{c}\right)\left(\frac{\Delta\lambda}{\lambda}\right)\left(\lambda^2 \frac{d^2n}{d\lambda^2}\right) \tag{11.8.2}$$

Substituting the value of Δt_{mat} in Eq. (11.8.1)

$$\left(\frac{L}{c}\right)\left(\frac{\Delta\lambda}{\lambda}\right)\left(\lambda^2 \frac{d^2n}{d\lambda^2}\right) \leq 0.35\, T_B \tag{11.8.3}$$

where L is the length of the link, c is the velocity of light, $\Delta\lambda$ and λ are the spectral width and wavelength of the optical signal, n is the refractive index of the fiber core.

In the limiting case $L = L_{max}$ for *RZ* code and from Eq. (11.8.3)

$$L_{max} = (0.35\, T_B c)\left(\frac{\lambda}{\Delta\lambda}\right)\left(\frac{1}{\lambda^2 \frac{d^2n}{d\lambda^2}}\right) \tag{11.8.4a}$$

when *NRZ* code is used

$$L_{max} = (0.7\, T_B c)\frac{\lambda}{\Delta\lambda}\left(\frac{1}{\lambda^2 \frac{d^2n}{d\lambda^2}}\right) \tag{11.8.4b}$$

Example 11.1: *A silica fiber link has a bit rate of 50 MbS^{-1}. The fiber is operating at a wavelength of 850 nm with a spectral width of 45 nm. The value of the parameter ($\lambda^2\, d^2n/d\lambda^2$) is found to be 0.023. Calculate the material dispersion delay Δt_{mat} when the length of the link is 1.25 km.*

Solution: From Eq. (11.8.2)

$$\Delta t_{mat} = \left(\frac{L}{c}\right)\left(\frac{\Delta\lambda}{\lambda}\right)\left(\lambda^2 \frac{d^2n}{d\lambda^2}\right)$$

Data supplied

$L = 1.25$ km, $\Delta\lambda = 45$ nm, $\lambda = 850$ nm, $c = 3 \times 10^8$ m/s and $\left(\lambda^2 \frac{d^2n}{d\lambda^2}\right) = 0.023$

Substituting the values

$$\Delta t_{mat} = \left(\frac{1.25 \times 10^3}{3 \times 10^8}\right)\left(\frac{45}{850}\right)(0.023) = 5.07 \text{ ns}.$$

Example 11.2: *A silica fiber link has a bit rate of 10 Mb s^{-1}. The fiber is operating at wavelength of 850 nm with spectral width of 45 nm. The value of ($\lambda^2 d^2n/d\lambda^2$) is 0.023. Estimate the material dispersion limited transmission distance.*

Solution: Material dispersion limited transmission distance L_{max} is

$$\Delta t_{mat} \leq 0.35\, T_B \text{ and } \Delta t_{mat} = \frac{L}{c}\left(\frac{\Delta\lambda}{\lambda}\right)\left(\lambda^2 \frac{d^2n}{d\lambda^2}\right)$$

Thus, $0.35\, T_B \geq \frac{L_{max}}{c}\left(\frac{\Delta\lambda}{\lambda}\right)\left(\lambda^2 \frac{d^2n}{d\lambda^2}\right)$

Rearranging proper by

$$L_{max} = (0.35\, T_B c)\left(\frac{\lambda}{\Delta\lambda}\right)\left(\frac{1}{\lambda^2 \frac{d^2n}{d\lambda^2}}\right)$$

$$T_B = \frac{1}{B_R} = \frac{1}{10 \times 10^6},\ c = 3 \times 10^8 \text{ m/s},\ \lambda = 850 \text{ nm},\ \Delta\lambda = 45 \text{ nm and } \lambda^2 \frac{d^2n}{d\lambda^2} = 0.023$$

Substituting the values

$$L_{max} = \frac{0.35 \times 3 \times 10^8}{10 \times 10^6}\left(\frac{850}{45}\right)\left(\frac{1}{0.023}\right) = 86.2 \times 10^2 \text{ m}$$

$$L_{max} = 0.862 \text{ km}.$$

11.9 MODAL DISPERSION LIMITED TRANSMISSION DISTANCE

For modal dispersion in a step index fiber using Eq. (3.14.9) we have

$$L_{max} = \frac{0.35c\, T_B}{n_1 - n_2} = \frac{0.35c}{(n_1 - n_2)B_R} = \frac{0.70c\, n_1}{(NA)^2\, B_R} \tag{11.9.1}$$

Where T_B is the bit period and B_R is the bit rate so that $T_B = 1/B_R$, n_1 and n_2 are the refractive index of the fiber core and cladding respectively, and (NA) is numerical aperture of the fiber.

When *NRZ* code is used

$$L_{max} = \frac{0.70c}{B_R(n_1 - n_2)} = \frac{1.4c\, n_1}{(NA)^2 B_R} \quad (11.9.2)$$

For the modal dispersion limited case in a graded index fiber using *RZ* code

$$L_{max} = \frac{2.8T_B c\, n_1^2}{[NA(o)]^2} = \frac{2.8c\, n_1^2}{[NA(o)]^2 R_B} = \frac{1.4c\, n_1}{(n_1 - n_2)B_R} \quad (11.9.3)$$

where n_1 is the refractive index at the centre of the core.

Example 11.3: *Calculate the modal dispersion limited transmission distance for 50 Mbs^{-1} data link in the case of a step index fiber and a graded index fiber where refractive index of the core n_1 = 1.45, $\Delta = n_1 - n_2$ = 1%. The code of the signal is RZ code.*

Solution: $n_1 - n_2 = \Delta n_1 = 0.01 \times 1.45$

Data rate are $B_R = 50 \times 10^6$.

In the case of the step index fiber;

$$L_{max_{SI}} = \frac{0.35c}{(n_1 - n_2)B_R} = \frac{(0.35)\times 3\times 10^8}{(0.0145)\times(50\times 10^6)} = 1.448\times 10^2 \text{ m}$$

For a graded index fiber

$$L_{max_{GI}} = \frac{2.8cn_1^2}{[NA(a)]^2 B_R} = \frac{2.8cn_1^2}{2n(n_1 - n_2)B_R} = \frac{1.4cn_1^2}{n_1(n_1 - n_2)B_R} = \frac{1.4cn_1}{(n_1 - n_2)B_R}$$

$$L_{max} = \frac{1.4\times(3\times 10^8)\times(1.45)}{(0.0145)\times(50\times 10^6)} = 840 \text{ m}.$$

11.10 ATTENUATION LIMITED TRANSMISSION DISTANCE

For making comparison, we like to calculate the maximum link distance for a system only by the fiber attenuation (loss). The equation for this is

$$L_{max} = \frac{P_T(dBm) - P_R(dBm)}{\alpha_{fiber}} \quad (11.10.1)$$

where P_T is the power of the transmitter (usually dependent on data except for a high data rates). P_R is the power of the receiver required to maintain the bit-error rate or signal to noise ratio and α is the attenuation constant of the fiber. In order to overcome the limits of attenuation we must have stronger signal. Two techniques are (1) optoelectronic repeater and (2) optical amplifier (discussed at the end of this chapter).

11.11 PLOTTING CURVES BETWEEN MAXIMUM TRANSMISSION DISTANCE VS. DATA RATE (DETERMINED BY DISPERSION LIMITS AND ATTENUATION LIMIT)

Example 11.4 will be given and from that example it will be possible to plot the curve.

Example 11.4: *A graded index fiber has n = 1.45, Δ = 1% and a loss of 1 dB per km. It is operated with a source of 850 nm with a line width of 60 nm and it produces power in fiber of –10 dBm. The receiver is a p-i-n diode that requires a power given by*

$$P_R(dBm) = -65.0 + 20 \log DR' \; [Mbs^{-1}]$$

to maintain a bit error rate of 1 × 10^{-9} where DR′ is the data rate in Mbs^{-1}. The value of the parameter ($\lambda^2 d^2n/d\lambda^2$) is found to be 0.022. The coding is RZ coding and

(a) *Calculate the material-dispersion limited distance.*

Solution: The equation for the material dispersion limited distance is

$$L_{max} = \frac{0.35c}{B_R} \cdot \left(\frac{\lambda}{\Delta\lambda}\right)\left(\frac{1}{\lambda^2 \dfrac{d^2n}{d\lambda^2}}\right)$$

Thus substituting the numerical values

$$L_{max} = \frac{(0.35)(3\times10^8)(850\times10^{-9})}{B_R(60\times10^{-9})(0.022)} = \frac{6.757\times10^{10}}{B_R}$$

(*b*) Calculate the modal limited distance

Solution: The distance is

$$L_{max} = \frac{2.8cn_1^2}{[NA]^2 B_R} = \frac{1.4cn_1^2}{n_1^2 B_R\Delta} = \frac{1.4c}{B_R\Delta}$$

$$= \frac{(1.4)\times(3\times10^8)}{B_R(0.01)} = \frac{4.2\times10^{10}}{B_R}$$

(*c*) Calculate the attenuation limited distance as a function of the data rate

Solution: The distance is

$$L_{max} = \frac{P_T(dBm) - P_B(dBm)}{\alpha} = \frac{-10-[(-65.0+20\log(B_R))]}{1}$$

$$= -55 - 20 \log B_R$$

(*d*) Draw the result in a graph for link length versus data rate extending from 1 Kbs^{-1} to 10 Kbs^{-1} (Fig. 11.11).

11.12 WAVELENGTH-DIVISION MULTIPLEXING (WDM) SYSTEMS

We have observed that an optical fiber link transmits a single wavelength. In order to increase the transmission capacity of a link we should utilize the transmission of several wavelengths simultaneously through the same fiber in a link. Fig. 11.12 shows a curve for the variation of attenuation (loss) with wavelength for an optical source. The source uses a small part of the available spectral transmission band of the fiber although optical fibers have a considerably broad region of low attenuation of the signal. Thus, the fiber can support the operation of the link with more than one source. We are able to control the operating wavelength of a semiconductor emitter such as a laser or LED through the alloy composition or through the spacing of the grating in the distributed feedback (DFB) laser diode or distributed feedback reflector (DBR) laser. It is thus, possible to design a link that would carry several wavelengths signals through a single fiber simultaneously.

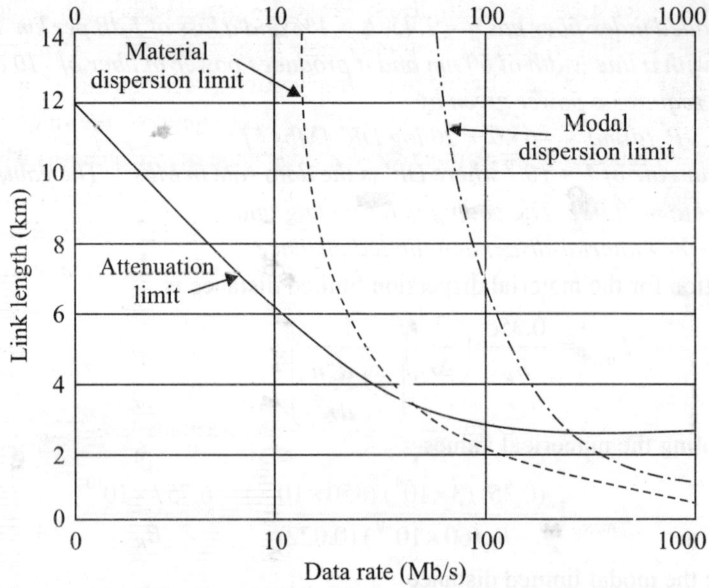

Fig. 11.11 A graph between the link length versus data rate for BER = 10^{-9}

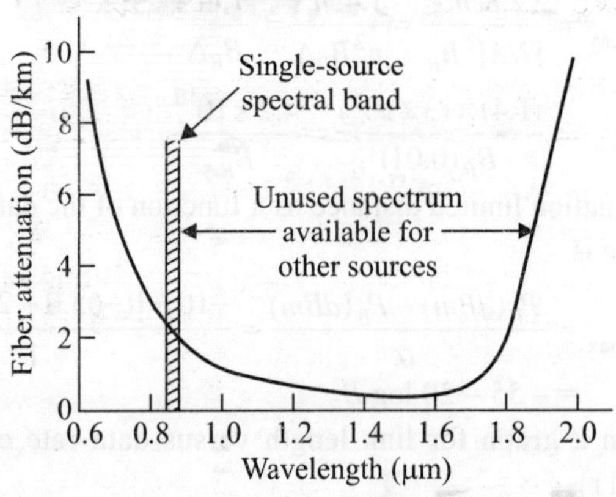

Fig. 11.12 A curve showing the variation of attenuation (or loss) as a function of wavelength for an optical source used in fiber communications

11.13 UNIDIRECTIONAL WAVELENGTH DIVISION MULTIPLEXER (WDM) SYSTEM

In implementing a wavelength-division multiplexer (WDM) system, a multiplexer in the transmitting end is required to combine the signals of different wavelengths from several light sources into a single fiber. At the receiving end a demultiplexer is used to separate the signals into different appropriate channels. The optical signals that are combined pass through the demultiplexer usually do not emit significant amount of optical power outside the channel. Fig. 11.13 shows the schematic diagram of the system using the wavelength-division multiplexing where the input signals of wavelength λ_1, λ_2, λ_3 and

λ_4, are passed through the fiber by multiplexer. From the optical fiber the combined signal passes through the demultiplexer that divides all the four signals through their appropriate channels. Cross talk between two adjacent channels is negligible at the transmitter end. The main design problem is that the multiplexer should provide a low-loss path for each optical signal to the output of the multiplexer system. Photodetectors are usually sensitive over a broad range of wavelengths which may include all WDM channels. Thus, to prevent a significant amount of wrong signals from entering into the receiving channel either multiplexer must be carefully designed or a very stable optical **filters** with wavelength cut off must be used. This will provide a good channel isolation of the different wavelengths being used.

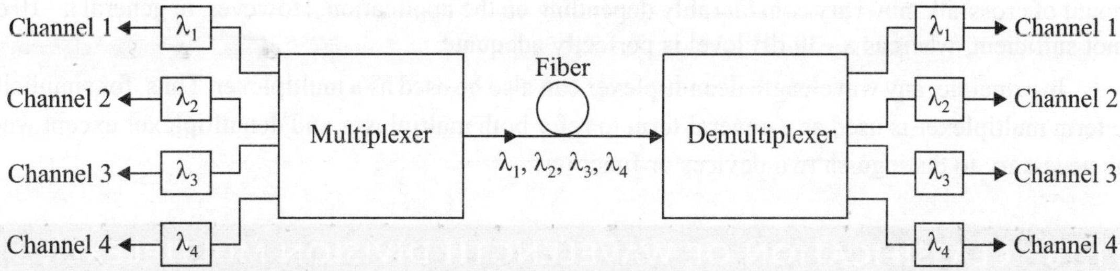

Fig. 11.13 A schematic diagram using wavelength division unidirectional multiplexing with output signals of wavelengths λ_1, λ_2, λ_3, and λ_4 and the multiplexer combines four independent input signals for transmission over a single fiber

11.14 BIDIRECTIONAL WDM LINKS

The bidirectional links are also possible in WDM system by intermixing sources and receivers at each end. Fig. 11.14 shows a bidirectional WDM link. It involves sending information in one direction at a wavelength λ_1 from left to right and simultaneously sending transmitting information data in the opposite direction from right to left in the figure at a wavelength λ_2. The transmission in both directions is accomplished by the same fiber.

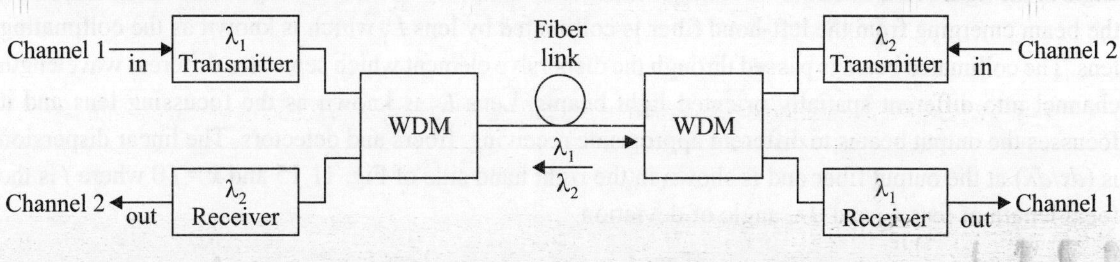

Fig. 11.14 Schematic representation of a bidirectional WDM system in which two or more wavelengths are transmitted simultaneously in opposite directions over the same fiber

11.15 INSERTION LOSS, CHANNEL WIDTH AND CROSS TALK

Insertion loss is defined as the amount of optical power loss that occurs in the optical fiber line due to the addition of WDM coupling devices. This includes the losses occurring in the connection points of WDM elements to the fiber line and any intrinsic losses within the multiplexing elements itself. In practice the designer can tolerate a loss of few decibels in each end.

Channel width is the wavelength range that is allocated to a particular source. If the laser diodes are used, the channel width is several tens of a nanometer (nm) so that it ensures that no inter channel interference takes place resulting from the source instability. We have seen in chapter IV and V that the change of peak of the operating wavelength of the source (such as a laser or LED) takes place with the change of temperature. When the LED sources are used, the required channel widths are to be 10 or 20 times larger than that with a laser source, because the widths of the spectral output of LED sources are more than those of the laser sources.

Cross talk is defined by the amount of signal coupling from one channel to another. The tolerable amount of crosstalk may vary considerably depending on the application. However, in general a –10 dB is not sufficient, whereas a –30 dB level is perfectly adequate.

In principle, any wavelength demultiplexer can also be used as a multiplexer. Thus, for simplicity, the term multiplexer is used as a general term to refer both multiplexer and demultiplexer except when it is necessary to distinguish two devices or functions.

11.16 CLASSIFICATION OF WAVELENGTH DIVISION MULTIPLEXER (WDM)

The wavelength division multiplexer (WDM) are most widely classified into two classes:

1. Angularly dispersive devices, such as prisms and gratings
2. Filter based devices, such as multilayer thin film interference filters or single mode integrated optical devices.

11.17 ANGULARLY DISPERSIVE WDM SYSTEM

Fig. 11.15 shows a diagram to explain the basic principle of an angularly dispersive multiplexer for a three element system. Here $(d\theta/d\lambda)$ is the angular dispersion of the device which is equal to the deviation angle of the light beam when the wavelength changes by unity. When the device is used as a demultiplexer, the beam emerging from the left-hand fiber is collimated by lens L_1 which is known as the collimating lens. The collimated beam is passed through the dispersive element which separates different wavelength channel into different spatially oriented light beams. Lens L_2 is known as the focussing lens and it focusses the output beams to different appropriate receiving, fibers and detectors. The linear dispersion is $(dx/d\lambda)$ at the output fiber and is shown in the right hand side of Fig. 11.15 and $x = f\theta$ where f is the focal length of lens L_2 and θ = angle of deviation

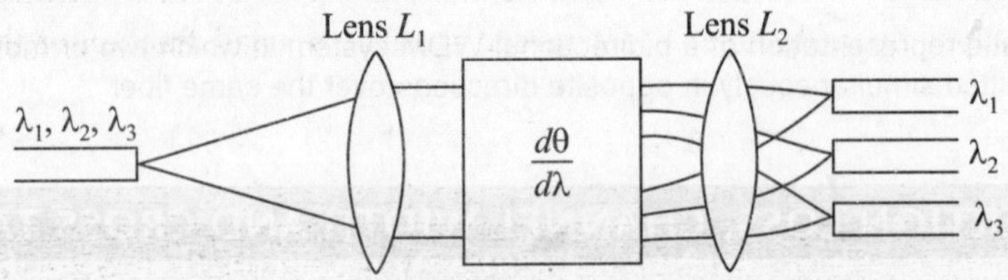

Fig. 11.15 Schematic representation of an angularly dispersive WDM element showing three wavelengths λ_1, λ_2 and λ_3 for explaining the basic principle

$$\frac{dx}{d\lambda} = \frac{d(f\theta)}{d\lambda} = f\frac{d\theta}{d\lambda} \tag{11.17.1}$$

where f is the focal length of lens L_2. When we consider an ideal case of aberration less optics and zero source spectral-width, the intrinsic insertion loss and cross talk will be zero provided the output signals are separated by more than their diameter d, that is if

$$\frac{dx}{d\lambda}\Delta\lambda \geqq d \tag{11.17.2}$$

where $\Delta\lambda$ is the spectral separation between channels. We shall assume here that all input and output fibers have the same diameter d and numerical aperture (NA). In the limiting case Eq. (11.17.1) becomes

$$f\frac{d\theta}{d\lambda}\Delta\lambda = d$$

or

$$f = \frac{d}{\frac{d\theta}{d\lambda}\Delta\lambda} \tag{11.17.3}$$

To collect all light from input fiber, the collimating lens L_1 must have a diameter b satisfying the condition

$$b > 2f\frac{NA}{n'} \tag{11.17.4}$$

where n' is the refractive index of the medium between the dispersive device and lens L_1.

On combining Eq. (11.17.3) and Eq. (11.17.4) yields

$$b > \frac{2(NA)d}{n'\left(\frac{d\theta}{d\lambda}\right)\Delta\lambda} \tag{11.17.5}$$

In any practical system the output beam spreads out because of the finite size of the light source and the angular dispersion resulting from the wavelength spread of the source. The fractional increase S in the beam diameter is given by

$$S = \frac{b' - b}{b} = (1+m)\frac{Wd\,(NA)}{b^2 n'} \tag{11.17.6}$$

where m is the number of wavelength channels, b' is the diameter of lens L_2 and W is the total path length from the output of lens L_2 to the input of the lens L_1. To avoid overfilling of the numerical aperture of the output fiber, the total beam spread must be a small fraction of the collimating lens diameter, that is

$$S << 1 \tag{11.17.7}$$

11.18 WDM SYSTEM USING MULTI-LAYER THIN FILM REFLECTOR

The filters are designed to transmit light in specific wavelength channel and either to absorb or to reflect all other wavelengths which are not required. Reflection types of filters are usually used, because the losses of optical power in absorption type filters tend to be high (greater than 1 dB). The reflection types of filter consist of a flat glass substrate and over this substrate a multiple layers of different dielectric films are deposited for wavelength selectivity.

Fig. 11.16 shows the diagram to explain the operation of filter type multiplexing system for two wavelength operation. Two light signals of wavelength λ_1 and λ_2 from the end of an optical fiber falls on lens L_1 from its focus. The rays pass as parallel rays to fall on a multi-layer reflector made of thin film. The device is transparent at the wavelength λ_2 so that a parallel ray signal at the wavelength λ_2 falls on the lens L_2 to converge finally at the focus of lens L_2. The other signal at wavelength λ_1 is reflected from the filter to fall on lens L_3. These filters can be used in series to separate additional wavelength channels. The complexity involved in stacking the filters in series is to cause the increase in signal loss that occurs with the addition of successive multiplexers. These double losses generally limit operation to two or three filters (that is three or four channels).

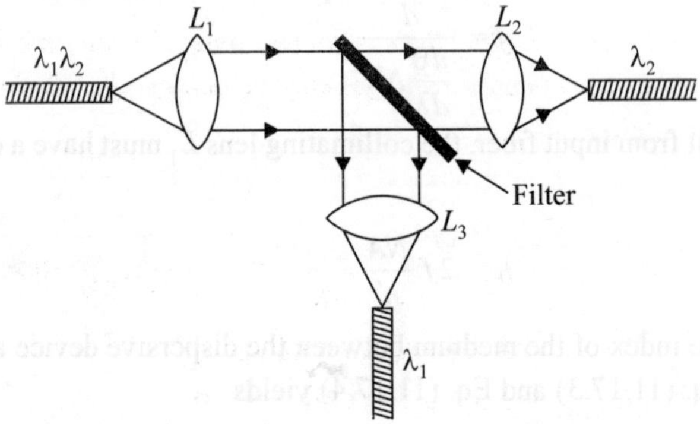

Fig. 11.16 Schematic diagram to explain the operation of filter type multiplexing system for two wavelength operation

11.19 REFLECTING GRATING WDM RECEIVER

Fig. 11.17 shows an angularly dispersive system. This system uses a grating as a device which is combined with a lens. A grating is a reflecting element or transmitting element. It has a series of close parallel lines engraved or etched into the surface.

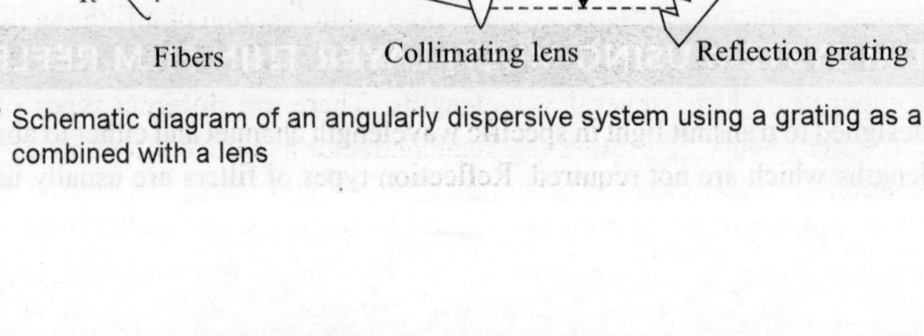

Fig. 11.17 Schematic diagram of an angularly dispersive system using a grating as a device combined with a lens

In the figure the top most line is the transmitter line where wavelengths λ_1, λ_2 and λ_3 are multiplexed. From the transmitter line multiplexed light signal enters into the collimating lens L. The signal containing wavelength λ_1, λ_2 and λ_3 moves as a parallel ray from the lens L to the grating. It is incident on grating causing different wavelengths to be reflected at different angle. After reflection from the grating the signal return back to the lens L and each wavelength is focussed on different points depending on their wavelength. The signals fall on fibers λ_1, λ_2 and λ_3 receiver lines. The loss suffered by the system ranges from 1 dB to 3 dB. The typical values of channel separations vary from 1 to 10 nm.

Example 11.5: *The LED used in 100 Mbs^{-1} silica fiber link has a rise time of 8 ns. The wavelength emitted by the LED is 830 nm with a spectral width of 40 nm. The p-i-n photo diode has a rise time of 10 ns. For the silica fiber the value of the parameter* $\left(\lambda^2 \dfrac{d^2 n}{d\lambda^2}\right)$ *is 0.024.*

The length of link is 2.5 km calculate the system rise time. The inter modal dispersion of the fiber is 3.5 ns/km and estimate the maximum bit rate that may be achieved in the link using RZ and NRZ for solution: The material dispersion delay time.

Solution:

$$\Delta t_{\text{mat}} = \left(-\frac{L}{c}\right)\left(\frac{\Delta\lambda}{\lambda}\right)\left(\lambda^2 \frac{d^2 n}{d\lambda^2}\right) = \frac{2.5\times10^3}{3\times10^8}\,\frac{40}{830}\,(0.024)$$

$$= -9.64 \times 10^{-9}\text{s} = -9.64 \text{ ns}$$

For a 2.5 km link has $\Delta t_{\text{modal}} = 2.5 \times 3.5 = 8.8$ ns

$$\Delta t_{\text{sys}} = 1.1\sqrt{(\Delta t_s)^2 + (\Delta t_r)^2 + (\Delta t_{\text{mat}})^2 + (\Delta t_{\text{modal}})^2}$$

$$= 1.1\sqrt{8^2 + 10^2 + 9.6^2 + 8.8^2} \text{ ns} = 19.8\times10^{-9} = 19.8 \text{ ns}$$

(*a*) $$B_T(\max) = \frac{0.7}{\Delta t_{\text{sys}}} = \frac{0.7}{19.8\times10^{-9}} = 34\times10^6$$

With *RZ* format $B_{T\max} = 34 \times 10^6$

(*b*) When *NRZ* format is used $B_T(\max) = \dfrac{0.35}{19.8\times10^{-9}} = 17\times10^6$

11.20 FILTERING DEVICES

In this section we shall discuss the basic property of a filtering device which finds application in wavelength division multiplexing (WDM) systems. The main function of a filtering device is either to reflect or transmit light depending on the wavelength of the light. Different varieties of filter devices can be used to separate light of several wavelengths. There are different types of filter, such as (*a*) Interference filter (*b*) Fabry Perot filter (*c*) Semiconductor amplifier filter and (*d*) Electronic filter for coherent detection.

(a) Interference Filters

Most optical filters which find applications in WDM system consist of multilayer dielectric stacks. They exhibit a sharp change in reflectivity at the desired wavelength and at the same time the insertion loss of signal is reduced to a small value. The filter passes one wavelength with typical response and stops all other wavelengths more than about 10% away from the centre wavelength of the filter. With more complex design of the filter, the spacing can be made as small as 1 to 0.1 per cent of centre wavelength. When it is desired to operate with more than two channels, a series arrangement of several filters becomes necessary. But the combination of the filters has a considerable insertion loss. For practical cases only a cascade of two or three filters is used, otherwise there will be a large insertion loss. As a result, broadband system operation will be restricted to five or six operation wavelength channels.

The typical channel bandwidth of a interference filter is of 5 nm wavelength.

(b) Fabry Perot Resonator Filter

The Fabry Perot resonator filter is a band pass optical filter. It is capable to transmit a narrow band of wavelength and to reflect other undesirable wavelengths. The transmission frequencies of the filter occur every $c/2L$ length in the resonator where c is the velocity of light and L is the distance between the two mirrors in the resonator. The period is referred to as Free Spectral Range (abbreviated as FSR) of the interferometer. The 3 dB frequency width of the filter pass band (*i.e.*, full width of the half maximum points) is Δf. The relationship between Δf and FSR is given by

$$\Delta f = \frac{(FSR)}{F} \tag{11.20.1}$$

where F is the fitness of the filter and $F = \pi\sqrt{R}/(1-R)$, where R is the reflectivity of the mirrors of the resonator (assumed that the same value of F for both mirrors of the resonator filter). The minimum transmission between the peaks is $(\pi/2F)^2$. The typical channel bandwidth ranges from 1 to 10 nm for a Fabry Perot filter.

(c) Semiconductor Amplifier Filters

A laser diode is pumped below the threshold voltage and it means that the voltage applied across the diode is not sufficient for laser action to occur. In that case the laser diode acts as a tuned bandpass amplifier with a narrow passband. The centre frequency of the pass band is determined by the composition of the semiconductor material used for fabricating the laser diode. When a distributed feedback (DFB) laser or a distributed feedback reflector (DFR) laser with multiple electrodes (Fig. 11.18) is used, it is possible to adjust the amplified wavelength. In the case of a distributed feedback amplifier (Fig. 11.18*a*) one current input is used as the pump, while the other is used to control the nominal centre wavelength of the amplifier. In the case of a distributed feedback-reflector-amplifier, the current I_1 varies the optical periodicity of the reflector and hence the centre wavelength of the reflector whereas current I_2 controls the phase of the amplifier and I_3 controls the pump power or gain of the amplifier (Fig. 11.18*b*). The typical channel bandwidth for a semiconductor amplifier filter ranges from 1 to 10 GHz.

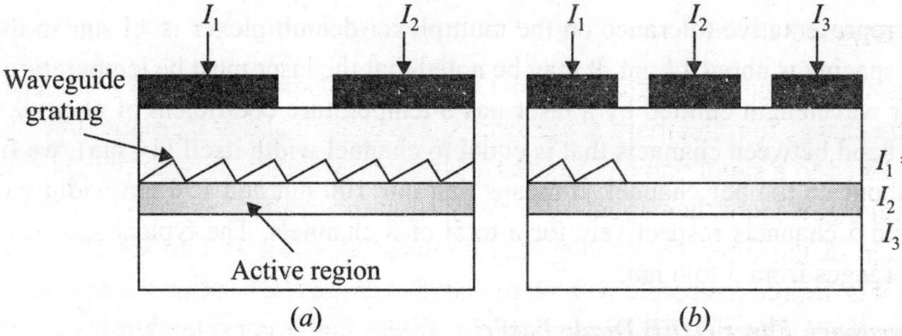

Fig. 11.18 (*a*) A distributed-feedback (DFB) amplifier (shares its feedback region with the current pump) (*b*) A distributed-reflector (DER) amplifier with multiple-control electrodes (separates the reflector region from the current pump region)

(d) Coherent Detection with Electronic Filters

When it is possible to fabricate a laser diode having a bandwidth smaller than the signal bandwidth, it becomes possible to use coherent detection techniques. In these techniques the signal is mixed with a frequency-shifted reference signal from a frequency-locked laser oscillator. The signals mix (or *i.e.*, interfere) coherently on the detector, producing a difference frequency which is called the "intermediate frequency". The channel spacing of the detected signals is determined by the bandwidth of the modulated signal depending on the type of modulation used and the data rate and the receiver arrangement used. This detection method allows the use of narrow channels that are separated in electronic processing subsystem of the receiver. In this system precision electronic filters can be used rather than filtering in the optical system where the separation filters are still crude and old in terms of the minimum channel width.

The minimum channel bandwidths for heterodyne receiver ranges from 1 to 10 GHz.

In order to make WDM successful we should have a source with good stability and tunability so that the desired wavelength can be selected rapidly.

11.21 SOURCES IN WDM SYSTEM

(1) LED Sources

Although the bandwidth of a LED source is considerably wide, this wide bandwidth is undesirable and disadvantageous, because large value of signal dispersion is encountered for this large bandwidth. However, it is possible to pass the output of a LED through a spectral filter to make the spectral bandwidth narrow. The technical term of this method is known as "*spectral splicing*". The advantage of this technique is that it is possible to utilize several inexpensive LEDs as sources and several appropriate filters are used to select operating wavelengths. This selective filtering further decreases the already attained small optical power level from the LED source. The reduction in power is proportional to the bandwidth reduction. This system is suitable only for short-distance and low-data-rate links. The typical spectral width ranges from 50 to 100 nm (or more).

(2) Fabry Perot Diode Lasers

The Fabry Perot (FP) laser will have 6 to 8 modes oscillating with a spacing of *c*/2L Hertz between two mirrors of the laser. For this type of laser the centre wavelength can vary by ±3 nm and the typical line width of 6 nm. A representative tolerance on the multiplexer/demultiplexer is ±1 nm so that the best available channel spacing is about 14 nm. It may be noted that the laser must be temperature controlled because the center wavelength emitted by a laser has a temperature coefficient of about 0.4 nm °C^{-1}. Allowing a guard band between channels that is equal to channel width itself (14 nm), we find that we need to allocate about 28 nm per channel. It means that our 100 nm and 150 nm width windows can accommodate 3 and 5 channels respectively for a total of 8 channels. The typical spectral width of a Fabry-Perot laser ranges from 3 to 6 nm.

(3) Distributed Feedback Laser (DFB Diode Laser)

The distributed feedback laser allows the single mode operation of the diode and has a lower temperature sensitivity of 0.08 nm °C^{-1}. The typical value of spectral width of the output signal is 10 to 100 MHz. This is smaller than the spectral width of the signal for high data rate or a wavelength chirp (unwanted high frequency oscillations) which accompanies the output of the laser when pulses are applied. The chirp can be as much as several tens of GHz (or several tenth of nm). This undesirable "Chirp" can be removed by using C.W. laser source and using an external modulator to modulate the signal data into light pulses. The channel width can be determined from a center wavelength tolerance of a source of value ±0.5 nm. It is assumed that the laser is temperature controlled, and a chirp tolerance of ±0.2 nm with a filter tolerance of ±1 nm. The actual channel width is about 3.5 nm and assuming a guard band equal to the channel width, the allocated channel width would be about 7 nm. Therefore, it is possible to have about 14 and 21 channels in respective fiber transmission windows. The typical spectral width of a DFB laser is less than 0.01 nm.

(4) Frequency Locked Laser

A laser can be made phase locked to an atomic resonance and with temperature and current feedback controls it can achieve a line-width as small as 5 MHz. This small line width allows the channel bandwidth to be determined only by the signal bandwidth, independent of the line width of the source. In this case the demultiplexing cannot be achieved by fixed filters, but must be done by a tunable, narrow-bandwidth element to ensure matching the element precisely to the wavelength of the source. Once frequency-locked, the laser cannot be easily tuned. Other techniques can be used to offset a second laser from the reference-locked laser by an adjustable amount. The external cavity laser has a typical spectral width less than 1 MHz.

11.22 TUNING OF OPTICAL SOURCES

The rapid tuning (changing wavelength slightly) mechanism of a source like a laser diode is discussed below.

Temperature Current Tuning

The emitting wavelength of an optical source like a laser diode can be changed by a small amount by changing the temperature of the device. In the case of distributed feedback laser a change of wavelength of about 0.08 nm can be achieved by changing its temperature by 1° centigrade. Therefore, it is not difficult to make a change of wavelength by 1 nm with the help of this technique of temperature-Tuning.

In the case of a distributed feedback (DFB) laser when the bias current is changed by 1 mA, it is possible to get a change of its frequency by 200 MHz-mA^{-1} for switching rates higher than 10 MHz.

DFB and DBR tuning

When a comparatively larger tuning ranges of wavelengths are desired, it can be achieved by combining a DFB laser with a phase-control system and an active section. The wavelength can also be achieved by combining a DFB laser with a fabricated integrating structure. The integrating structure can controlably combines the power of different laser structures oscillating at various wavelengths.

External Element Tuning

The external element tuning is possible by using a Fabry Perot laser with an external cavity. The cavity has a movable grating which can be controlled by a piezoelectric pusher. It is reported that the tuning with this technique has been demonstrated over 55 nm.

Receiver Tuning

In order to make receiver tuning, tunable filters can be based on technology using multilayer filters, Fabry Perot interferometers or integrated optical devices.

1. A tunable multilayer filter device can be fabricated whose transmission properties vary spatially (from point-to-point) across the face of the filter and it is achieved by varying layer thickness. Tuning is done by translating the filter (*i.e.*, by changing the position). A filter that uses this technique has been reported a coverage of 270 nm with a pass band of 7 nm. Fig. 11.19(*a*) shows tunable optical filter characteristics: channel bandwidth, channel spacing and tuning range.
2. Fabry Perot interferometer filters can be tuned by changing the length of the cavity using a piezoelectric pusher. Devices with 30 channel operation with a tuning time of few microseconds have been demonstrated in the laboratory.

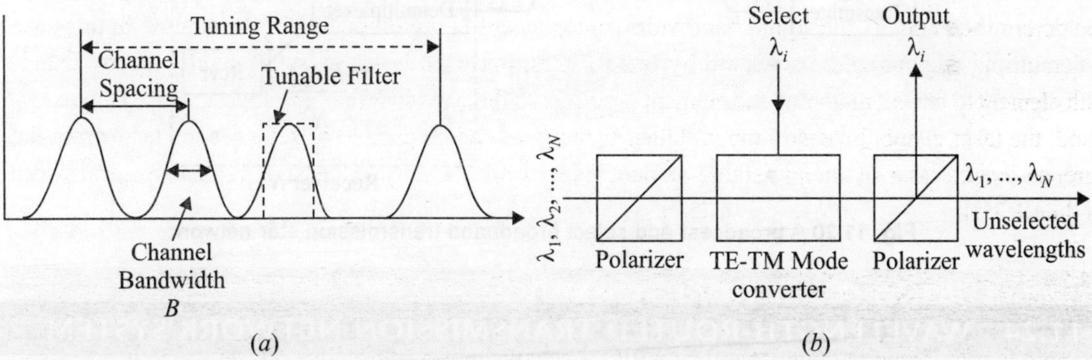

Fig. 11.19 (*a*) Tunable optical fiber characteristics, (*b*) Integrated optics tunable filter that is used as TE to TM mode converter

3. Integrated optic tunable filters that use. TE-to-TM mode converters (that converts transverse electric modes to transverse magnetic modes) have been demonstrated in the laboratory (Fig. 11.19*b*).
4. Distributed Feedback lasers operated below threshold (it means that operated as amplifiers rather than oscillators) exhibit a narrow pass bands. This device can be tuned by incorporating a phase-

controlled section. The response times are fast on the order of nanosecond switching time and tuning ranges for ten channel operation have been demonstrated as 71 GHz.

11.23 MULTI POINT, MULTI WAVELENGTH NETWORKS

Broadband Transmission

Broadband transmission network which is also known as broadcast and select network is shown in Fig. 11.20 where any frequency transmitted by a fixed frequency transmitter is broad casted to all receivers (providing the maximum in network flexibility). The receivers tune to the desired transmitter.

If N is the number of stations in the array, then the power is split into $\frac{1}{2^N}$ and the overall splitting loss is 10 log 2^N dB. The configurations of the network may be either a linear tree or as a star as shown in Fig. 11.20 and 11.21.

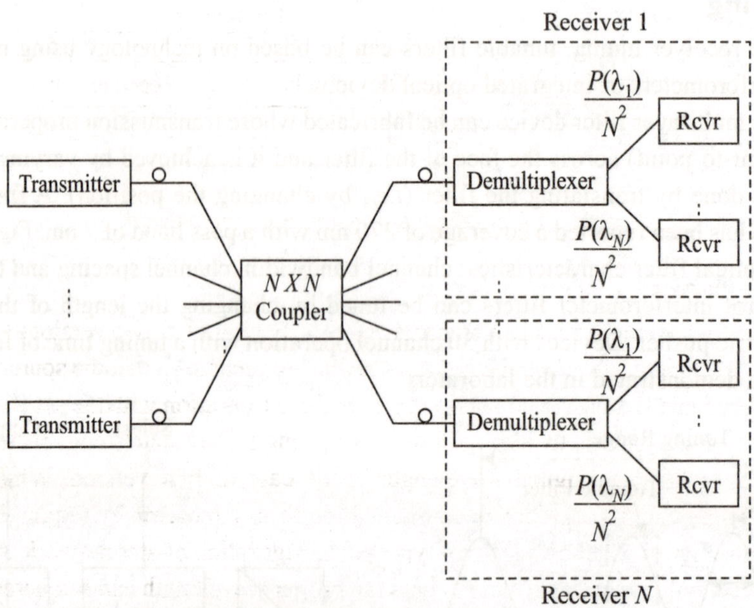

Fig. 11.20 A broadcast and select broadband transmission star network

11.24 WAVELENGTH-ROUTED TRANSMISSION NETWORK SYSTEM

These networks combine wavelength division multiplexer systems in several techniques. The optical power transmitted in a particular wavelength from a transmitter is routed to a particular receiver through the network with the use of wavelength selective elements. Once a wavelength is dropped (sent) to the desired receiver, it can be reutilized to add another signal to the mix, and thereby the number of signals is conserved. Some networks for the use of wavelength conversion devices to change wavelengths of the signal within the network routing device. This increases the efficiency of the network. Both the wavelength spectrum and the optical power can be utilized more efficiently, but at a cost of network flexibility.

These types of networks are (1) the Star network (2) the Chain network and (3) the ring network. A broadcast and select broadband transmission star network is shown in Fig. 11.20 and Fig. 11.21 shows broadband transmission tree network.

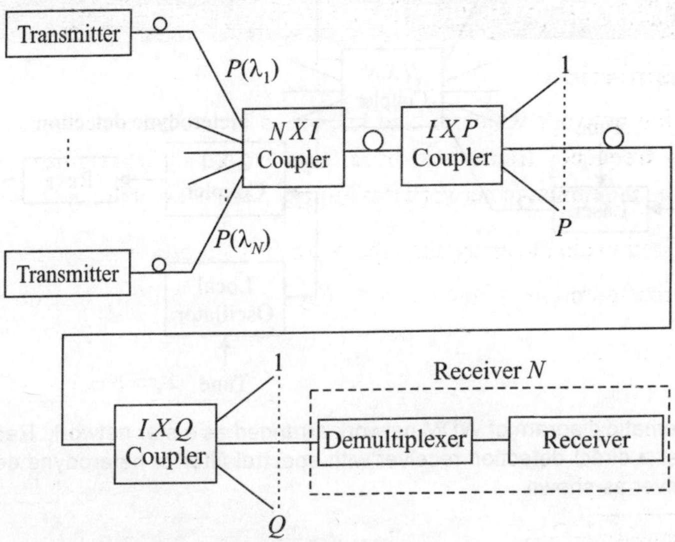

Fig. 11.21 A broadband transmission tree network

WDM Star Network

Fig. 11.20 shows the wavelength selective star network. The network has two versions (1) in one version the receivers have a set (fixed) wavelength but the sources are tunable so that the source wavelength can be changed as desired and (2) In second version there is a configuration with the set (fixed) wavelength of the source and the receivers are tunable to desired wavelength. Any source can talk to any receiver by tuning the receiver in the preassigned wavelength in the case of first version. While in the case of second version also any source can have communication to any receiver by tuning the source to any desired wavelength. Fig. 11.22 shows the two types of configuration of star network. An optical WDM direct-detection receiver uses optical filters to pass the proper wavelength and a coherent FDM receiver that uses "Heterodyne detection" to tune to the proper frequency of the receiver. There are NXN couplers in this arrangement. This is a wide band device that combines source power and splits (divides) equally among all receivers.

WDM Chain Network

Fig. 11.23 shows the schematic for a wavelength selective chain network. In this network each node has a wavelength splitter that separates the wavelengths. In this add/drop configuration certain wavelengths are routed to the receivers (*i.e.*, dropped from the network to their receivers). New signals are added to the network at these wavelengths. Additional wavelengths can also be added to the network provided the output coupler of the nodes has input for them. Data can be added or dropped at each wavelength channel as the information moves through the network. In this figure we notice that nodes 1 and 5 should be co-located to have transmit/receive capabilities.

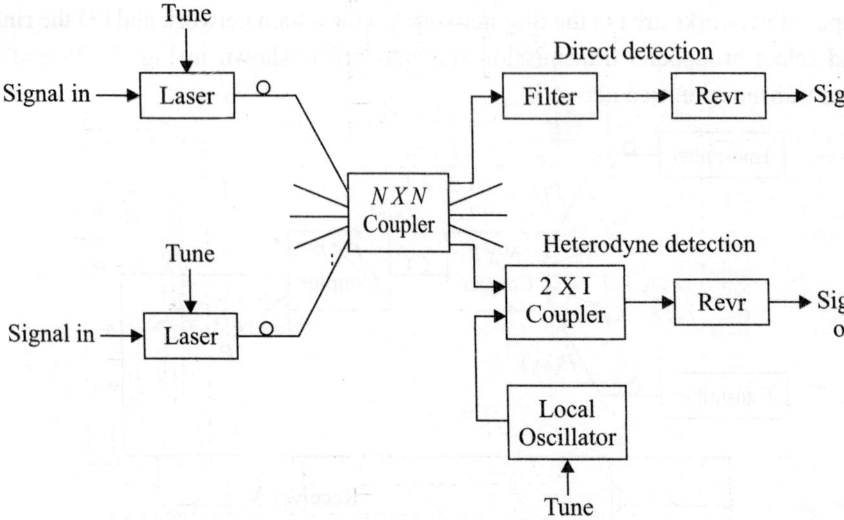

Fig. 11.22 Schematic diagram of WDM network arranged as a star network. Receiver can be either a direct detection receiver with spectral filter or heterodyne detection FDM receiver as shown

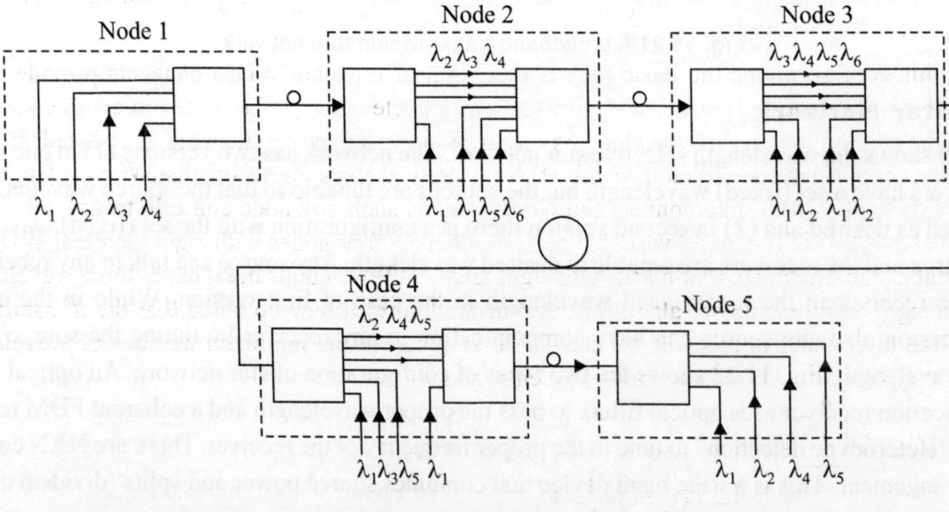

Fig. 11.23 A wavelength-division multiplexed chain network illustrating wavelength dependent routing to accomplish signal add/drop functions. Nodes 1 and 5 should be co-located to have full transmission and reception capabilities in the unidirectional link

WDM Ring Network

In the wavelength selective ring network as shown in Fig. 11.24 the configuration is identical to the chain network with its first and last ends tied joined together. In the simplest form the number of wavelengths added are equal to the number received and as such all the couplers and splitters are the same in number.

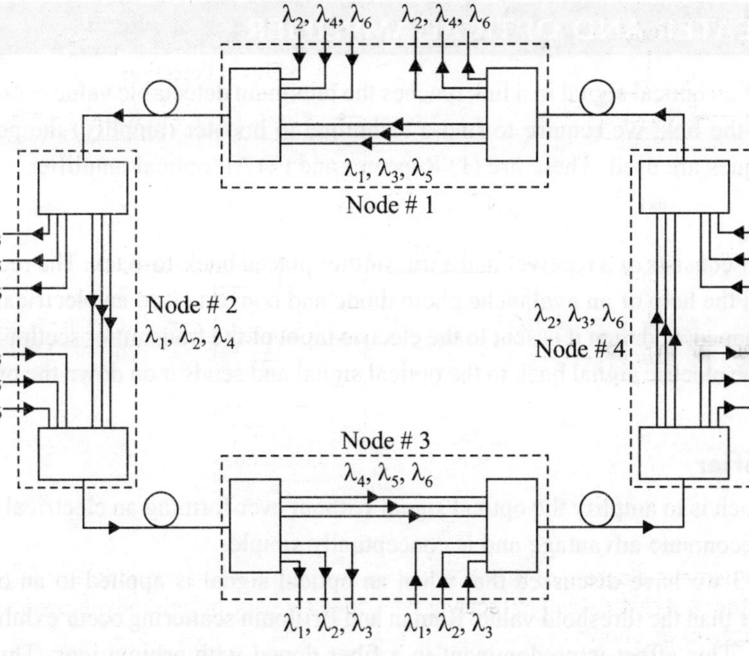

Fig. 11.24 A wavelength division multiplexed ring network with wavelength routed add/drop functionability

In this configurations the basic idea is that a set of identical WDM elements provide a fixed optical path between the transmitter and receiver. The wavelengths can be reused in another section of the network, and the number of wavelengths required to implement the network is reduced. It can be shown that a set of N modes will require only $N(N-1)$ optical channels, rather than N^2 wavelengths that might be expected to fully interconnect the Nodes and as such any node can talk to any other node. Fig. 11.25 shows a WDM star network that can interconnect four nodes with eight identical WDM star network that can interconnect four nodes with eight identical WDM units but it needs three wavelengths. The wavelength assignments are also illustrated in the figure. It may be noted that the transmitter #1 does not like to connect to receiver #1 etc. Table 11.1 shows the minimum number of wavelengths required to implement a WDM network between N nodes.

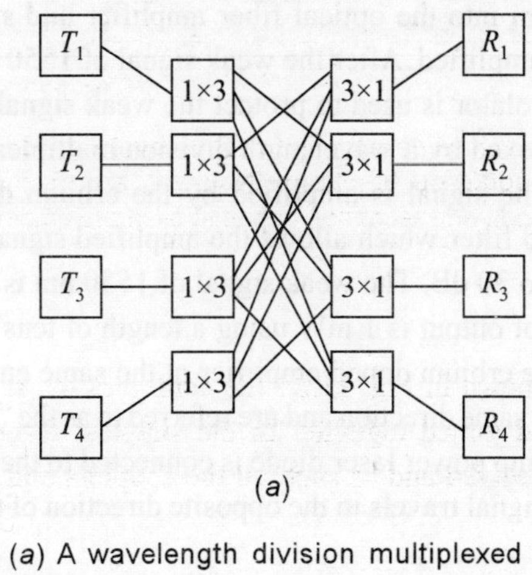

(a)

	R_1	R_2	R_3	R_4
T_1	—	λ_1	λ_2	λ_3
T_2	λ_3	—	λ_1	λ_2
T_3	λ_2	λ_3	—	λ_1
T_4	λ_1	λ_2	λ_3	—

(b)

Fig. 11.25 (*a*) A wavelength division multiplexed star network with four nodes and three wavelengths (*b*) wavelength assignment table

11.25 REPEATER AND OPTICAL AMPLIFIER

When the power of an optical signal in a link reaches the minimum detectable value even before reaching the destination of the link, we require to find a technique to booster (amplify) the power to a higher level. Two techniques are used. These are (1) Repeater and (2) All optical amplifier.

Repeater

An optical repeater consists of a receiver and a transmitter placed back-to-back. The receiver detects the optical signal with the help of an avalanche photo diode and converts it to an electrical signal which is amplified and reshaped, and then it is sent to the electric input of the transmitter section. The transmitter section converts the electric signal back to the optical signal and sends it on down the optical fiber wave guide.

Optical Amplifier

Alternative approach is to amplify the optical signal without ever forming an electrical equivalent. This technique has an economic advantage and is conceptually simple.

In chapter 3 we have discussed that when an optical signal is applied to an optical fiber at a power level greater than the threshold value, Raman and Brillouin scattering occur exhibiting a nonlinear effect in the fiber. This effect is predominant in a fiber doped with erbium ions. This is the basis of optical fiber amplifier.

11.26 ERBIUM-DOPED FIBER AMPLIFIER

An optical fiber doped with erbium ions has become a potential all optical amplifier. It is known as "erbium doped amplifier" which operates in the 1520 to 1550 nm window of the fiber loss characteristics. This is the region of attenuation minimum of the fiber loss characteristics. A laser diode pump source operating at 950 to 1480 nm serves as a pump. Most optical amplifiers use the wavelength of 1450 nm as a pump power source. Gains of 20 to 30 dB with power level of l mW can be attained in these fiber amplifiers with a length of fiber of tens of meters. The pump power required for such gains is a few tens of mW which can be achieved with modern laser sources. Fig. 11.26 illustrates the coupling of the pump source laser working at 980 or 1480 nm into the optical fiber amplifier and simultaneously a weak signal of 1550 nm is applied to it to get amplified. After the weak signal of 1550 nm is applied to it to get amplified. After the weak signal an isolator is used to protect the weak signal generator from damage. These two optical signals are multiplexed by a wavelength division multiplexer (WDM) and are passed through an erbium doped fiber. The signal is amplified by the erbium doped fiber. The amplified signal passes through a narrow band filter which allows the amplified signal of 1550 nm to develop at the output and the gain may be 20 to 30 dB. The weak signal of 1550 nm is amplified at the expense of the pump power. The power level of output is 1 mW using a length of tens of meters.

When the pump power is coupled to the erbium doped amplifier at the same end of the fiber as the optical signal, the two beam propagate in the same direction and are referred to as the "copropagating". A second version is also possible where the pump power laser diode is connected to the other end of the signal applied to the amplifier and the optical signal travels in the opposite direction of the pump signal. Two beams are then "Counter propagating".

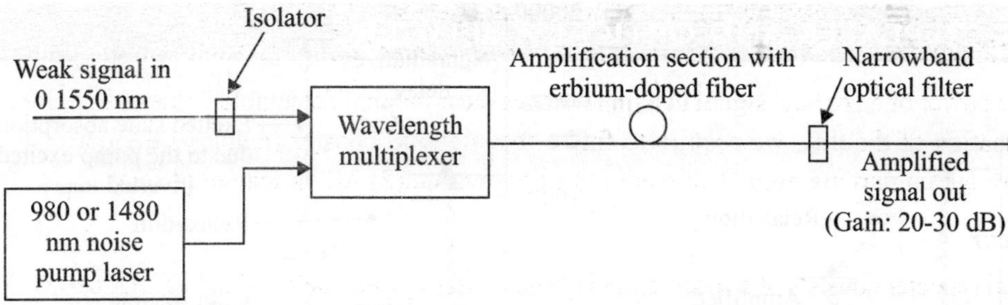

Fig. 11.26 Schematic arrangement for coupling of the pump source laser at 980 or 1480 nm wavelength in the optical fiber amplifier and simultaneously a weak signal of 1550 nm is applied to the input of erbium doped fiber to get the signal amplified

The insertion loss when the amplifier is spliced (connected to the link) is as low as less than 1 dB.

Wave Forms of Output Signals

Fig. 11.27(*a*) shows the waveform of the weak signal at the input and Fig. 11.27(*b*) illustrates the amplified optical signal at the output. At the output of the amplifier the narrow band signal is superimposed on the broadband amplified spontaneous emission (ASE) noise. The shortcomings of these optical amplifiers are that (1) their operation is limited near the wavelength of 1550 nm although the response is moderately broadband in the vicinity of wavelength, (2) they require a non-trivial (not small amount) pump power of 40 to 50 mW at 1480 nm, (3) they require a length of fiber that is greater than several metres because a short lengths do not provide high gain and (4) they add noise to the optical signal.

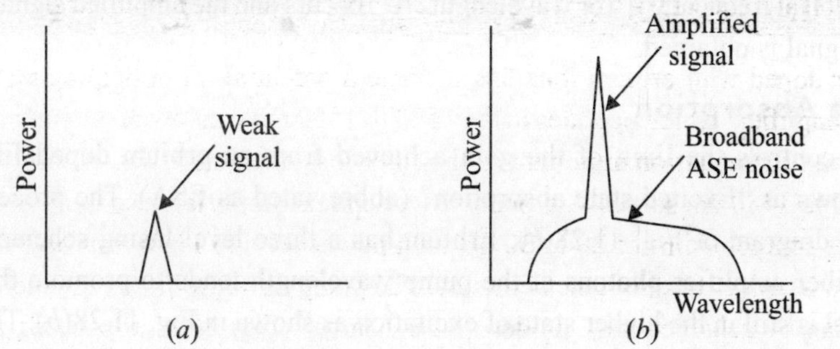

Fig. 11.27 Waveforms of (*a*) the weak signal at the input to be amplified and (*b*) amplified optical signal at the output

Energy Level Diagram

Practical pump bands exist at wavelengths 532 nm, 670 nm, 807 nm, 980 nm and 1480 nm. However, it is found that wavelengths of 807 nm, 980 nm and 1480 nm comprise the most important pump bands. The amplification depends on the material gain of a relatively short section of the fiber (1 to 100 nm). Aluminium coding can be used to broaden the spectral bandwidth to around 40 nm. It should be noted, however the spectral dependence on gain is not constant as that illustrated in Fig. 11.28.

The spectrum bandwidth for erbium doped silica fibers may be restricted to around 300 GHz (2.4 nm).

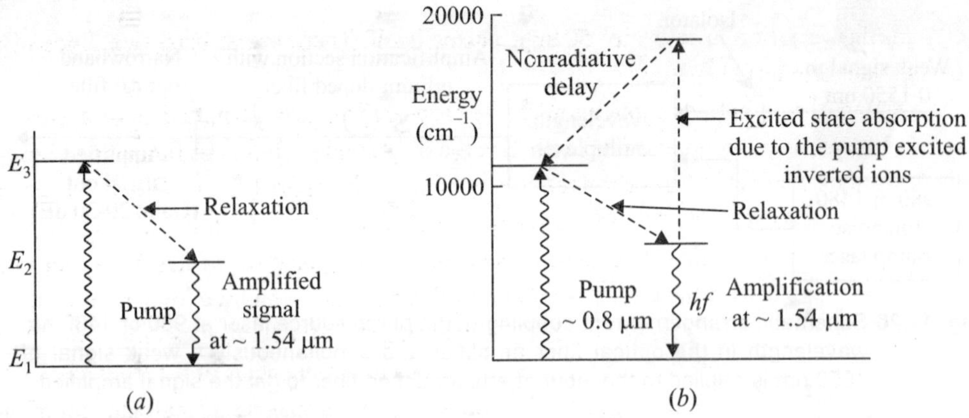

Fig. 11.28 Energy level diagrams for erbium doped silica fiber laser: (*a*) three level lasing scheme provided by E_r^{+3} doping (*b*) an illustration of excited state absorption which is a major limitation

The gain available from an erbium doped fiber amplifier is illustrated in Fig. 11.28(*a*). The process of obtaining gain is illustrated in the energy level diagrams for an erbium doped fiber system in Fig. 11.28(*a*). Erbium provides a three level lasing scheme. However in the erbium fiber amplifier photons at the pump wavelength elevates large number of electrons from level E_1 to the level E_3. So that $E_3 - E_1 = h\upsilon_p$ where υ_p is the pump frequency. Then some electrons from E_3 fall on intermediate level E_2 to relax at E_2. Now an optical signal of frequency υ_s is applied to the fiber so that $E_2 - E_1 = h\upsilon_s$. Now stimulated emission at frequency υ_s (or wavelength c/υ_s) occurs and the amplified signal at the wavelength of the original signal is obtained.

Excited State Absorption

A factor which controls the limit of the gain achieved from an erbium doped fiber amplifier is a phenomenon known as "Excited state absorption" (abbreviated as ESA). The process is illustrated in the energy level diagram of Fig. 11.28(*b*). Erbium has a three level lasing scheme. However, in the Erbium doped fiber amplifier photons at the pump wavelength tends to promote the electrons in the upper lasing level is still in the higher state of excitation as shown in Fig. 11.28(*b*). The electrons decay nonradiatively for intermediate level such as pump band and eventually back to the upper lasing level. Hence ESA reduces the pumping efficiency of the device and thus it is necessary to pump at a higher power to get specific gain.

11.27 SOLITON TRANSMISSION

The most important recent discovery in the optical fiber communications is probably the transmission of soliton. Although the method of transmission of soliton has been theoretically known for several years, it was only in 1989 when the optical components such as erbium doped amplifier became readily available that the efforts to achieve soliton transmission succeeded.

Now we shall give the idea about what are "solitons". These are regarded as the dispersionless pulse of light (optical signal). They only exist in a non-linear medium, where the refractive index changes with the variation in intensity of light pulses itself. There are at least two kinds of solitons. They are (1) temporal or (2) spatial.

The temporal soliton finds applications in the ultra-long distance communication system. A soliton is a transmission mode of an optical fiber which is the only stable solution of fundamental propagation equation. Solutions cancels out the chromatic dispersion of the optical fibers. When it operates at the zero dispersion shifted value of 1.55 μm wavelength, the pulse will be spectrally broadened because the nonlinearity is exhibited in silica. The mere presence of a light pulse which has a varying intensity of light over its envelope causes the refractive index of the material to attain a higher value at the pulses peak than at its lower intensity tails. Fig. 11.29 illustrates the formation of soliton. Fig. 11.29(*a*) the peak wave is retarded compared to the tails, which reduces the frequencies of the front half of pulse and increases the frequencies in the trailing half. For the soliton to propagate a small amount of chromatic dispersion is necessary. If the system operates at a wavelength slightly lower than the zero dispersion wavelength, there will be a tendency for the pulse spreading to occur with the higher frequencies of the pulse leading and the lower frequencies being retarded as shown in Fig. 11.29(*b*). It is apparent that the effects of the nonlinearity of the fiber are balanced by the chromatic dispersion of the fiber and as a result, the soliton is produced as shown in Fig. 11.29(*c*). Thus, the soliton is truly non-dispersive in both frequency and time domains. Furthermore, it has already been demonstrated that the soliton remains stable after passing through several amplifiers provided the distance between two consecutive amplifiers is short enough so that the shape of the pulse is not disturbed. A practical demonstration of this theory was successfully completed over a 9000 km long optical fiber with a 4-GHz signal in a circulating loop incorporating Er-doped amplifiers spaced 25 to 30 km apart.

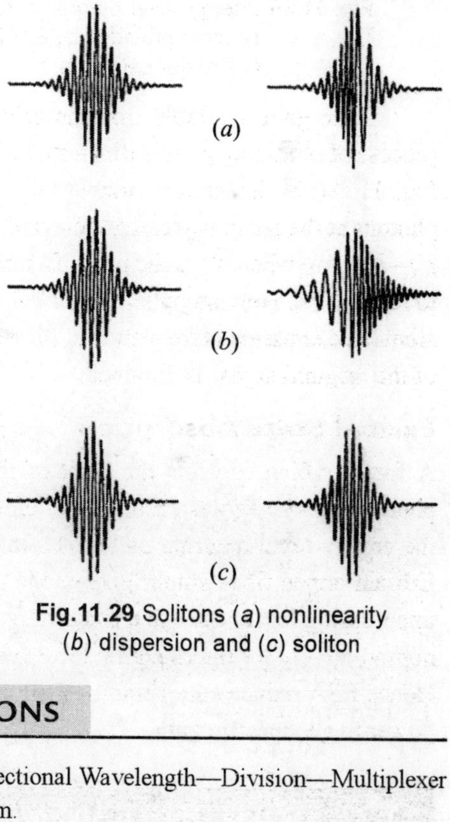

Fig.11.29 Solitons (*a*) nonlinearity (*b*) dispersion and (*c*) soliton

REVIEW QUESTIONS

11.1 Describe with diagram to explain the operation of a unidirectional Wavelength—Division—Multiplexer (WDM) system. Discuss about a bidirectional WDM system.

11.2 Give the definition of following parameters (1) Insersion loss (2) Channel width and (3) Cross talk of a WDM system.

11.3 Discuss about the choice of different components in designing an optical fiber link. Discuss the shortcomings of decency of each components.

11.4 Define the rise time of pulse in a passive RC circuit and determine the relationship between the rise time and bandwidth of the circuit.

11.5 An optical fiber link is designed to operate 8 km length of the link without using a repeater. The rise time of different individual components are:

The rise time of LED source = 5 ns

The rise time of *p-i-n* photodetector = 8 ns

Rise time due to modal dispersion = 5 ns/km

Rise time due to material dispersion = 1 ns/km

Calculate the system rise time and maximum bit rate that can be achieved in the link using (1) *RZ* format and (2) NRZ format.

11.6 The bit rate of a silica fiber link is 40 Mbs^{-1}. The fiber is operating at a wavelength of 840 nm with a spectral width of 42 nm. The value of the parameter ($\lambda^2 d^2n/d\lambda^2$) is 0.022.

Calculate the material dispersion limited transmission distance.

CHAPTER 12

Electrooptic Effect

CHAPTER FOCUS

- *Introduction*
- *Introduction to modulation techniques in optical communication system*
- *An introduction to optical OFDM*
- *References*

12.1 INTRODUCTION

In this section we consider the problem of propagation of optical radiation in crystals in the presence of an applied electric field. We find that in certain types of crystals it is possible to effect a change in the index of refraction that is proportional to the applied electric field. This is the linear electrooptic effect. It affords a convenient and widely used means of controlling the intensity or phase of the propagating radiation. This modulation is used in an ever expanding number of applications, including the impression of information onto optical beams.

In the optoelectronic devices, information is coded as a variation of some characteristics of the optical output from the LED or the LASER diode. The modulation property depends upon the application. It may involve variation of amplitude, phase, pulse width etc.

The propagation of optical radiation in a crystal can be described completely in terms of the dielectric tensor. We know that, for a direction of propagation in a crystal, in general two possible linearly polarized modes exist-the so-called normal modes of propagation. Each mode possesses a unique direction of polarization (*i.e.*, direction of D or E) and a corresponding index of refraction (*i.e.*, a velocity of propagation). The mutually orthogonal polarization directions (D or E vectors) and the indices of the two modes are found most conveniently by using the index ellipsoid, which assumes its simplest form in the principal coordinate system:

$\frac{x^2}{n_x^2}+\frac{y^2}{n_y^2}+\frac{z^2}{n_z^2}=1$ where the directions x, y, and z are the principal dielectric axes that is, the directions in the crystal along which the D and E field vectors are parallel. We note that $\left(\frac{1}{n_x^2}\right)$, $\left(\frac{1}{n_y^2}\right)$, and $\left(\frac{1}{n_z^2}\right)$ are the principal values of the impermeability tensor, which is defined as $n=\frac{\varepsilon_0}{\varepsilon}$; where ε is the dielectric tensor.

According to the quantum theory of solids, the optical dielectric tensor depends on the distribution of charges in the crystal. The application of an applied electric field can cause a redistribution of the charges and possibly a slight deformation of the crystal lattice. The net result is a change in the optical impermeability tensor. This is known as the electrooptic effect.

How electrooptic effect can be employed to obtain phase or amplitude modulation?

We know that an applied electric field can change the index ellipsoid of certain crystals. We also know that the propagation characteristics are governed by the index ellipsoid. Consequently, we can employ the electro-optic effect of these crystals to manipulate the propagation of optical waves, fundamentally their phase and polarization state. This modulation can, if desired, be converted into amplitude modulation.

Factors to be considered for high speed optical network

High-speed and high-capacity optical network are driven by several key technologies.

- Low-loss optical components (including transmission fiber, dispersion-compensating devices, and optical switching/routing elements) minimize the need for optical amplification and reduce the associated amplification noise.
- Low-noise optical amplifiers (such as distributed Raman amplifiers) lower the noise accumulated along transmission lines.
- Advanced optical fibers reduce nonlinear signal distortions and enable higher signal launch powers.
- Forward Error correcting codes allow for operation at poorer channel BER, which relaxes the requirements on the OSNR at the receiver.
- Advanced modulation formats are used to trade off noise resilience, fiber propagation characteristics.

12.2 MODULATION TECHNIQUES IN OPTICAL COMMUNICATION SYSTEM

Choice of the right modulation format is key to building flexible and cost-effective high-capacity optically routed wavelength division multiplexed fiber networks. To achieve a communication system which is noise tolerant and distortion tolerant, the modulation format must be properly designed.

There are two basic types of modulation for optical fiber communications systems: analog and digital. The analog method is preferred for relatively short distances; from 10 km to 20 km. Transmission of multiple signals through the same channel is now possible using sub-carrier modulation (SCM) technique. Voice or video channels are frequency-division multiplexed (FDM). The output current of such multiplexers modulates the optical source. Analog modulation presents certain disadvantages compared to digital method. For example, the high SNR required by the analog system limits the usable system bandwidth, thus reducing the link information capacity as well as the path length between repeaters. Another disadvantage of analog transmission is increased chromatic distortion due to optical source nonlinearities generated by the intensity modulation.

Digital modulation is much preferable to analog, especially for long distance transmission, because of its increased carrying capacity and resistance to various link impairments. Although a binary signal travelling through the optical fiber suffers a considerable loss due to signal dispersions, a BER of 10^{-9} can easily be maintained. Optical sources like LED, LASER diodes exhibits switching frequencies in the order of several MHz to GHz. These enhanced switching capabilities also increase the system available bandwidth and overall system capacity. To improve SNR and to minimize signal distortion, now-a-days, the digital system is chosen for fiber optic communication system. Modern digital Optoelectronic Integrated Circuits (OEICs) provide a reliable and economic method of transmitting both voice and data signals.

In this section we will come across different modulation formats. Fiber-optic communication systems form the high-capacity transport infrastructure that enables global broadband data services and advanced Internet applications. The desire for higher per-fiber transport capacities and, at the same time, the drive for lower costs per end-to-end transmitted information bit has led to optically routed networks with high spectral efficiencies. Among other enabling technologies, advanced optical modulation formats have become key to the design of modern wavelength division multiplexed (WDM) fiber systems. So, we review different optical modulation formats in the broader context of high speed optical network.

On-Off Keying (OOK)

Optical direct detection effectively measures the energy in the optical signal impinging on the detector. Since direct detection does not respond to phase, but is capable of distinguishing only between different intensity levels, intensity modulation is required. The most common forms of digital intensity modulation are on-off keying (OOK) and pulse-position modulation (PPM). Most of the common modulation formats suitable for direct detection can be decomposed into a simple binary on-off "pulsed" form of modulation. At a given wavelength or during a given period of time an optical pulse is either present or absent. The simplest of these is binary OOK, for which each modulated bit consists of one of these on-off pulses.

More complex pulsed modulations, such as PPM, can be conceptually regarded as coded versions of binary OOK. Coherent reception can be used with any pulsed modulation such as PPM, but additionally there are many non-pulsed modulations suitable for coherent reception. Examples include M-ary phase shift keying (M-PSK) and M-ary quadrature amplitude modulation (M-QAM), *i.e.*, the whole suite of modulations used for coherent RF communication. Pulsed modulations, such as OOK and PPM, share a common characteristic that the laser is either "on" or "off" during every slot time interval.

With on-off keying, binary data is represented by the presence or absence of a single light pulse in each T-second symbol interval. The binary information sequence can be mapped directly to a sequence of light pulses at the transmitter according to the rule: if the information bit is 1, transmit a laser pulse; if it is 0, transmit nothing. Therefore, there is a one-to-one correspondence between 1's in the data-stream, and the occurrence of light pulses emanating from the transmitter. This is illustrated in Figure 12.1 below.

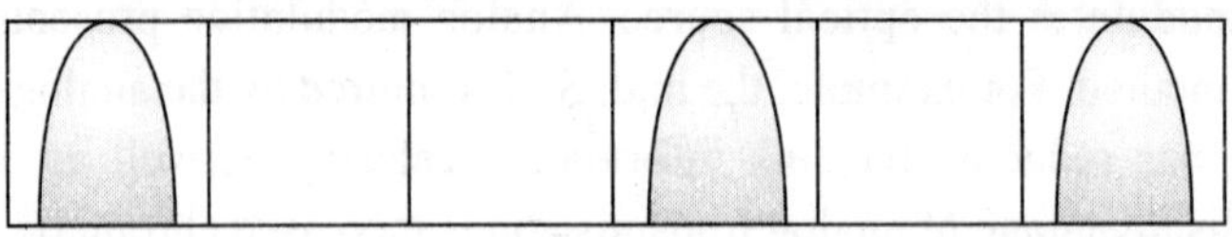

Fig. 12.1 OOK uses one slot per bit. Modulation of information 100101 is shown

The Amplitude shift keying (ASK) format is also called on-off keying (OOK) and is identical with the modulation scheme commonly used for noncoherent digital lightwave systems.

The implementation of ASK for coherent systems differs from the case of the direct-detection systems in one important aspect. Whereas the optical bit stream for direct-detection systems can be generated by modulating a light emitting diode (LED) or a semiconductor laser directly, external modulation is necessary for coherent communication systems. The reason behind this necessity is related to phase changes that invariably occur when the amplitude or the power is changed by modulating the current applied to a semiconductor laser. For noncoherent digital lightwave systems, such unintentional phase changes are not seen by the detector (as the detector responds only to the optical power). The situation is entirely different in the case of coherent systems, where the detector response depends on the phase of the received signal. The implementation of ASK format for coherent systems requires the phase to remain nearly constant. This is achieved by operating the semiconductor LASER continuously at a constant current and modulating its output by using an external modulator.

Analog Optical Fiber Communication Systems

Voice or video analog signals transmitted through an optical fiber communication link are first modulated, then frequency-division multiplexed, finally the optical source is modulated. Figure 12.2 below shows the block diagram of such a system.

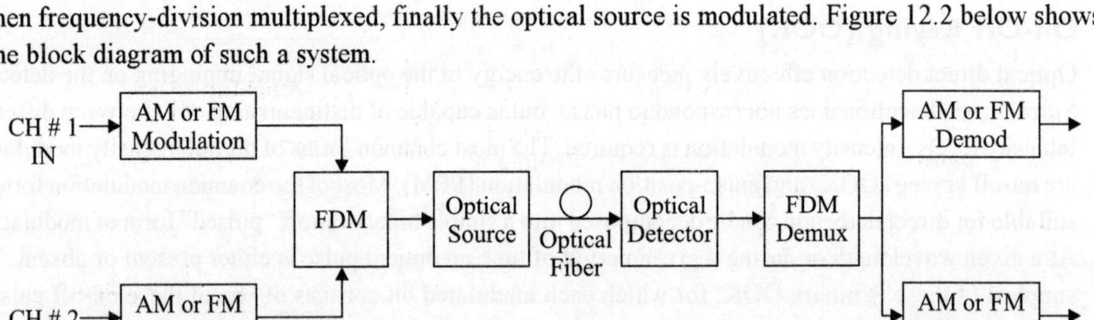

Fig. 12.2 Block diagram of an optical fiber communication link

The modulation techniques employed in an optical fiber communication system are amplitude modulation or frequency modulation. The use of subcarrier frequencies for voice or video signal transmission is made possible through the utilization of single-mode optical fibers and laser diodes.

There are several analog modulation techniques in optical fiber communication.

Intensity modulation with direct detection systems

Direct intensity modulation is one of the simple analog modulation techniques. In this type of modulation current flowing through the optical source is proportional to amplitude of the message signal.

The overall performance of such a system is a direct function of the optical index of modulation (m_o).

$$m_o = \frac{P_{max} - P_o}{P} \quad \ldots(12.2.1)$$

where

m_o = optical index of modulation

P_{max} = peak optical power at the input of the optical detector

P_o = average optical power at the input of the optical detector.

The higher the index of modulation, the higher the received optical power, results in higher SNR. This increase of received optical power due to increase of the modulation index results in a proportional increase in inter-modulation distortion as optical source is operating in the non-linear region of the operating characteristic curve. This distortion, coupled with cochannel interference, results in a substantial degradation of overall system performance.

The transmitted optical power in the direct intensity modulation technique is given by

$$P_t = P_a[1 + s(t)] \quad \ldots(12.2.2)$$

where P_a is the unmodulated optical power, $s(t)$ is the base band signal can be written as

$$s(t) = m \cos\omega t \quad \ldots(12.2.3)$$

where m is the modulation index.

A photodiode is used for direct detection of the transmitted signal. The photodiode converts the received input optical power into an electrical current. The current is converted into voltage, employing an amplifier, as shown in Fig. 12.3 below.

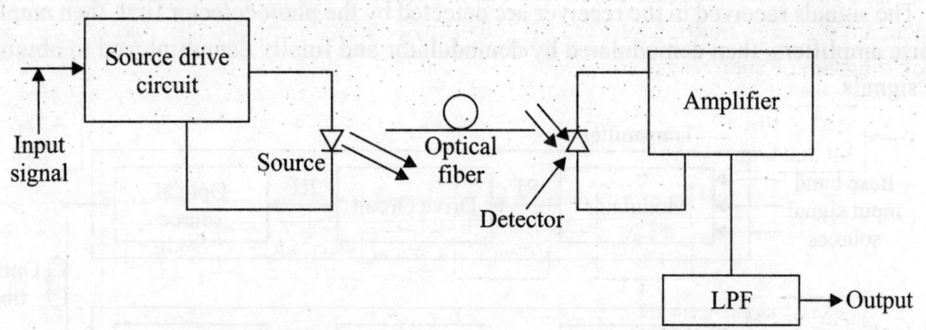

Fig. 12.3 Block diagram of direct intensity modulation

The photocurrent generated by the APD detector can be written as $I(t) = I_p M(1 + m \cos\omega t)$ if dispersion is neglected

From equation (6.7.3*b*) of chapter VI of this book the average photo current I_p generated by a steady average optical power P_o incident on the photo detector is given by

$$I_p = \frac{\eta e P_o}{hf} \quad \ldots(12.2.4)$$

where η is the quantum efficiency.

Therefore the photocurrent generated in APD can be written as

$$I(t) = \frac{\eta e P_o}{hf} M(1 + \cos \omega t) \qquad \ldots(12.2.5)$$

Thus output current of the detector obtained is directly proportional to the base band signal $s(t)$.

Subcarrier intensity modulation

These schemes are employed to utilize the large bandwidth of optical fiber. The basic concept behind subcarrier multiplexing (SCM) is borrowed from microwave technology, which employs multiple microwave carriers for transmission of multiple channels (electrical FDM) over coaxial cables or free space. The total bandwidth is limited to well below 1 GHz when coaxial cables are used to transmit a multichannel microwave signal. However, if the microwave signal is transmitted optically by using optical fibers, the signal bandwidth can easily exceed 10 GHz for a single optical carrier. Such a scheme is referred to as SCM, since multiplexing is done by using microwave subcarriers rather than the optical carrier. The sub-carrier signals are intensity modulated by the base band signals prior to the intensity modulation and are multiplexed in a single fiber.

Standard modulation techniques like AM, FM or PM can be utilized. Pulse modulations like PAM, PPM or PWM can also be used. Thus modulated signal can be transmitted by intensity modulation of the optical source.

A block diagram of a sub-carrier intensity modulation technique for analog modulation is shown in Fig. 12.4 the transmitter and receiver shown separately.

The drive circuit amplifies the combined signal using broadband amplifier and passes the same to the optical source. The amplified signal then used to intensity modulate by optical source. The output power from the source is then coupled to optical fiber for transmission.

The signals received in the receiver are detected by the photodetector first, then amplified by the low noise amplifiers, then demodulated by demodulator and finally demultiplexed to obtain base band output signals.

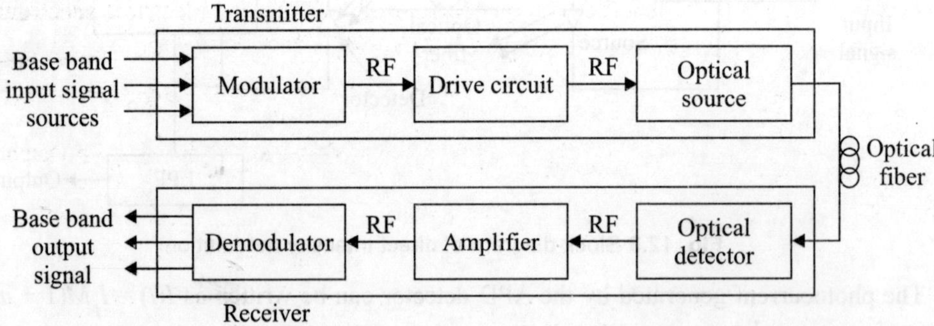

Fig. 12.4 Block diagram of sub-carrier intensity modulation system

A modulation system with large bandwidth utilization, SNR improvement, low-cost, simple with higher system capacity can be obtained as a result of sub-carrier intensity modulation. The disadvantages are, shorter distance between repeaters, more susceptible to noise and cochannel interference.

Optical Coherent Detection

In optical fiber communication 'coherent system' means any system or technique which employs nonlinear mixing of two light waves. In coherent optical receiver the input signal is added with the locally generated optical wave before detection of optical signal. The basic idea behind coherent detection consists of combining the optical signal coherently with a continuous-wave (CW) optical field before it falls on the photodetector (see Fig. 12.5). The CW field is generated locally at the receiver using a narrow line width laser, called the local oscillator (LO), a term borrowed from the radio and microwave literature. The schematic block diagram of a coherent optical detector is shown in the Fig. 12.6.

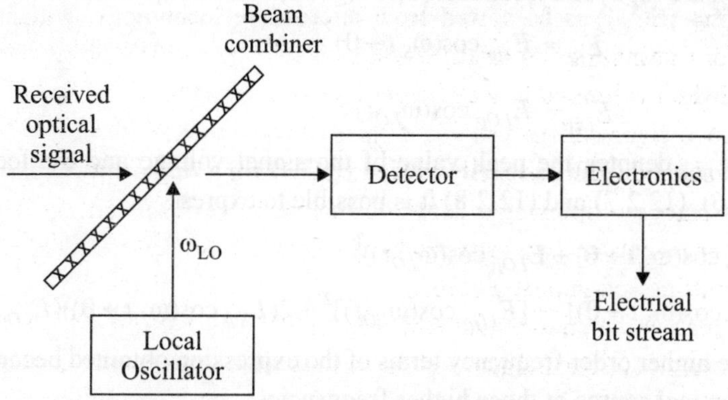

Fig. 12.5 Block diagram of a coherent optical receiver

In this detector, the input received optical wave is mixed with the local optical wave generated from a local oscillator. The mixed signal is applied to the photo detector for direct detection. The oscillator consists of a laser diode and its driving circuitry. The function of the mixer is performed by a single mode fiber fused biconical coupler. If the frequency of the input signal is same with the frequency of the local oscillator then the receiver is said to be of homodyne type. In homodyne receiver the output signal is the base band signal. When the local oscillator frequency is different from the input signal frequency then the receiver is said to be of heterodyne type. In heterodyne type, the electrical spectrum of the output of the detector is called intermediate frequency (IF).

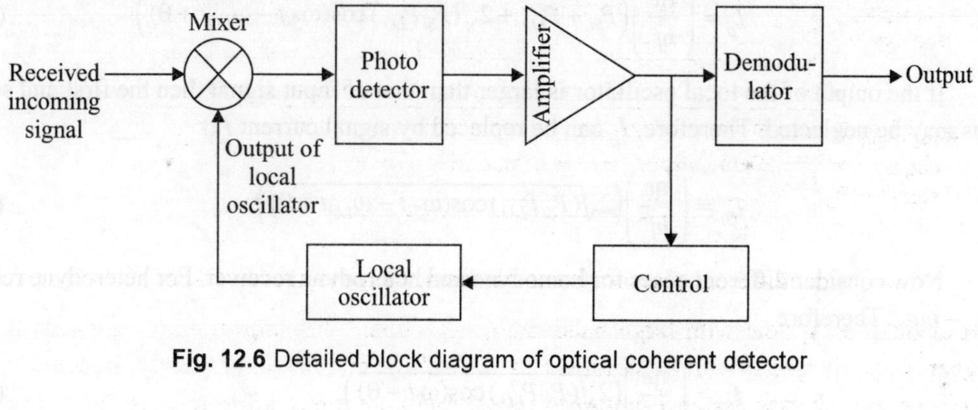

Fig. 12.6 Detailed block diagram of optical coherent detector

The IF is the difference of frequencies of the two signals. It depends on the information transmission rate and the characteristics of the modulation. The IF is chosen in such a way so that, it is possible to write

$$\omega_i = \omega_{si} - \omega_{LO} \quad \text{...(12.2.6)}$$

where ω_{si} and ω_{LO} denote the signal frequency and local oscillator frequency.

On the other hand, the signal photocurrent generated by the photodiodes is directly proportional to the intensity of incoming received light. Therefore it is possible to write that

$$I_P \propto (E_{Si} + E_{LO})^2 \quad \text{...(12.2.7)}$$

where, E_{Si} and E_{LO} are input and local oscillator signals, can be expressed as

$$E_{Si} = E_{Sip} \cos(\omega_{Si} t + \theta)$$

$$E_{LO} = E_{LOp} \cos(\omega_{LO} t) \quad \text{...(12.2.8)}$$

where, E_{Sip} and E_{LOp} denotes the peak value of the signal voltage and the local oscillator voltage. Combining (12.2.6), (12.2.7) and (12.2.8) it is possible to express

$$I_p \propto (E_{Sip} \cos(\omega_{Si} t + \theta) + E_{LOp} \cos(\omega_{LO} t))^2$$

$$I_p \propto [E_{Sip} \cos(\omega_{Si} t + \theta)]^2 + [E_{LOp} \cos(\omega_{LO} t)]^2 + 2(E_{Sip} \cos(\omega_{Si} t + \theta))(E_{LOp} \cos(\omega_{LO} t))$$

we can neglect the higher order frequency terms of the expression obtained because it is not possible to response for the optical source at those higher frequencies.

Therefore neglecting higher order frequencies we can write

$$I_p \propto \frac{1}{2}[E_{Sip}]^2 + \frac{1}{2}[E_{LOp}]^2 + 2E_{Sip} E_{LOp} \cos(\omega_{Si} t - \omega_{LO} t + \theta)$$

The above expression can be also written as

$$I_p \propto P_{Si} + P_{LO} + 2\sqrt{(P_{Si} P_{LO})} \cos(\omega_{Si} t - \omega_{LO} t + \theta) \quad \text{...(12.2.9)}$$

where P_{Si} and P_{LO} denote the signal power of input received signal and the local oscillator respectively.

Using (3.7.3(*b*)) of chapter 6 of this book in above expression, it is possible to write that

$$I_p = \left(\frac{\eta e}{hf}\right)\left(P_{Si} + P_{LO} + 2\sqrt{(P_{Si} P_{LO})} \cos(\omega_{Si} t - \omega_{LO} t + \theta)\right) \quad \text{...(12.2.10)}$$

If the output of the local oscillator is larger than that of input signal then the first and second dc terms may be neglected. Therefore, I_p can be replaced by signal current I_{Si}.

$$I_{Si} = \left(\frac{\eta e}{hf}\right)\left(2\sqrt{(P_{Si} P_{LO})} \cos(\omega_{Si} t - \omega_{LO} t + \theta)\right) \quad \text{...(12.2.11)}$$

Now consider different cases for homodyne and heterodyne receiver. For heterodyne receiver $\omega_i = \omega_{Si} - \omega_{LO}$. Therefore

$$I_{Si} = \left(\frac{\eta e}{hf}\right)\left(2\sqrt{(P_{Si} P_{LO})} \cos(\omega_i t + \theta)\right) \quad \text{...(12.2.12)}$$

For homodyne receiver, $\omega_{Si} = \omega_{LO}$, therefore, in homodyne receiver the output of the photodetector is the desired base band signal.

$$I_{Si} = \left(\frac{\eta e}{hf}\right)\left(2\sqrt{(P_{Si}P_{LO})}\cos\theta\right) \quad \ldots(12.2.13)$$

The advantage of coherent detection is evident from equation (12.2.13). Because the last term in this equation contains the signal phase explicitly, it is possible to transmit information by modulating the phase or frequency of the optical carrier. Direct detection does not allow phase or frequency modulation, as all information about the signal phase is lost. Since P_{LO} can be made much larger than P_{Si} the power enhancement can exceed 20 dB. Although shot noise is also enhanced, it can be shown that homodyne detection improves the signal-to-noise ratio (SNR) by a large factor because, the local oscillator still amplifies the received signal by a large factor, thereby improving the SNR.

The advantages obtained from coherent receiver are

- Improved receiver sensitivity.
- Higher wavelength selectivity.
- Larger repeater spacing.
- Higher transmission rates.

Digital communication systems

The advantages of digital transmission over analog transmission are listed below in brief:

- Increase of information capacity
- Resistance to link impairments
- BER in the range of 10^{-9} can be easily achievable.

The most important application of optical fiber communication is digital telephone exchange link. Digital optical communication system incorporates the same basic components in their design as analog systems. For example, they must require

Optical Sources: LED or laser diode
Optical Fiber: single-mode or multi mode
Optical wavelength: 800-900 nm or 1300-1600 nm
Optical Detector: APD or PIN diode
Modulation: Digital (two-level or multi-level)
Multiplexing: TDM or WDM.

To achieve acceptable signal to noise ratio base band digital transmission is suitable for optical fiber communication. Base band digital signal transmission using intensity modulation for high capacity optical fiber communication can mitigate the problem of nonlinearities associated with optical source. Base band fiber optic network requires dedicated Integrated Circuits. Optical OEICs already developed provides the transmission reliability with cost-effectiveness.

Fiber optic base band PCM system

Pulse Code Modulation is an efficient digitization technique to digitize the analog signals. A simple digital communication system can also employ PCM in an optical fiber communication link.

The analog signal is converted into discrete time signals by performing sampling at a rate higher than the Nyquist rate. The sampled signals are quantized to obtain discrete level of voltages. These voltage levels are coded and hence PCM signal is generated.

PCM requires a large bandwidth which optical fiber can provide. In an 8 bit PCM 256 levels are required where 7 bits are used for coding and one bit for polarity. For voice signal transmission, nonlinear encoding is used by using companding mechanism. Companding technique involves compression of input signal before transmission and expansion in the receiver terminal after decoding.

Figure 12.7 below shows a block diagram of a base band PCM system. The PCM transmission system consists of sampler, qunatizer, encoder and optical source. The PCM receiver system consists of optical detector quantizer, decoder and low pass filter. The analog information base band signal is directly fed to the sampler. The output of the sampler is fed to quantizer for quantization of the sampled signal. The error introduced during quantization process is known as quantization error arises as a result of truncation and rounding-off. A compressor or companding process may be introduced to minimize this quantization error at the transmission end. The quantized signal is then encoded by the encoder which provides PCM signal output. The PCM signal is then applied to an optical source for intensity modulation. However a drive circuit may be used to amplify the signal prior to intensity modulation of the optical source. The optical digital signal is then coupled to optical fiber for transmission.

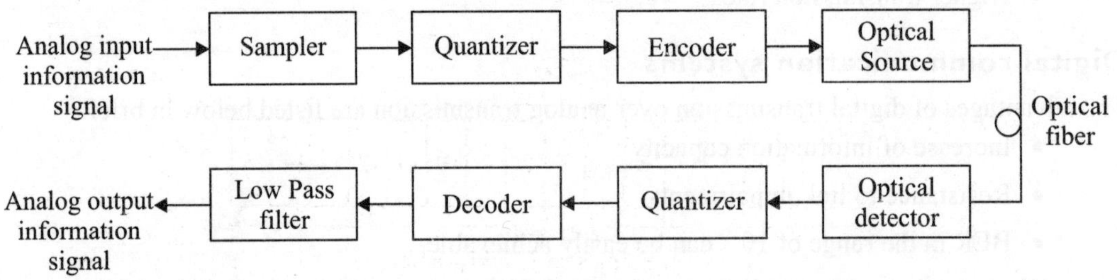

Fig. 12.7 Block diagram of PCM system for fiber optic communication

In the receiving system, a photodetector is used to generate electrical signal from the corresponding incoming optical signals. The detected electrical signal is quantized by the quantizer and fed to decoder for decoding. The decoded signal is finally fed to a Low pass filter (LPF) to eliminate noise and provide smoothness in the analog output signal.

Time division multiplexing (TDM) can also be introduced to achieve simultaneous transmission of several channels over a single optical fiber link. In this process the encoded PCM signals are interleaved by TDM to obtain composite signal for transmission. This composite signal is then intensity modulated by an optical source which converts the electrical signal into light signal and couples the light power to an optical fiber. In receiver side, the signals are demultiplexed and original base band signal is recovered by the photodetector followed by PCM receiver.

Optical pulse format of PCM system

It is possible to allow communication for a large number of voice channels simultaneously using the large bandwidth of a single fiber optic link. Figure 12.8 below shows the signalling and frame structure of a conventional 30 channel PCM system used in Europe.

Figure 12.8 shows that the time slots from respective channels are multiplexed into a frame. The extra two time slots are used for signalling and synchronization.

Figure 12.8(*c*) shows a multi-frame consisting of sixteen frames. Thus 10 Mb/sec digital fiber optic links can transmit approximately 120 telephone channel and 10 Gb/sec link can transmit around 10000 telephone channels.

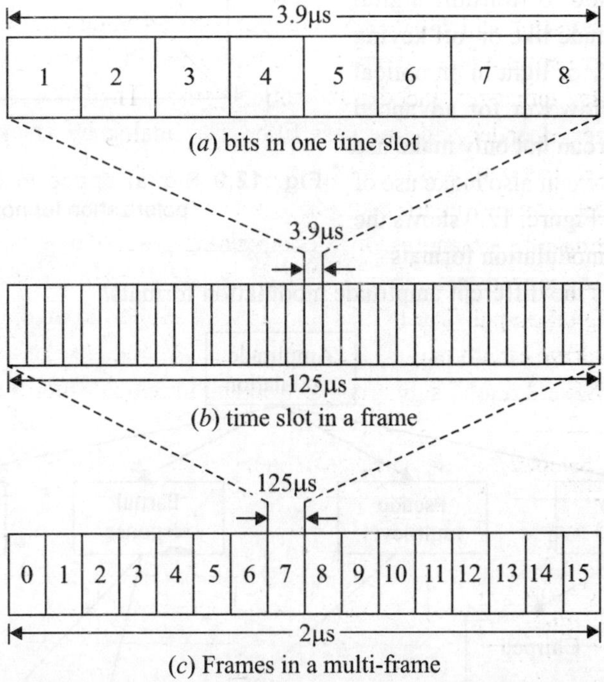

Fig. 12.8 Signal and frame structure of a thirty-channel PCM system

Optical pulse format for digital communication is an important factor to be considered in designing an optical fiber link. The decision-making electronic circuitry in receiver must get accurate timing information from the input optical signal for following reasons.

1. To provide sufficient time for sampling the signal.
2. To maintain proper pulse spacing.
3. To indicate start and stop of each timing interval.

Most importantly, the optical signal must have the capability of error detection and correction. Different line code formats are used in optical fiber transmission links.

1. Non-return-to-zero (NRZ) format.
2. Return-to-zero (RZ) format.
3. Phase ended (PE) format or the optical Manchester Code format.

Section 11.5 of chapter 11 of this book contains the description about the line codes, their relative advantages and disadvantages.

Different modulation types in fiber optic communication

In this section, we classify optical modulation formats according to their key attributes. In single-mode optical fibers, the optical field has three physical attributes that can be used to carry information: intensity, phase (including frequency), and polarization. Depending on which of the three quantities is used for information transport, we distinguish between intensity, phase (or frequency), and polarization data modulation formats.

The simplest method to transmit digital data is by changing amplitude like on-off keying *i.e.*, to change the intensity of light in an optical communication system. However for advanced modulation techniques we can not only make use of the light intensity, but we can also make use of its phase and polarization. Figure. 12.9 shows the signal space for different modulation formats.

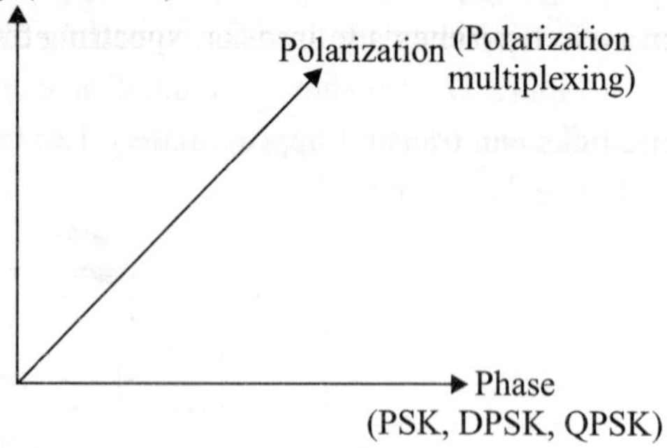

Fig. 12.9 Signal space in intensity, phase and polarization for novel modulation formats

Figure 12.10 shows the different amplitude modulation formats.

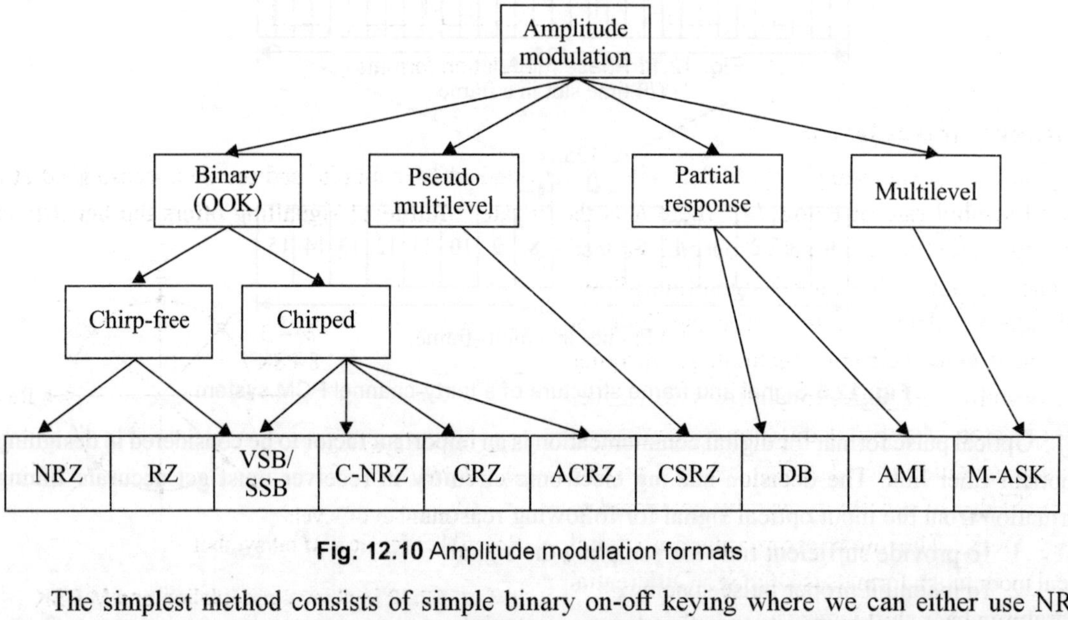

Fig. 12.10 Amplitude modulation formats

The simplest method consists of simple binary on-off keying where we can either use NRZ (non-return-to-zero) or RZ (return-to-zero) CSRZ (carrier-suppressed RZ), duobinary (DB) or alternate-mark-inversion (AMI) formats, are discussed in section 11.5 of chapter 11 of this book.

As a broad level, modulation formats can be classified into two types as phase and amplitude modulation. Thus as an alternative to amplitude modulation, phase modulations are also of high interest. Different phase modulation schemes are listed in Fig. 12.11 below.

While intensity and phase modulation formats have been widely used in high-speed optical communications, encoding information onto the polarization of light (Pol-SK) has received comparatively little attention. For intensity- or phase-modulated DMFs and direct-detection receivers, polarization management is only required in the presence of substantial polarization mode dispersion be acceptable if Pol-SK offered a significant receiver sensitivity improvement over intensity modulation, which it does not. However, the polarization degree of freedom can be used to improve the propagation properties of a format by pseudo-multilevel or correlative coding. Furthermore, polarization is sometimes used in research experiments to increase spectral efficiency.

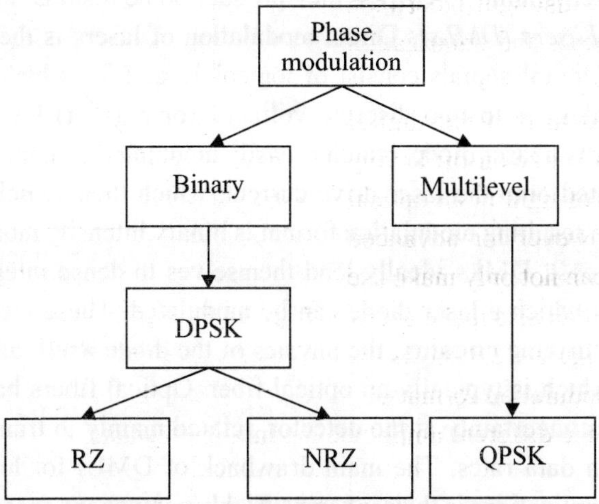

Fig. 12.11 Phase modulation formats

Multilevel modulation

Using multilevel signalling, $\log_2 M$ data bits are encoded on M symbols, and are then transmitted at a reduced symbol rate of $R/\log_2 M$, where R is the bitrate. Multilevel signalling offers the benefits of higher spectral efficiencies at the cost of a reduced tolerance to noise. Multilevel signalling allows single-channel data rates to exceed the limits of high-speed optoelectronics technology. In other words, multilevel signalling allows for lower symbol rates at a fixed data rate. Multilevel intensity modulation (M-ASK) has not proven beneficial due to sensitivity penalty compared to binary OOK. The most promising multilevel optical modulation formats is DQPSK (Differential quadrature phase shift keying).

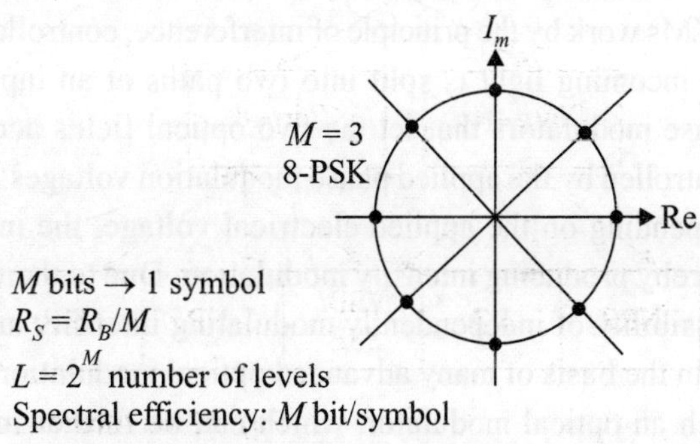

Fig. 12.12 Multilevel modulation and M-PSK

Different types of modulators (DML, EML and MZM)

It is of utmost importance to always consider practical aspects of modulation and detection hardware when designing optical modulation formats. Three basic modulator technologies are widely in use today: Directly Modulated Lasers (DMLs), Electroabsorption Modulators (EAMs), and Mach Zehnder Modulator (MZMs).

Electroabsorption Modulators (EAMs, EMLs): In this scheme, usually a semiconductor material that can be switched between transmissive and opaque states to provide the on-off transitions. Materials used in these modulators can be conveniently integrated with the laser diode. As with direct modulation, the driving circuitry uses microwave design and layout to couple high speed data streams to the modulating devices. EAMs are P-I-N semiconductor structures whose bandgap can be modulated by applying an external voltage, thus changing the device's absorption properties. EAMs feature relatively low drive voltages (typically 2 V), and are cost-effective in volume production. They are available for modulation up to 40 Gb/s today. However, similar to DMLs, they produce some residual chirp.

Directly Modulated Lasers (DMLs): Direct modulation of lasers is the easiest way to imprint data on an optical carrier. Digital signals consist of logical 1s and 0s, which readily corresponds to electrical ON and OFF states, or to two discrete voltages (or current) levels. The typical optical communications light source is a laser diode, which is easily modulated by controlling its current. Here, the transmit data is modulated onto the laser drive current, which then switches on and off the light emerging from the laser. The resulting modulation format is binary intensity modulation (OOK). Due to their compactness and low cost, DMLs ideally lend themselves to dense integration. Several factors limit the upper frequency at which a laser diode can be modulated. These include the time constants (frequency response) of the driving circuitry, the physics of the diode itself, and the characteristics of the transmission medium, which is typically an optical fiber. Optical fibers have a certain amount of dispersion, which introduces uncertainty at the detector, related mainly to transmission distance. This uncertainty is worse at high data rates. The main drawback of DMLs for high-bitrate transmission beyond short-reach access applications is their inherent, highly component-specific chirp, *i.e.*, a residual phase modulation accompanying the desired intensity modulation; laser chirp broadens the optical spectrum, which impedes dense WDM channel packing and can lead to increased signal distortions caused by the interaction with fiber chromatic dispersion. Ultimately, direct modulation is limited by the characteristics of the diode itself.

Mach–Zehnder Modulators (MZMs): Unlike EAMs, which work by the principle of absorption, MZMs work by the principle of interference, controlled by modulating the optical phase. In this modulator: the incoming light is split into two paths at an input coupler. One (or both) paths are equipped with phase modulators that let the two optical fields acquire some phase difference relative to each other, controlled by the applied phase modulation voltages. Finally, the two fields interfere at an output coupler. Depending on the applied electrical voltage, the interference varies from destructive to constructive, thereby producing intensity modulation. Due to their well-controllable modulation performance and the possibility of independently modulating intensity and phase of the optical field, $LiNbO_3$ based MZMs form the basis of many advanced optical modulation formats. Figure 12.13 below is a block diagram of such an optical modulator, which can be further integrated with a detector onto a microprocessor or other device, as on-chip optical interface.

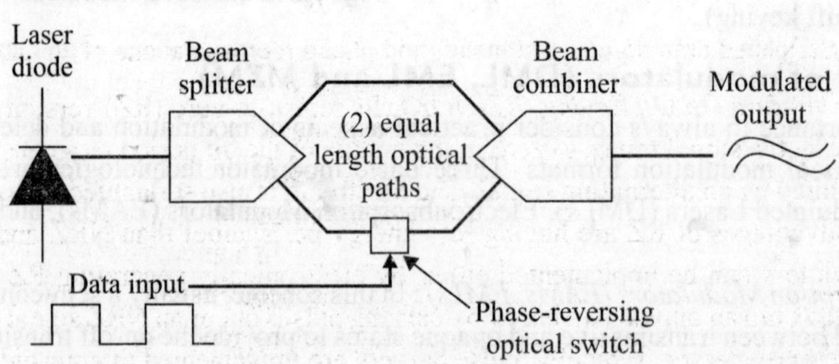

Fig. 12.13 Functional diagram of an MZM optical modulator

Different intensity modulation formats (NRZ-OOK, RZ-OOK, CSRZ, CRZ, C-NRZ)

A. Nonreturn-to-Zero On/Off Keying (NRZ-OOK): The simplest way to generate optical modulation format is NRZ-OOK, often just referred to as NRZ. At bitrates of 10 Gb/s and above, NRZ is most conveniently generated using DMLs or EAMs (for short-reach and intermediate reach transponders) or chirp-free MZMs (for long-haul applications).

The schematic diagram of NRZ transmitter, with an external electro-optic intensity modulator, is shown in Fig. 12.14 below. The intensity modulator can either be Mach-Zehnder type or electro-absorption type, which converts an OOK electric signal with data rate of R_b into an OOK optical signal at the same data rate. For a long time, NRZ has been the dominant modulation format in optical fiber communication systems. In recent years, as optical communication is advancing to higher data-rates, long distance with optical amplifiers, NRZ modulation format may not be the best choice for high capacity optical systems. However, due to its simplicity, and its historic dominance, NRZ can be used as a good reference for the purpose of comparison.

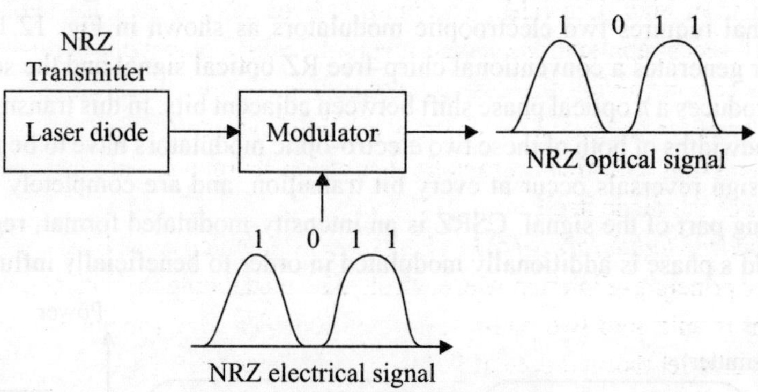

Fig. 12.14 NRZ OOK transmitter

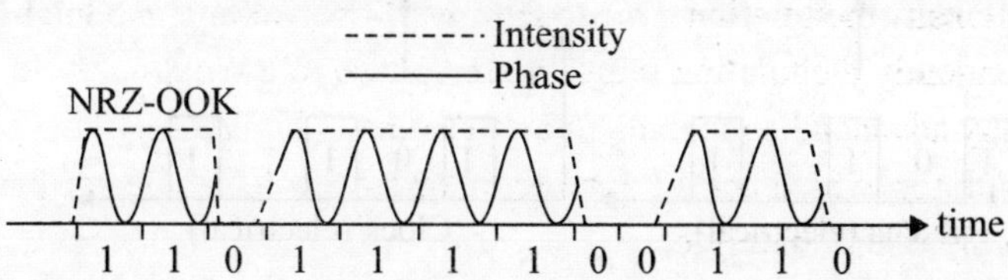

Fig. 12.15 Simplified time domain (intensity) and phase representations of an OOK NRZ signal

B. Return-to-Zero On/Off Keying (RZ-OOK): Its return-to-zero (RZ) version improves the clock recovery by making the signal return to zero in the middle of the bit period so a constant stream of ones would be represented by an alternating signal. The bit stream is also scrambled to avoid long streams of zeroes. The disadvantages of RZ are having less energy per symbol than NRZ and higher frequency. RZ-OOK transmitters can be implemented either by electronically generating RZ waveforms, which are then modulated onto an optical carrier, or by carving pulses out of an NRZ signal using an additional modulator, called pulse carver. Typically, pulse carvers are implemented as sinusoidal driven EAMs or MZMs.

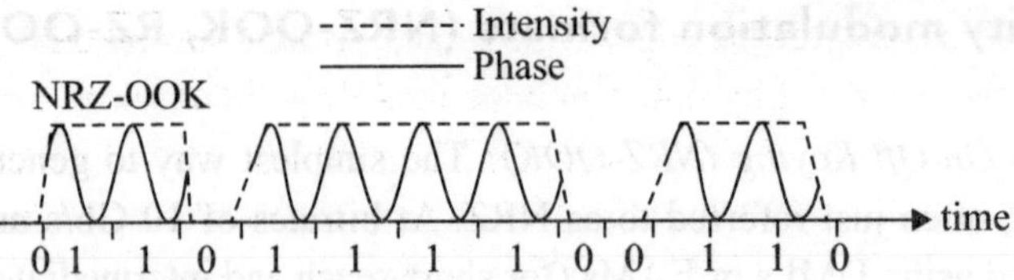

Fig. 12.16 Simplified time domain (intensity) and phase representations of an OOK RZ signal

C. Carrier-Suppressed Return-to-Zero (CSRZ): CSRZ is a pseudo-multilevel modulation format, characterized by reversing the sign of the optical field at each bit transition. The most widespread (since easiest to generate) pseudo-multilevel format is CSRZ, where the information is encoded on the intensity levels {0,1}, but the phase is changed by π every bit, regardless of the data information. The major difference between a CS-RZ and a conventional RZ is that in CS-RZ optical signal there is a π phase shift between adjacent bits.

Therefore, the average optical field in a CS-RZ signal is zero (no DC component). In the frequency domain. this translates into a carrier-suppression in the optical spectrum. In general, the generation of a CS-RZ optical signal requires two electrooptic modulators as shown in Fig. 12.17 below. The first intensity modulator generates a conventional chirp-free RZ optical signal and the second electro-optic phase modulator produces a π optical phase shift between adjacent bits. In this transmitter configuration, the modulation bandwidths of both of these two electro-optic modulators have to be in the same level as the data rate. The sign reversals occur at every bit transition, and are completely independent of the information-carrying part of the signal. CSRZ is an intensity-modulated format, regardless of the fact that the optical field's phase is additionally modulated in order to beneficially influence the spectrum.

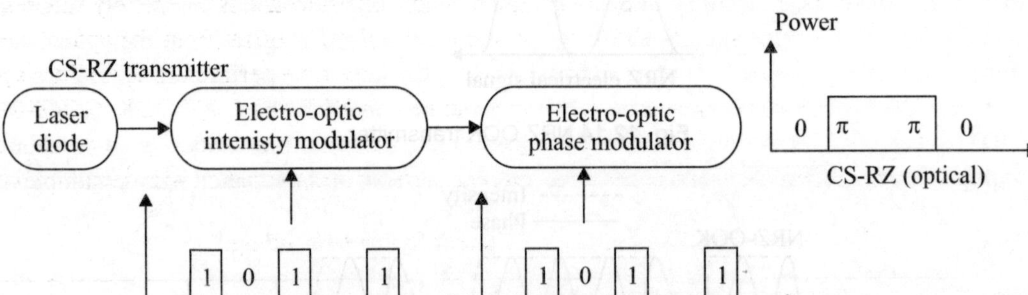

Fig. 12.17 Schematics of CSRZ transmitters

D. Chirped Return-to-Zero (CRZ): Auxiliary modulation features such as chirp, can be used to improve a format's transmission properties. If a controlled amount of analog phase modulation is applied to a Digital Modulator, the qualifier chirped is added to describe the format. In the case of CRZ, or ACRZ, bit-synchronous periodic chirp spectrally broadens the signal bandwidth. Although this reduces the format's suitability for high spectral efficiency WDM systems, it generally increases its robustness to fiber nonlinearity and also enhances its resilience to multipath interference. CRZ is predominantly used in ultra long haul point-to-point fiber communications, as found in transoceanic (submarine) systems.

E. Chirped NRZ (C-NRZ): To counteract effects caused by chromatic dispersion, appropriate amount of chirp across each optical pulse: lowering the frequency across the pulse's leading edge, and raising the frequency across the trailing edge is introduced. Among several other techniques, prechirping of optical pulses can be performed by imbalancing a dual-drive MZM used for NRZ modulation

F. Correlative Coding and Partial Response Formats(DB, AMI): Optical Duo Binary (DB) and alternate mark inversion (AMI) belong to the general class of correlative coding formats, a subclass of which being referred to as partial response signalling. Correlative coding introduces a distinct correlation between the optical phase and the data information: in optical DB, a phase change occurs whenever there is an odd number of 0-bits between two successive 1-bits, whereas for AMI the phase changes for each 1-bit (even for adjacent 1-bits), independent of the number of 0-bits in between. The time-domain explanation considers a 1 0 1 bit pattern. If, due to dispersion or optical filtering, the two 1-bit pulses spread into the 0-bit between them, duobinary encoding lets them interfere destructively, thus maintaining a low 0-bit level. Conventional OOK formats, on the other hand, let the pulses interfere constructively, which raises the 0-bit level. The frequency-domain explanation is based on the narrower spectral extent of properly filtered DB signals, which reduces dispersion induced signal distortions. AMI, sometimes also classified as partial response due to the way it is generated, does not share the bandwidth limitation characteristic of other partial response formats. In optics, AMI is typically implemented in RZ form (RZ-AMI). Like other RZ-OOK formats using auxiliary phase coding, the additional phase modulation of AMI can help to reduce the effects of some fiber nonlinearities

G. Vestigial Sideband (VSB) and Single Sideband (SSB): Apart from shaping (and compressing) the optical signal spectrum by means of (pseudo)-multilevel signalling or correlative coding, it is possible for some modulation formats to additionally suppress half of their spectral content by appropriate optical filtering. Since the spectrum of real-valued baseband signals is symmetric around zero frequency, filtering out the redundant half of the spectrum (*i.e.*, one of the two spectral sidebands) preserves the full information content. This is exploited in SSB signalling, where one sideband is completely suppressed, and in VSB signalling, where an optical filter with a gradual roll-off is offset from the optical carrier frequency to suppress major parts of one sideband, while at the same time performing some filter action on the other. Optical VSB has been successfully demonstrated on NRZ-OOK, RZ-OOK, CSRZ-OOK. In a WDM system, VSB filtering can either be done at the transmitter (*i.e.*, before or in combination with multiplexing the WDM channels) or at the receiver (*i.e.*, after or in combination with demultiplexing).

Table 12.1 Examples for Different Symbol Encodings

Data Sequence	0	0	1	0	1	1	1	0	0	1	0	1
Pseudo-multilevel (CSRZ)	0	0	+1	0	+1	–1	+1	0	0	–1	0	–1
Partial-response (DB)	0	0	+1	0	–1	–1	–1	0	0	–1	0	+1
Alternate Mark Inversion (AMI)	0	0	+1	0	–1	+1	–1	0	0	+1	0	–1

12.3 AN INTRODUCTION TO OPTICAL ORTHOGONAL FREQUENCY DIVISION MULTIPLEXING (OFDM)

Digital modulation techniques are broadly classified into two categories.

1. *Single carrier modulation (SCM):* Where the data are carried on a single main carrier. It is the conventional modulation techniques for last few decades with many variants like non-return-to-

zero (NRZ), return-to-zero (RZ), Duobinary, differential phase-shift keying (DPSK), and coherent quaternary phase-shift keying (QPSK), etc.

2. *Multicarrier modulation (MCM):* Where the data are carried through many closely spaced subcarriers.

Orthogonal frequency division multiplexing (OFDM) is a special class of MCM system that has only recently gained attention in the optical communication community because, it is also possible to detect this multi-subcarrier optical signal using narrowband optical filters. Because the frequency separation between subcarriers is very narrow, which is comparable to the datarate, this type of optical signal structure is commonly referred to as optical frequency-division multiplexing (OFDM). Experiments on coherent optical OFDM (CO-OFDM) transmission at 100 Gb/s by various groups have put the optical OFDM in the race for the next generation of 100 Gb/s Ethernet transport. OFDM has already emerged as a leading modulation technique for RF domain. Recently all new major future communication standards like wireless LAN (IEEE 802.11 also known as Wi-Fi), digital video and audio standards (DAV/DAB), and digital subscriber loop (DSL) etc. has adopted OFDM. Therefore, it is not surprising that the two competing fourth-generation (4G) mobile network standards, Worldwide Interoperability for Microwave Access (WiMAX, or IEEE 802.16) from the computing community and Long-Term Evolution (LTE) from the telecommunication community, both have adopted OFDM as the core of their physical interface. Recent technological advancements in narrowband optical filters and optical interleavers make OFDM a feasible approach to achieve high bandwidth efficiency and flexible datarate granularity. Conventionally, an OFDM transmitter uses a laser diode and an external optical modulator for each wavelength. Because of the narrow channel spacing of an OFDM system, the requirement on the wavelength control and stabilization of each transmitter laser is stringent. A number of very narrowly spaced OFDM channels can be generated using a single laser diode and an external optical modulator. In addition to using conventional on-off key modulation, advanced optical modulation schemes, such as duobinary and optical DPSK, can be also applied through RF domain signal processing.

Single carrier or Multicarrier transmission

There is a debate about the supremacy of single carrier or multicarrier transmission.

There is no single answer, rather than a split about this debate. For example, the United States has chosen single-carrier 8-level vestigial sideband modulation (8-VSB) as the digital TV standard, whereas Europe, Japan, and most other countries have elected OFDM.

Again the debate may produce different meaning when applicable to optical domain rather than OFDM in RF domain.

Figure 12.18 below shows the transmitter architectures for single carrier systems and CO-OFDM systems.

We can observe two differences:

1. Single-carrier systems employ a relatively conventional and simpler architecture, in which discrete digital-level modulation is fed into the two arms of the QPSK modulator. With regard to generating the I and Q component, the QPSK modulation resembles that of conventional BPSK or DPSK modulation. In contrast, the CO-OFDM architecture includes drastic modification from the conventional single-carrier system, in which the electronic DSP module and DAC are required for complex OFDM signal generation at the transmit end. The OFDM transmitter strictly enforces linearity in each component associated with the CO-OFDM transmitter.

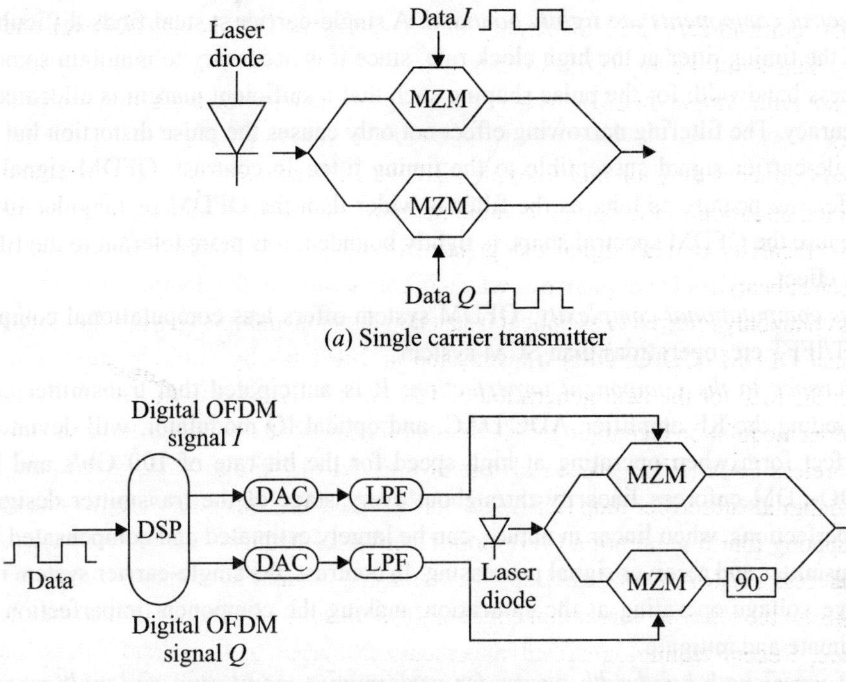

(*a*) Single carrier transmitter

(*b*) Multi carrier transmitter

Fig. 12.18 Transmitter architecture for (*a*) single-carrier systems and (*b*) multicarrier systems

2. In the single-carrier systems, the information is coded in the time domain, whereas in CO-OFDM, the information is encoded in the frequency domain, more precisely onto each individual subcarrier.

Main Features of OFDM

Advantages:

(i) *Ease of signal processing:* Signal processing can be coupled to OFDM system in a more easier way than SCM system since OFDM system places the signal processing capability in the transmitter side to get the advantage of transmitter adaptability. In CO-OFDM-based systems, by using pilot symbols or pilot subcarriers, the signal processing procedures like channel estimation and phase estimation are made relatively straightforward. In contrast, in a SCM system employs blind equalization for channel estimation.

(ii) *Easier higher order modulation:* More complicated optical modulation format can be configured using OFDM technique with more ease. For example, for a QPSK transmitter it seems that the transmitter complexity is higher. But once the modulation goes beyond 2 bits per symbol, such as 8-PSK or 8-QAM, the CO-OFDM has lower complexity than the single-carrier system, which subsequently reduces the system cost. A CO-OFDM system can be changed to a higher order modulation system without changing optical hardware. The only change from 4-QPSK to 8-QAM modulation is enabled via the software to reconfigure the DSP and DAC.

(iii) *Spectral components are tightly bounded:* A single-carrier system finds difficulty in reducing the timing jitter at the high clock rate, since it is necessary to maintain some additional excess bandwidth for the pulse shaping such that a sufficient margin is allocated for timing accuracy. The filtering narrowing effect not only causes the pulse distortion but also makes single-carrier signal susceptible to the timing jitter. In contrast, OFDM signal practically suffers no penalty as long as the filter is wider than the OFDM rectangular-like spectrum because the OFDM spectral shape is tightly bounded, it is more tolerant to the filter narrowing effect.

(iv) *Less computational complexity:* OFDM system offers less computational complexity (like FFT/IFFT etc. operation) than SCM system.

(v) *Tolerance to the component imperfection:* It is anticipated that transmitter components, including the RF amplifier, ADC/DAC, and optical IQ modulator, will deviate from their perfect form when operating at high speed for the bit rate of 100 Gb/s and beyond. As COO-FDM enforces linearity throughout every stage of the transmitter design; thus, the imperfections, when linear in nature, can be largely estimated and compensated through the transmitter and receiver signal processing. In contrast, the single-carrier system relies on the drive voltage operating at the saturation, making the component imperfection difficult to estimate and mitigate.

(vi) *Sub-wavelength bandwidth access for performance monitoring and multiaccess/multicast networks:* It is a great advantage to place the DSP in the transmitter for CO-OFDM systems. The ability and flexibility to allocate a certain number of subcarriers for channel estimation and performance monitoring will prove to be an attractive feature for CO-OFDM. For instance, this leaves the option of grouping a band of subcarriers for monitoring, which can be easily detected without processing the entire spectrum. Similarly, sub-banding of OFDM allows for the dynamic bandwidth allocation for multiaccess networks using the orthogonal frequency domain multiple-access scheme. All these are difficult to achieve with the single-carrier system.

(vii) *Bit and power loading:* Bit and power loading can be done in a OFDM system which is a great advantage. In RF communication when the channel can be in deep fading or some part of the spectrum is notched out due to severe multipath interference, this operation can be done. It is a matter of research that how this can be employed in optical communication.

In view of the previous comparisons, we can conclude that CO-OFDM system is a future communication system which is advantageous for high data rate and transrecevier adaptability.

Disadvantage of OFDM system: The CO-OFDM unavoidably inherits the two main problems intrinsic to OFDM: (1) high PAPR that suggests the CO-OFDM is more susceptible to nonlinearity, and (2) sensitivity to the frequency and phase noise.

Difference between RF OFDM and Optical OFDM Systems

Optical OFDM bears both similarities to and differences from the RF counterpart.

A typical wireless channel can be modeled as summarization of the multiple paths that each undergoes a Rayleigh process. An optical SMF can be more conveniently modeled in the frequency domain for the two polarization components in the fiber.

Table 12.2 Comparison between wireless and optical channels

	Mathematical Model	Nonlinearity	Speed
Wireless OFDM	Time domain multiple discrete Rayleigh fading	none	Can be fast for mobile environment
Optical OFDM	Continuous frequency domain dispersion	significant	medium

Wireless channel is in free space and therefore does not possess any nonlinearity. On the other hand, optical fiber is fairly nonlinear. Coupled with the fiber dispersion, polarization effects make the optical channel more complicated than wireless channel. At a first glance, OFDM suffered with high PAPR would not fit for the optical fiber with high nonlinearity. However, fortunately, the fiber chromatic dispersion saves OFDM in optical domain that tends to mitigate against the nonlinearity, and experiments have shown successful transmission of 100 Gb/s CO-OFDM over 1000 km SSMF fiber.

OFDM in wireless domain has been studied extensively for last few years. But from the above discussion it is clear that the translation from wireless to optical domain is not effortless and it needs clear understanding of optical channel to get the most out of it.

REFERENCES

[1] Peter J. Winzer, R.J. Essiambre "*Advanced Optical Modulation Formats*" Proceedings of the IEEE | Vol. 94, No. 5, May 2006.

[2] G. P. Agrawal, "*Fiber-Optic Communication Systems*", 3rd ed., New York: Wiley, 2002.

Index

J

L

M

N

O

P

R

S

T

V

W